DEPLOYING RFID – CHALLENGES, SOLUTIONS, AND OPEN ISSUES

Edited by **Cristina Turcu**

Deploying RFID – Challenges, Solutions, and Open Issues
Edited by Cristina Turcu

Published by InTech
Janeza Trdine 9, 51000 Rijeka, Croatia

Publishing Process Manager Davor Vidic

Technical Editor Teodora Smiljanic

Cover Designer InTech Design Team

Additional hard copies can be obtained from orders@intechopen.com

Deploying RFID – Challenges, Solutions, and Open Issues, Edited by Cristina Turcu
p. cm.
ISBN 978-953-307-380-4

For EU product safety concerns:
IN TECH d.o.o., Prolaz Marije Krucifikse Kozulić 3, 51000 Rijeka, Croatia,
info@intechopen.com or visit our website at intechopen.com.

Meet the editor

Cristina Turcu is an Associate Professor of Software Engineering and Artificial Intelligence at Stefan cel Mare University of Suceava, Romania. She received her Diploma (M.Sc.) in Automatics and Computers and her Doctorate (Ph.D.) in Automatics, in 1991 and 2000, respectively, both from the Gheorghe Asachi Technical University of Iasi, Romania where she has been Head of Computer Department since 2004. Her research interests include software engineering, RFID applications for the end-consumer, and intelligent systems. Dr. Turcu is an Editor of four books and has served on various program committees of conferences in computing and RFID systems. She also has served as a reviewer for numerous referred journals and conferences. She is the Editor in Chief of the International Journal of Radio Frequency Identification & Wireless Sensor Networks. Dr. Turcu has published over 80 publications in books or book chapters, refereed journals, technical reports, and refereed conference/workshop/seminar proceedings.

Contents

Preface

Radio frequency identification (RFID) is a technology that is rapidly gaining popularity due to its several benefits in a wide area of applications like inventory tracking, supply chain management, automated manufacturing, healthcare, etc. The benefits of implementing RFID technologies can be seen in terms of efficiency (increased speed in production, reduced shrinkage, lower error rates, improved asset tracking etc.) or effectiveness (services that companies provide to the customers). Despite these numerous benefits, the technology has limited widespread implementation, due to the insufficient robustness and reliability of the RFID technology, cheaper alternatives to RFID (bar-codes), the costs-benefits balance of RFID implementation and the absence of common standards and interoperability.

In these conditions, it is the researchers' goal to contribute to the improvement of this technology and the providing of new and valuable solutions to the industry personnel. The book includes interesting research studies from experienced scientists in the RFID domain.

In chapter 1, the authors present the importance of RFID and the shortcomings of the current approaches designed to correct some issues among the integration of systems including security, privacy, and data abnormalities. They also recommend solutions to these issues.

Chapter 2 deals with the study of the RFID components such as antenna and reader. It also discusses the RFID active and passive tags, and compares these tags, considering both advantages and disadvantages of RFID system. RFID applications are explored and a technical model is analyzed. The chapter also considers the healthcare perspectives and RFID use within healthcare settings. This study outlines a model for connected RFID applications, which provides quick support for various healthcare functions and enhances flexibility for different systems' components integration.

Chapter 3 outlines the experience and achievements attained in a project carried out by the National University of Colombia. This project was intended to design and implement RFID-based tools for training students in medical and nursing techniques applied on neonatal patients.

The authors of chapter 4 propose an RFID-based solution to reduce the human factor in the preparation and administration of cytostatic infusions.

In chapter 5 the authors propose an RFID-based system that integrates RFID and multi-agent technologies in health care in order to make patient emergency care as efficient and risk-free as possible, by providing doctors with as much information about a patient and as quickly as possible. Also they describe a general purpose architecture and data model that is designed for both collecting ambulatory data from various existing devices and systems, and storing clinically significant information in order to be accessed by the emergency care physician.

In chapter 6 the authors propose an RFID-based multi-agent system, that facilitates the integration of data from heterogeneous sources in order to achieve a complete patient electronic medical record. The adoption of this system does not require major changes in terms of the software resources existing in the medical units.

In chapter 7 the authors propose a farm operation monitoring system using wearable sensor devices with RFID readers and various sensing devices such as motion sensors, cameras, and a GPS. This system recognizes detailed farming operations automatically in various situations by analyzing the data from sensors and detected RFID tags. The tags and sensors are attached to relevant objects such as farming materials, machinery, facilities, and so on. In this chapter, the authors, based on their research, describe the concept and features of this system and the results of several experiments conducted on a prototype system. The major applications and extensions of the current systems are also outlined.

In chapter 8, the authors introduce the application of RFID in day-to day activity in cow industry, regarding the use of RFID technology in automatic cow feeding machine.

Chapter 9 focuses on the cow-calf sector, with an overview of U.S. agriculture and the beef-cattle sector. Finally, the author presents what can be called "the National Animal Identification System (NAIS) Pushback".

In chapter 10 the author presents the new potential of RFID-applications in mine planning. An RFID-based system can be used to visualize the placement of machines inside roadways; to monitor miners with personal transponders; to prevent non-permitted control of machines; to prioritize the control of machines; to evaluate the productivity of both machines and mining areas; to evaluate fuel consumption and machine resources, etc. After being gathered, this information is used for the mine management.

In chapter 11, the authors investigate the applicability of the RFID technology in location sensing, the main design, and environmental factors that should be considered before developing an RFID-based localization scheme. The authors present a scenario according to which the location of multiple reader-enabled terminals need to be estimated based on the information retrieved from low cost passive tags, which are deployed in a particular area. Also, the authors propose a mathematical model for taking into account all implicating factors which affect the accuracy performance of the system, like types of collisions among its components, interference of materials, and temporal environmental changes.

Chapter 12 analyzes an indoor object localizing method by using active RFID tags and simple switch sensors embedded in the environment. The authors focus in their work on object's "location" in the environment (e.g. Table, Bed, Sofa, etc.) instead of object's 3-dimensional position, the only object location allowing the achievement of their application.

In chapter 13, an approach for developing an RFID sensor model is presented. The authors examine recent progresses in fuzzy logic-based RFID sensor modeling using an autonomous robot. Constructing a reliable sensor model is very important for successive applications such as tag localization, robot localization, just to mention a few.

Chapter 14 deals with optimizing distributed robotic control systems, considering as example an intelligent cart system designed to be used in common airports. The presented framework employs an RFID-based localization algorithm and control methods using mobile software agents.

In chapter 15 the authors present the services, use cases and the future challenges of Near Field Communication, which is the most customer-oriented one among RFID technologies.

In chapter 16 the authors study a cyber-physical system based on RFID technology. They compare the proposed RFID system with a traditional wireless sensor network system and discuss the applicability of the first one. Finally, the authors present the design, methodology, and development of an active RFID-based relative positioning system, also showing the experimental results.

Chapter 17 gives an overview of SAW-based RFID transponders made for extreme conditions like temperatures up to 400°C or cryogenic temperatures down to –196°C. The authors give an explanatory outline of SAW transponder function principles and system performance, and also some application examples from steel and automotive industries.

The chapter 18 discusses the RFID/WSN technology from a networking perspective. The authors outline the development needed to integrate RFID systems with Internet of Things and present the evolution from today's connection of objects to the future networking of objects.

Chapter 19 deals with the application of RFID technology to improve user interaction in novel environments. The authors describe the development and implementation of three different use cases, actually implementing the concept of context-awareness, location awareness, and Internet of Things.

Chapter 20 provides a historical and conceptual introduction to the IoT topic. The work also outlines key aspects in the process of moving from the current state of the art of IoT, where objects have digital identities, towards a network of objects having digital personalities and being able to interact with each other and with the environment. Finally, a selection of the possible impacts of the IoT is analyzed.

Chapter 21 address issues of the protecting privacy of RFID tag carriers in a "privacy by design" model, which is described on four different levels: legal aspects, policy services, technical specifications and security services.

The final chapter of this book goes beyond identification matters and addresses aspects referring to the ethics of RFID technology. The author focuses on the ethical approach that must be concerned with the subjectification of people through the use of the technology.

Leading to considerable operational and strategic benefits, RFID technology continues to bring new levels of intelligence and information, strengthening the experience of all participants in this research domain, and serving as a valuable authentication technology. We hope this book will be useful for engineers, researchers and industry personnel, and provide them with some new ideas to address current and future issues they might be facing.

Cristina TURCU
Stefan cel Mare University of Suceava
Romania

1

The Challenges and Issues Facing the Deployment of RFID Technology

Peter Darcy, Prapassara Pupunwiwat and Bela Stantic
Institute of Integrated and Intelligent Systems, Griffith University
Australia

1. Introduction

Radio Frequency Identification refers to wireless technology that uses radio waves to automatically identify items within certain proximity. This process involves tagging items with a transmitter which will emit bursts of information including, but not limited to, the identification of the tag. There are three main varieties of tags: Active, Semi-active and Passive. Active tags rely solely on a battery for its power source resulting in the maximum integrity rate and reading range but, also, a limited lifespan and higher cost. Semi-Active tags use batteries to extend the range of the tag only resulting in a higher reading rate than passive tags, a longer lifespan than the active tags, but also higher cost. The passive tag uses the electromagnetic pulse from readers as a power source to transmit its identifier. Due to its lack of a battery, passive tags are the most cost effective and theoretically have an unlimited lifespan. However, due to their lack of the power source, passive tags also have a limited range and produce the largest amount of data anomalies. The RFID Reader is used to interrogate the zone to discover tags within proximity of the reader range. If a tag is discovered, its identification along with the reader's ID and the timestamp of the observation are recorded. This information is then passed through the Middleware where initial filtration is done to avoid data anomalies being recorded. Finally, the information will then be stored within a database ready to be queried for future analysis.

Due to the benefits of the technology, RFID is currently employed in various commercial sectors to provide automated assistance for mundane tasks. There are hospitals which have employed tagged bracelets to ensure maximum care is given to surgical patients. At various airports around the world, RFID is being utilised to track passengers' bags to ensure that the location of the luggage will be known at all times. In various cities around the world, pets have had RFID chips implanted to ensure that, when lost, the authorities can find their owners' information by simple scanning the tag. Various countries have also introduced the RFID-enabled toll system designed for cars at RFID-enabled toll booths which allow drivers to continue on their journey and avoid the necessity of stopping to pay.

Despite the advantages gained from RFID technology integration, various drawbacks prevent the wide-scale adoption into the majority of the commercial sector. There are three main issues concerning the integration of the architecture. The first issue is security when using the technology as tags are prone to various physical and virtual attacks upon the system. The second concern stems from the need of privacy surrounding the data collected as the observations recorded can be used for breaches in privacy. The third issue is that the

data collected among systems, in particular where passive tags are utilised, produces data characteristics that make the systems harder to use.
With regard to the data characteristics issue of RFID, there are four main problems. The first is that the data collected only contains two identifiers and a timestamp making the low-level data useless without context of other information. The large amounts of data gained in short periods of time is the second complication that arises from the use of RFID technology resulting in the database storing massive amounts of observations, some of which are useless. The third obstacle found among the integration of RFID systems is the complex spatial and temporal dimensions resulting from handheld readers and other advanced devices. The final difficulty is the tags generating ambiguous and incorrect observations resulting in duplicate, wrong and missing anomalies.
Various methodologies have been mentioned in literature to address the current problems with RFID data anomalies. We have categorised these solutions into three main groups: Physical, Middleware and Deferred approaches. Various physical solutions have been proposed in past studies to avoid missed readings in particular such as metallic-proof tag pads, tag orientation and multiple tagging. Smoothing Filers and Anti-Collision Protocols are Middleware solutions proposed to correct anomalies found within the Reader at the point of scanning. Finally, there have been several rule-based and classification algorithms proposed in past methodologies to be utilised at a deferred stage of the scanning cycle to correct various anomalies already stored in the database.
Unfortunately, each of the proposed solutions has drawbacks that prevent it from eliminating all problems found within RFID systems. With regard to the physical solutions, most have been designed to eliminate a specific problem (i.e. the metallic padding) or it will generate additional and unforeseen complications (multiple tags introducing duplicate reads). Middleware solutions have been intended to be applied at the edge of the device when the scanning is conducted which results in a limited amount of analytical information for correction allowing ambiguous anomalies to persist. The Deferred approaches have the advantage of having access to additional information in the database. However, they cannot be applied in real-time and rely on user-specified rules or probabilistic algorithms that may result in additional artificial anomalies.
We have examined RFID technology and its current uses in various applications. We have also examined three core issues stopping the mass integration of RFID in the systems including security, privacy and problematic data characteristics. We have further explored the data characteristics issue to find that it contains low-level nature, large data gathering, complex spatial and temporal aspects, and data anomalies. There have been various methodologies proposed in the past to cope with the various data anomalies which we have categorised into physical, middleware and deferred solutions. Unfortunately, due the various drawbacks such as application-specified solutions, lack of analytical information or reliance on user-specified/probabilistic algorithms, current approaches do not provide the adequate support needed in RFID systems to be adopted in commercial sectors. In this work, we have identified the importance of RFID, the shortcomings of existing approaches designed to correct its issues, and have recommended solutions to these methodologies.

2. Radio Frequency Identification

Radio Frequency Identification (RFID) has had a long history commencing with its utilisation during the Second World War to its modern usage. The basic architecture of RFID itself

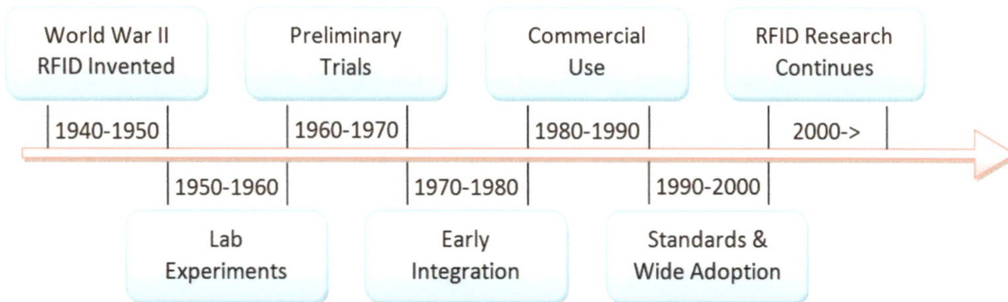

Fig. 1. The timeline of recent RFID history from the 1940s through to the present day (Landt, 2001).

consists of a tag, reader and middleware to perform advanced analysis on the data which makes it practical for use in many applications with beneficial outcomes. There are several problems which arise when using the passive tags due to the nature of the system, in particular, the amount of unreliable readings in the raw data.

2.1 History of RFID

For a general overview of RFID's historical achievements, please see the timeline illustrated in Figure 1. The physical birth of RFID would not come until the fusion of two technologies was achieved approximately around the era of the World Wars. The first technology was the Continuous Wave radio generation which was created in 1906 by Ernst F. W. Alexanderson. The second technology was the Radar device which is thought to have been developed in 1922 and was utilised extensively in World War II (Landt, 2001). The combination of these two devices resulted in the concept of RFID which was first academically proposed in theory by Harry Stockman in 1948. During this time, RFID was employed as a means to distinguish between enemy and allied aircrafts in the war. Unfortunately, as Stockman notes, technology had not progressed to the point that the complete potential of RFID technology could be realised (Stockman, 1948).

RFID research continued to be pursued in both the academic community and the military aircrafts' division who were attempting to develop "Identification Friend or Foe" (IFF) technology throughout the 1950s. It was not until the late 1960s that a Sensormatic and Checkpoint developed the first commercial RFID product in the form of EAS or "Electronic Article Surveillance" which consisted of a security system incorporating RFID tags that only stored an "on or off" command to prevent theft in stores. RFID's focus throughout the 1970s was in the tracking of animals and vehicles and, also, within the automation of factories. This adoption of the technology eventually led to the first RFID integrated road toll which was established in Norway in 1978. It was employed later in various other locations world-wide, the second notable one having been set up in 1989 at the Dallas North Turnpike in America (Landt, 2005).

In the 1990s, RFID had been integrated into people's daily activities. An example of this includes the utilisation of RFID key cards for enhanced security to enable a higher level of integrity for secure locations (Chawathe et al., 2004). In its most recent history from 2000-2010 and onwards, RFID has received the majority of its attention from various commercial sectors adopting its technology (Derakhshan et al., 2007). Some of these industries include Wal*Mart (Engels, 2005) where it has been used to enhance the supply chain, the US Department of

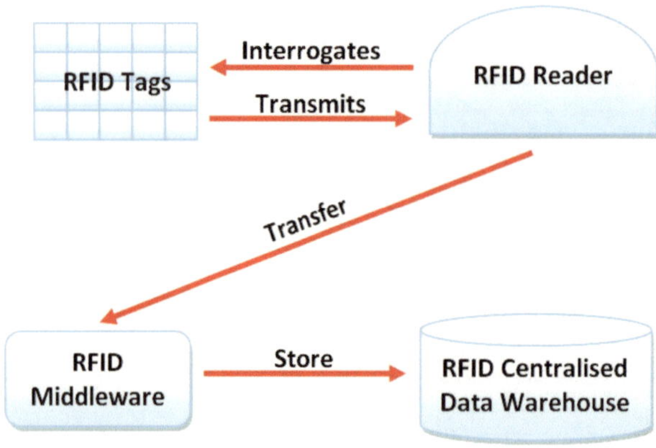

Fig. 2. The flow of information between the different components of the RFID System Architecture

Defence which has developed smarter tags (Collins, 2005) and the Aviation Industry which attaches tags to identify different parts when shipping out items (Collins, 2004). For a more comprehensive analysis of current RFID applications please see Section 3.

2.2 System Architecture

The System Architecture of an RFID system contains four important components (Chawathe et al., 2004): an RFID Tag, an RFID Reader, the RFID Middleware and the Database Storage. For a diagram representing the flow of information in this System architecture, please see Figure 2.

The RFID Tag is the simplest, lowest level component of the RFID System Architecture. These tags come in three types - Passive, Semi-Passive and Active. The Tag itself is made up of three different parts: the Chip which holds the information the tag is to dispense, the Antenna which is used to transmit the signal out and the Packaging which houses the Chip and Antenna and may be applied to the surface of other items. The Passive Tags are the most error-prone, but due to not needing a battery, also the most cost-effective and long-lasting. Electromagnetic pulses emitted from the Readers allow the Passive Tag enough energy to transmit its identification back. In comparison, the Semi-Passive Tag has a battery. However, it is only utilised to extend the readability scan resulting in a shorter life-span but increased observation integrity. The final tag is the Active Tag which utilises a battery to, not only extend its range, but also to transmit its identification number. From its heavy reliance of the battery, the Active Tag has the highest cost and shortest life-span of all the tags currently available (Chawathe et al., 2004). Even today, there are novel and emerging technologies to reduce the production cost even further such as the *Chipless RFID System Tags* and Readers (Preradovic et al., 2008; Preradovic & Karmakar, 2009).

The RFID Readers are the machines used to record the Tag identifiers and attach a timestamp of the observation. It does this by emitting a wave of electromagnetic energy which then interrogates the Tags until they have responded. These devices have a much greater purpose when needing to interrogate Passive and Semi-Passive Tags as they also provide the power necessary to transmit the information back. Readers, like the Tags, come in a variety of types such as the Hand-held reader and the Mounted Reader. The mobile hand-held tags are used for mainly determining which objects are present within a group, for example, when needing

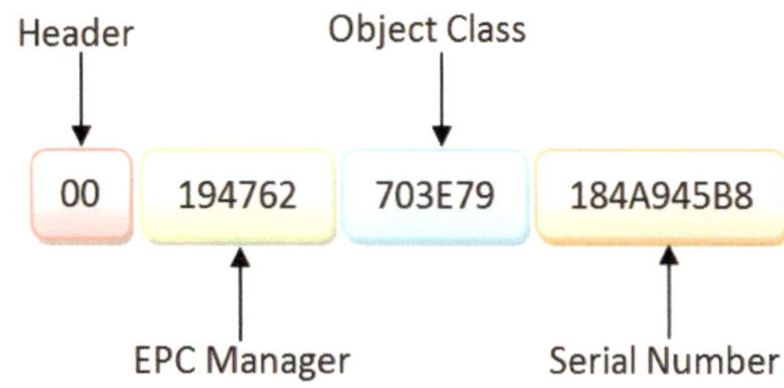

Fig. 3. The various parts of a Electronic Product Code (EPC) stored on RFID Tags.

to stocktake several items within a supermarket. In comparison, the Mounted Readers are static in geographical locations and used primarily to track items moving through their zones such as mounted readers to observe all items on a conveyer belt.

The Middleware, also commonly known as the Savant or Edge Systems, is the layer at which the raw RFID readings are cleaned and filtered to make the data more application-friendly. It receives information passed into it from the Readers and then applies techniques such as Anti-Collision and Smoothing Algorithms to correct simple missing and duplicate anomalies (Jeffery et al., 2006; Shih et al., 2006). The filtrated observational records, including the Tag and Reader Identifiers along with the Timestamp the reading was taken, are then passed onto the Database Storage.

The final destination of all the observational records is to be placed within a collection of readings taken from all connected RFID Readers. This component is known as the Database Storage and is used to hold all information which is streamed from the Readers. In most cases, due to the massive amount of interrogation undertaken to read all Tags at all times, this can result in massive floods of data, for example, 7TB of data generated daily (Schuman, 2005). Having all information stored in a central database also allows for higher level processes such as data cleaning, data mining and analytical evaluations.

EPC	Reader	Timestamp
030000E500023C000431BA3	001	2008-07-29 14:05:08.002
030000E500023C000431BA3	003	2008-07-29 14:32:12.042
030000E500023C000431BA3	002	2008-07-29 14:45:54.028
030000E500023C000431BA3	004	2008-07-29 15:02:06.029
030000E500023C000431BA3	007	2008-07-29 15:18:49.016

Table 1. A table populated with sample RFID Data containing the information of EPC, Reader and Timestamp.

2.3 Format of observations

The format of the data recorded in the database after a tag has been read consists of three primary pieces of information: the Electronic Product Code, the Reader Identifier which made the observation, and the Timestamp which contains the time the reading occurred. Table 1 contains information typically found stored in the Database Storage.

The Electronic Product Code (EPC) is a unique identification number introduced by the Auto-ID Center and given to each RFID Tag which is made up of a 96 bit, 25 character-long code containing numbers and letters. The number itself, as seen in Figure 3, is made up of a Header for 8 bits, EPC Manager for 28 bits, Object Class for 24 bits and Serial Number for 36 bits (Ward et al., 2006). Ward and Kranenburg state that a possible alternative to using the

EPC is to employ IPv6 which is the advanced version of internet addresses. These will take over the current system which is IPv4 (Ward et al., 2006). It is estimated that, since IPv6 will have 430 quintillion internet addresses as opposed to the current 4 billion address limit, there will be enough addresses for all items being tracked with RFID.
The EPC Class 1 Generation 2 is widely used in the Ultra High Frequency (UHF) range for communications at 860-960MHz. The passive RFID tag is sometime referred to as EPC Gen-2 tag, where the standards have been created by EPCGlobal (EPCGlobal, 2006), (EPCGlobal, 2005), (EPCGlobal, 2008). The most common encoding scheme with 96 bits encoding currently used includes: the General Identifier (GID-96), the Serialised Global Trade Item Number (SGTIN-96), the Serialised Shipping Container Code (SSCC-96), the Serialised Global Location Number (SGLN-96), the Global Returnable Asset Identifier (GRAI-96), the Global Individual Asset Identifier (GIAI-96), and the DoD Identifier (DoD-96).
In order to manage and monitor the traffic of RFID data effectively, the *EPC pattern* is usually used to keep the unique identifier on each of the items arranged within a specific range. The EPC pattern does not represent a single tag encoding, but rather refers to a set of tag encodings. For instance, the General Identifier (GID-96) includes three fields in addition to the 'Header' with a total of 96-bits binary value. *25.1545.[3456-3478].[778-795]* is a sample of the EPC pattern in decimal, which later will be encoded to binary and embedded onto tags. Thus, within this sample pattern, the Header is fixed to *25* and the General Manager Number is *1545*, while the Object Class can be any number between *3456* and *3478* and the Serial Number can be anything between *778* and *795*.
Within each EPC, the Uniform Resource Identifier (URI) encoding complements the EPC Tag Encodings defined for use within RFID tags and other low-level architectural components. URIs provide an information for application software to influence EPC in a way that is independent of any specific tag-level representation. The URI forms are also provided for pure identities, which contain just the EPC fields which are used to distinguish one item from another. For instance, for the EPC GID-96, the pure identity URI representation is as follows:
urn:epc:id:gid:GeneralManagerNumber.ObjectClass.SerialNumber

In this representation, the three fields GeneralManagerNumber, ObjectClass, and SerialNumber correspond to the three components of an EPC General Identifier (EPCGlobal, 2008). There are also pure identity URI forms defined for identity types corresponding to certain encodings, the URI representations corresponding to these identifiers are as shown in Table 2.

Encoding Scheme	Uniform Resource Identifier
GID	urn:epc:id:gid:GeneralManagerNumber.ObjectClass.SerialNumber
SGTIN	urn:epc:id:sgtin:CompanyPrefix.ItemReference.SerialNumber
SSCC	urn:epc:id:sscc:CompanyPrefix.SerialReference
SGLN	urn:epc:id:sgln:CompanyPrefix.LocationReference.ExtensionComponent
GRAI	urn:epc:id:grai:CompanyPrefix.AssetType.SerialNumber
GIAI	urn:epc:id:giai:CompanyPrefix.IndividualAssetReference
DoD	urn:epc:id:usdod:CAGECodeOrDODAAC.serialNumber

Table 2. The Uniform Resource Identifier encoding complements the EPC Tag Encodings defined for use within RFID tags and other low-level architectural components

An example encoding of GRAI is demonstrates as follows:
urn:epc:id:grai:0652642.12345.1234

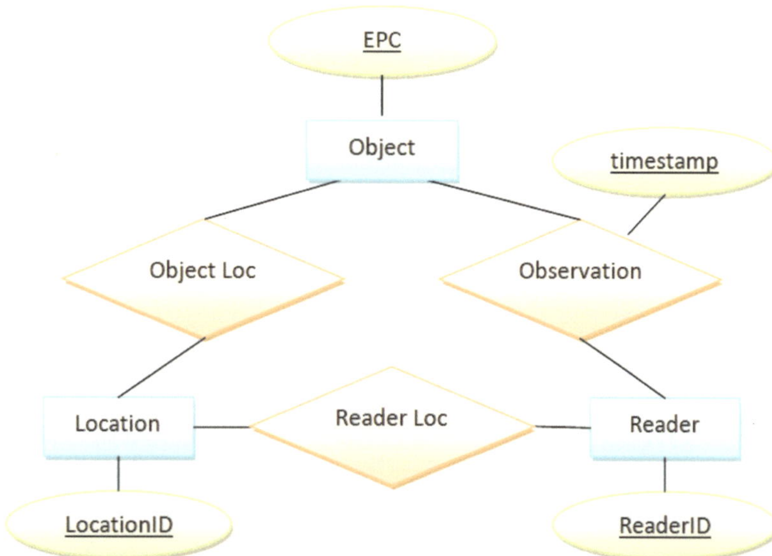

Fig. 4. An example RFID scheme which could be used to house the captured information generated from a RFID system.

From the above example, the corresponding GRAI is 06526421234581234. Refereing to Table 2, the CompanyPrefix, AssetType, and SerialNumber of GIAI are represent as 0652642, 12345, and 1234 respectively.
The Reader Identifier attribute is the unique identifier of the Reader so that the analyser will be informed of which reader took the EPC reading. If the Reader is static in its location as well, such a position of the reading may be derived from a simple query in the database later using this value. Knowledge of the geographical location of each unique Reader identifier may also provide additional information needed in future business processes.
The Timestamp contain a temporal reading used to identify the date and time that the Tag passed within vicinity of the Reader. For example, 2008-07-29 14:05:08.002 would be stored as a timestamp.

2.4 Storage of RFID data

In its rawest form, RFID data is recorded in a temporal stream of data consisting of EPC, Reader and Timestamp. After the burst of information is recorded from the reader, the RFID Savant or RFID Middleware modifies data to represent a higher level description of the events that took place. For example, the Siemens RFID Middleware extracts the data and loads it into a Dynamic Relationship Entity Relationship Model (Wang & Liu, 2005). Figure 4 depicts the Entity Relationship Diagram (ERD) used as a basic Database Storage for RFID events. As seen in the diagram, there are three prime entities that must be known, the Object, the Reader, and the Location of the Reader. Each entity has an identifying tuple attached including the Observation weak entity that also attaches the timestamp of a recorded event. Additionally, more advanced systems will only record the start and end time that an Object is within a Location, thereby saving memory so that observations are not recorded as frequently (Wang et al., 2010).

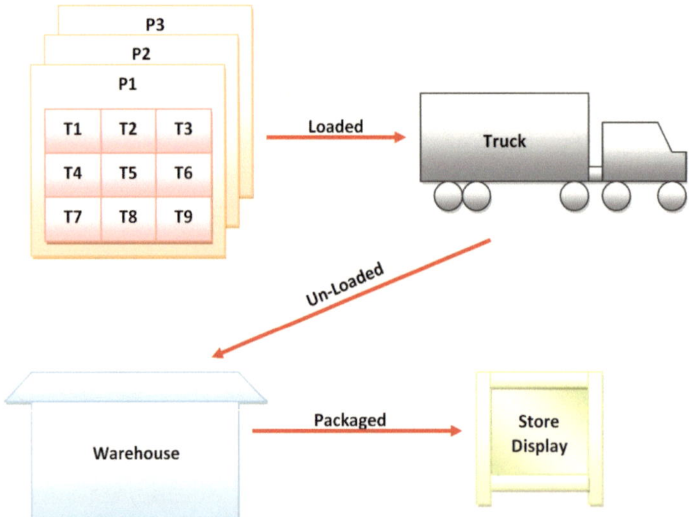

Fig. 5. The various stages taken when transporting various RFID-enabled items in a supply chain.

2.5 RFID advantages

The main advantage of RFID technology is that it is not necessary to have a line-of-sight between the object and the reading device (Derakhshan et al., 2007). In comparison to object scanners currently employed in various commercial sectors such as supermarkets, an object is needed to be taken out, place on a conveyor belt, rotated until the barcode is within the position and then placed back into the shopping trolley. If RFID is employed within this scenario, all items would automatically be recorded when the customer approaches the register and the cost tallied in one scan without the need of moving the items outside the trolley, thus saving the company time, money and physical labour. Specifically in relation to Passive Tags, there are two main advantages found when integrating RFID technology (Chawathe et al., 2004). The first is that the manufacture of the RFID Passive tag is extremely cheap. It is estimated that it only costs 5 cents per tag when bought in bulks of billions. The second advantage of the Passive RFID System is that, due to the ingenuity of the tag itself, it is not application-specific and may be applied to almost any domain. With regard to the variety of uses of RFID, as stated by Polniak - "Uses of automatic identification are manifold, limited only by one's imagination" (Polniak, 2007).

3. Current uses of RFID

From investigating the current uses of RFID, we have discovered that each utilisation may be placed into two different categories of RFID applications. The first, which we have labelled "RFID Integrated Applications", includes already existing systems which have been enhanced and made more effective and efficient using RFID technology. We have labelled the second category "RFID Specific Applications" in which prototype machines have been built from the bottom-up to incorporate RFID technology in its very make up.

3.1 Integrated RFID Applications

We have defined Integrated Applications as scenarios in which originally existing business operations have been augmented with the integration of RFID technology. The most common use of RFID integrated applications is the generic supply-chain example of RFID integration commonly employed by commercial stores such as Wal-Mart. In the example illustrated in Figure 5, tagged Objects (T1-T9) are added to specific Pallets (P1-P3), which are then loaded onto a Truck. The Truck will then transport the Pallets to their Warehouse destination at which point the items are then packaged for display at their retail stores. Additionally, as described by Derakhshan, Orlowska and Li, there are several other applications which have integrated RFID technology into their business models (Derakhshan et al., 2007) such as:

- **Defense and Military:** The US Department of Defence (DOD) is investigating a new active tag which has the ability to access and communicate via satellites. This new tag, known as the "Third Generation Radio Frequency Identification with Satellite Communications (3G RFID w/SATCOM)", is expected to be used to increase the visibility of the DOD's supply chain and, in turn, increase the confidence of shipments to various war-torn regions (Collins, 2005).
- **Postal Package Tracking:** The postal service has been found to incorporate RFID world-wide with the primary goal of increasing the effectiveness of tracking packages and parcels thereby increasing customers' property security (Harrop, 2005).
- **Aviation Industry:** Two major aircraft manufacturers, Boeing and Airbus, have started ensuring that the supplying factory parts for the aircraft use RFID tags for identifications resulting in an easier process to locate and identify needed parts (Collins, 2004).
- **Health Care:** The Taiwanese Chang-Gung Memorial Hospital has been monitoring surgical patients with RFID wristbands in order to ensure maximum care is given where needed. The features available in the wristbands include the ability to decrypt data, obtain read-only static fields (such as blood-types) and read/write dynamic fields which may be updated and modified by medical staff (Swedberg, 2005).
- **Baggage/Passanger Tracing:** The Boston Logan International Airport and the Boston Engineering Inkode Corporation have integrated RFID technology within the Secure Environment for Airport Terminal Systems (SEATS) which passengers and their baggage with passive RFID tags to track all movements from their arrival at the airport to boarding the flight (Ferguson, 2005). This technology ensures not only that passengers will be able to make their flight easier, but that their baggage location will always be known.

3.2 Specific RFID applications

We have categorised applications specifically designed and built with the integration of RFID technology as Specific Applications. Four such examples which have been developed in the recent years include the Magic Medicine Cabinet, the Multipurpose Smart Box, the Augmentation of Desktop Items and the Smart Shelves (Brusey et al., 2003; Floerkemeier, 2004).

The Magic Medicine Cabinet, as described in (Wan, 1999), is a bathroom cabinet which is used to assist in bridging the gap between the informational and physical aspects of the medical world. The Magic Medicine Cabinet will allow RFID based tracking systems to describe the content of what is being placed into and removed out of storage by the user. Through a combination of Facial Recognition, Vital Sign Monitors, Voice Synthesisers and

RFID technologies, the Cabinet can intelligently decide whether or not the person currently interacting with it should be taking the medicine. This, in turn, would being the action to the owner's attention if necessary.
As discussed in (Floerkemeier et al., 2003; Lampe & Floerkemeier, 2004), an automatic content monitoring application called the "Smart Box", similar to the Magic Medicine Cabinet, has been designed to monitor the RFID-enabled contents placed inside. The Smart Box may also be set up in different configurations to suit the context to which it will be applied such as a Smart Surgical Kit for hospitals and a Smart Toolbox for mechanics (Floerkemeier et al., 2003).
The Augmentation of Desktop Items is a means of combining physical objects with virtual interfaces using the inexpensive power of RFID tags and readers (Want et al., 1999). In a typical scenario, an office object such as a book would be tagged and then read by a Reader connected to a computer to allow the user additional functionality. For example, when someone scans a book by the reader, the computer would use stored information relating to the office to identify the book's title and would begin to provide additional internet-features such as summaries, discussions or would allow the user to order the book from Amazon.com.
The Smart Shelf is an RFID enabled device which tracks all items placed on it to accurately determine the location of the said object (Decker et al., 2003; TecO & SAP/CEC, 2003). The Smart Shelf was designed specifically with the secondary goal of obtaining the unobserved events of a person handling an item at retail outlets and, subsequently, returning it to the shelf thereby allowing business analysts further glimpses into the decision-making of the consumers. From this information, it would be possible to detect if a shopper mentally debates over the decision to purchase the product.

4. RFID issues

Before RFID can be utilised to its maximum potential, as opposed to the fraction in which it is currently employed, certain issues need to be understood by the users, and corrected if possible. The three core obstacles include the concerns of security, the problems surrounding the privacy of the data captured and the characteristics associated with the nature of RFID. Additionally, we will further examine the specific problems associated with anomalies present within the captured observational records which are regarded as a characteristics of RFID. When all of these issues are rectified to provide maximum security, privacy and integrity, RFID will be able to realise its full potential in massive wide-scale adoptions.

4.1 RFID security

The issues associated with RFID Security, also known as Intrusion Detection, refers to the discovery of foreign attacks upon the system usually utilising the tags that hinder the overall integrity of the data. The following five issues are some of the most dominant with regard to RFID security (Mitrokotsa et al., 2010; Thamilarasu & Sridhar, 2008):

- Eavesdropping: The act of setting up an additional reader to record tag data.
- Unauthorised Tag Cloning: Copying tag data onto an additional tag to gain the same privileges.
- Man-in-the-Middle (MIM) Attack: When an external object pretends to be either a tag or reader between actual tags and readers.
- Unauthorised Tag Disabling: When an external reader disables a tag not allowing it to be utilised again.

- Unauthorised Tag Manipulation: Manipulating the tag data using an external reader.

Until these security issues existing in the current architecture, it becomes difficult for facilities to employ RFID as a means of combatting unauthorised actions such as safe-guarding sensitive or expensive objects or restrict personnel access into various locations. Currently, there are techniques and approaches such as Tag Deactivation and Encryption (Karygiannis et al., 2007), Mutual Authentication (Konidala et al., 2007), Detections in Tag Ownership (Mirowski & Hartnett, 2007), Reader Analysers (Thamilarasu & Sridhar, 2008) and certain data cleaners (Darcy, Stantic, Mitrokotsa & Sattar, 2010) to reduce the difficulties associated with RFID Security.

4.2 RFID privacy

Privacy within the context of an RFID-enabled facility refers to either unknowingly releasing critical information (deriving specific knowledge or tracking meaningless data) (Langheinrich, 2009), or compiling a list of all items currently found on a person (Juels, 2006). There have been several methodologies proposed in the past to ensure maximum privacy of an individual, including the general approaches of Encrypting/Rewriting and Hiding/Blocking Tags (Langheinrich, 2009). In addition to these general solutions, there have been more specific and advanced approaches suggested such as killing/sleeping the Tags, carrying around a privacy-enforcing RFID device, releasing certain information based solely on distance from the reader and introducing Government Legislations (Juels, 2006).

4.3 RFID characteristics

There are certain characteristics associated with the nature of RFID technology (Cocci et al., 2008; Derakhshan et al., 2007). These challenges include Low Level Data, Error-Prone Data, High Data Volumes and its Spatial and Temporal Aspects. Low Level Data refers to the raw observational readings being taken by the RFID Reader; Error-Prone Data is the problem which RFID has with capturing the data; High Data Volumes refers to the ongoing obstacle with managing exponential RFID data streams and Spatial and Temporal Aspects alludes to the aspects of RFID's freedom in being capable of being used in all situations.

As previously discussed in Section 2.3, the format of the data at the time of scanning is very low level and lacks crucial information needed later for analysing the information captured. The core problem with these observations is the lack of associations between the readings and other information such as what the tags are attached to or the locations of the readers thereby making captured data useless on its own. Humans must find significant information extracted from these low level observations such as high level RFID Events (Khoussainova et al., 2007) which are the transformed state of the raw readings into meaningful milestones. For example, if a certain tag "202" is read at the reader "794" at timestamp "25/05/08 07:30:04", there is not enough information to comprehend the significance of the observation. By using relational information such as reader locations and tag information, these low level observations may be transformed into a high level event depicting the person named John being at the Front Door of location at 7:30:04 on the 25/05/08.

RFID Data integrity is constantly lowered to the point of questioning its authenticity especially when utilising passive tags due to errors captured within the observational data. These errors include Missed Reads in which a tagged item is present but not recorded, Wrong Reads in which data is captured where it should not resulting in the data set not reflecting events which are actually taking place, and duplicate reads in which a tagged item is stored twice in the

database where it should only be stored once. Section 4.4 further expands the error-prone nature of RFID where an analysis together with each of these errors are given.
Due to the continuous stream of information and the need to constantly interrogate tags, readers record massive amounts of data over long periods of time. It has been estimated that Wal*Mart currently generates about 7TB of information daily due to its RFID integration (Raskino et al., 2005). Additionally, it is estimated that by the year 2015, with a steady increase of RFID presence but lack of content management of the data generated, the information collected will be a serious problem for integrated systems. This may ultimately lead to a decrease of RFID usability and waste of information already gained unless either the management of data collected is properly attended to or the technology currently employed greatly increases its storage capacity.
As previously discussed in Derakhshan et al. (2007); Wang & Liu (2005), the exponential growth of smaller hardware RFID solutions coupled with the cost reduction in manufacturing these units results in RFID applications becoming increasingly dynamic in both spatial and temporal properties. For example, there are hand-held RFID Readers which are carried by people to scan groups of RFID tags in various locations. However, these scans will never be able to be placed into a geographical context thus limiting the potential of analytical processes that may be performed. Unless properly managed, the dynamic properties of RFID's spatial and temporal aspects may result in increasingly complex ambiguity ultimately resulting in the data losing significance, context and usability.

4.4 RFID anomalies

RFID observational data suffers from three main anomalies which are recorded with the correct RFID readings. The first is a Wrong Reading in which data is captured where it should not be. The second is Duplicate Readings in which a tag is observed twice rather than once. The third is the Missed Readings which occur when a tag is not read when and where the object it is attached to should have been physically within proximity. Figure 6 contains an example of a RFID-enabled shelf which has also generated the three anomalies, the recorded data may be seen in Table 3.

What is Recorded		
Tag EPC	**Timestamp**	**Reader ID**
T1	13/10/2010 14:31:05	R1
T2	13/10/2010 14:31:05	R3
T3	13/10/2010 14:31:05	R3
T3	13/10/2010 14:31:05	R4
T3	13/10/2010 14:31:05	R5
What is meant to be Recorded		
Tag EPC	**Timestamp**	**Reader ID**
T1	13/10/2010 14:31:05	R1
T2	13/10/2010 14:31:05	R2
T3	13/10/2010 14:31:05	R3
T4	13/10/2010 14:31:05	R5

Table 3. The recordings that took place from the example in Figure 6 and the observations that should have been recorded.

Wrong Readings, also known as Unreliable Readings or Ghost Reads falling into the False Positives category, refer to observations found in the data storage of tag which were not physically present in the location or time. These false readings may be produced when

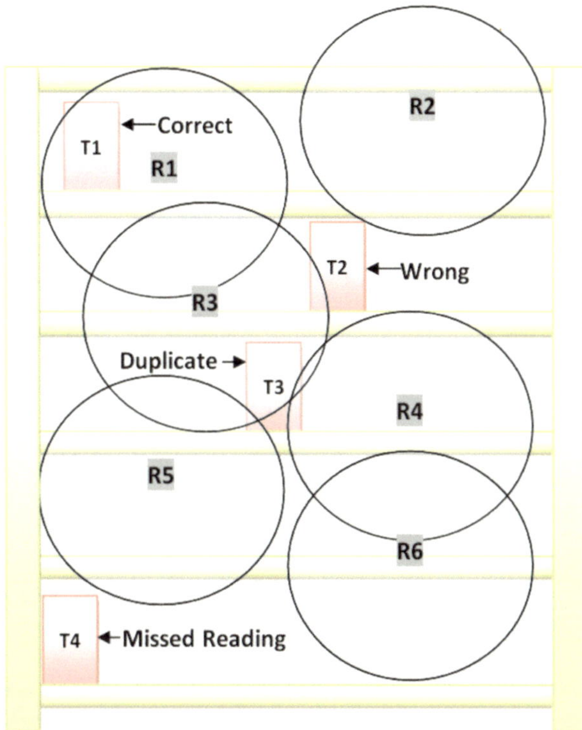

Fig. 6. A graphical representation of a RFID-enabled bookshelf with the data anomalies that may occur highlighted.

tags outside the normal Reader range are captured or where there is a problem with the environmental setup (Bai et al., 2006). As discussed previously (Embry, 2005; Engels, 2005), this problem has been identified as one of the two main technical problems with RFID. It may also result in additional unnecessary labor to continually monitor the objects where the locations of the tagged items is vital to the business process, for example, the tracking of livestock. Within the example in Figure 6 and Table 3, Tag T2 is read by Reader R3 when in reality it is closer to the area which should be scanned by Reader R2.

Duplicate Readings refer to an RFID tag which has been scanned twice in the database as opposed to just one scanning. Like the Wrong Readings, Duplicate anomalies also fall into the category of False Positive observations as they record the data which do not accurately depict reality. This may occur in several situations such as the situation in which there is more than one Reader covering an area and a tag happens to pass within overlapped region (Carbunar et al., 2005). This can be hazardous and redundant as the tag is represented in two areas during the same time period. Other duplicate reading situations occur when a scanned item stays in the reader range for a long period of time or when the owners of the RFID system attach multiple tags in order for an item to enhance its read rate (Bai et al., 2006). Ultimately, not only does this anomaly cause contradicting observations where tags may appear in two locations at the same time but it also leads to memory being wasted where it could be utilised to store factual information. In the sample scenario depicted in Figure 6 and Table 3, Tag

T3 is captured by not only the correct Reader R3, but also R4 and R5 resulting in T3 having duplicate entries in the recorded data set.
Missed Readings, also known as False Negative observations, refers to tagged objects not being scanned when, in actuality, they were present. The exact percentage of tags that are read remain only at 60%-70% under certain circumstances (Floerkemeier & Lampe, 2004). Reasons for these anomalies stem from problems such as Tag Collisions, Tag Detuning, Water/Metal Interference and Misalignment of the Tags. The missed reads anomaly has been identified as the second major problem in RFID deployment by an array of researchers (Engels, 2005; Floerkemeier & Lampe, 2004; Rahmati et al., 2007). The result of this anomaly may cause the users to believe that all items which are meant to be present are not, thereby hindering the overall process it was designed to make more efficient. Tag T4 in the example Figure 6 is shown to be a missed read due to it being placed slightly outside the scanning range of Reader R5 resulting in it not being recorded with the other tags in Table 3.

5. Current state-of-the-art approaches

In this section, we will provide a brief summary of all the current state-of-the-art approaches we have investigated to correct the RFID issues. We have divided the methods into three categories: Physical Approaches in which methods attempt to correct RFID anomalies by improving the environment around the scanners, Middleware Approaches in which algorithms attempt to correct the anomalies at the time of capturing and Deferred Approaches which attempt to correct RFID data when it is stored in the Database. Table 4 provides a list of each of the techniques examined in this section and the Corrected and Potentially Generated Anomalies.

	Methodology	Anomalies Corrected			Anomalies Generated		
		Wrong	Duplicate	Missed	Wrong	Duplicate	Missed
Physical	Tag Orientation	-	-	X	X	-	-
	Weighing	-	X	X	-	-	-
	Multiple Tags/Cycles	-	-	X	X	X	-
	Eccopad	-	-	X	-	-	-
Middleware	Edge Filtering	X	X	X	X	X	X
	Anti-Collision	-	-	X	-	-	-
	Thresholds	-	-	X[1]	-	-	-
	Statistical Approx.	X	X	X	X	X	X
Deferred	P2P Collaboration	X	X	X	-	-	-
	Proximity Detection	X	-	X	-	-	-
	Cost-Conscious Cleaning	X	X	X	X	X	X
	Data Mining Techniques	X	X	X	X	X	X
	Probabilistic Inference	-	-	X	X	-	-
	Event Transformation	X	X	X	X	X	X
	Intelligent Classifiers	X	X	X	X	X	X

Table 4. A table depicting which anomalies are corrected and generated by the various methodologies proposed. The 'X' denotes where the methodology either corrects or generates the anomaly. Note[1]: The Thresholds methodology does not actually correct the missing data but, rather, alerts the user to a False-Negative anomaly.

5.1 Physical approaches

One common solution to improve the tag reads in RFID systems is to employ Physical Approaches. This enhances the environment where the scanning is conducted. We define Physical Approaches as any solution which requires interaction with the equipment as opposed to virtual interaction used at the middleware or at a deferred warehouse stage to correct the captured data.

Rahhmati, Zhong, Hiltunen and Jana have conducted a study into the effects of reader rate when positioning the RFID tag in different positions (Rahmati et al., 2007). The research found that the Reader may scan the tags on an object most effectively when the Tag is positioned at the front.

Potdar, Hayati and Chang have formulated a novel and simple solution designed to detect missing RFID tags through the use of weights (Potdar et al., 2007). This method was created for applications in situations in which items are required to be tracked while being transported to various venues. It requires all tagged items to be weighed at the start of the transportation route. The items are then weighed at the end of the trip to determine any difference in the cargo weight. The system will determine if there are any missed reads coupled with an attempt to find any missing weight. If there are missing readings but a constant weight, the system will scan the cargo again until all items have been recorded.

As described by both (Bai et al., 2006; Vogt, 2002), a common solution to deal with RFID anomalies is to either install multiple readers or to attach multiple tags. Multiple readers are installed in the environment in an attempt to enhance the reader rate by covering a more substantial amount of ground. Another method of dealing with the enhancement of the read rate is to attach multiple tags housing identical EPC numbers to the same object in an effort for at least one of these tags to be read by the reader. Unfortunately, drawbacks arise from both methods in the form of duplicate readings and tag collision occurrences.

Emerson & Cuming Microwave Products (Emerson & Cuming Microwave Products, 2008) provides a device known as the Eccopad which is designed to enhance the reading rate of tags placed on metal. As described in (Floerkemeier & Lampe, 2004), metallic objects within a certain proximity will affect the reading potential of a passive RFID tag causing missed readings. The Eccopad insulates the RFID tag in a discrete manner which enables the maximum potential reading rate with little or no change in the spatial properties occurring.

5.2 Middleware approaches

Middleware Approaches refers to employing an algorithm to eliminate anomalies found in systems to correct the data before storing it. This can refer to any program used at the middleware stage of the RFID capture cycle to correct the raw incoming streamed data.

Edge Filtering refers to the cleaning being completed at the edge of the RFID system, that is, at the point of raw observations being read. Jeffery, Garofalakis and Franklin have proposed a method analysing the usage of an adaptive sliding window to correct unreliable readings within an RFID system (Jeffery et al., 2006). A sliding window is used to smooth out the raw data in order to accommodate both false positive and false negative readings. The problem associated with this technique is that the result of utilising a small sliding window will be false negatives whereas the large window may result in false positives being introduced. Thus, Jeffery et al, proposed a solution to create a declarative and adaptive smoothing window named SMURF (Statistical sMoothing for Unreliable RFid data) which they have continually improved (Jeffery et al., 2008).

RFID Anti-collision protocols are algorithms used at the edge to avoid missed readings. When an RFID scan is performed on several RFID tags, there are many relaying messages sent back and forth between the tag and the reader. If there are a large number of tags to be scanned in a certain read, these messages may collide in the air between their source and destination resulting in the information not arriving at the correct time if at all. Certain protocols are also designed to handle other forms of hazards such as instances where readers placed within a certain proximity interfere with each other's interrogation cycle causing collisions (Shih et al., 2006).

The various types of anti-collision methods for *collision* can be reduced to two basic types: *probabilistic* and *deterministic* methods. In a probabilistic method, tags respond at randomly generated times. If a collision occurs, colliding tags will have to identify themselves again after waiting for a random period time frame. From past literature, there have been several methods proposed such as: Basic Framed-Slotted ALOHA (Lee et al., 2008); Dynamic Framed-Slotted ALOHA (Ding & Liu, 2009); Enhanced Dynamic Framed-slotted ALOHA (Lee & Lee, 2006); and Probabilistic Cluster-Based Technique (Pupunwiwat & Stantic, 2010d), to enhance the performance efficiency of the data capturing process. In addition, several *Frame Estimation* approaches have been suggested to improve the accuracy of *frame-size* prediction including the Schoute method (Schoute, 1983), the Lowerbound method, the Chen1 and Chen2 methods (Chen, 2006), the Vogt method (Vogt, 2002), the Bayesian method (Floerkemeier, 2007), and the Precise Tag Estimation Scheme (Pupunwiwat & Stantic, 2010b), (Pupunwiwat & Stantic, 2010a).

The deterministic method operates by asking for the first EPC string of the tag until it gets matches for the tags, it will then continues to ask for additional characters until all tags within the region are found. There have been several methods proposed in literature in order to improved quality of the captured data such as: the Query Tree (Myung & Lee, 2006a); the Adaptive Splitting Tree (Myung & Lee, 2006b); the Hybrid Query Tree (Ryu et al., 2007); and the Joined Q-ary Tree (Pupunwiwat & Stantic, 2009), (Pupunwiwat & Stantic, 2010c).

Tan, Sheng and Li have proposed in their research the utilisation of a threshold to identify an excessive amount of missing RFID readings (Tan et al., 2008). By using two different protocols, the trusted reader and un-trusted reader protocols, the methodology analyses a RFID data set and finds missing data without the need for ascertaining tag identifiers. The system will then consult a threshold defined by the owner as to the number of missing tags which are tolerable in a given situation with the system alerting the user if this threshold is breached. It will not however replace the missed readings.

Statistical Approximations refer to the use of a Model-Based Querying system to return approximate readings found from the sensor networks (Deshpande et al., 2004; Deshplande et al., 2005). Although this method is not used primarily for RFID technology, the method is applied to wireless sensors which provide additional functionality that RFID tags do not (i.e. Temperature Sensors). This approach is designed to capture a query from the User, find the values from the sensor readings, and return approximate values to the User.

5.3 Deferred approaches

We have defined Deferred Approaches as methodologies applied at a deferred stage of the capturing cycle when the observational data is stored in the database. This includes P2P Networks, Probabilistic Tag Proximity Detection, Cost-Conscious Cleaning, Data Mining Techniques, Probabilistic Inference and Probabilistic Event Extraction.

The P2P Collaboration method, proposed by Peng, Ji, Luo, Wong and Tan (Peng et al., 2008), is an approach utilising Peer-to-Peer (P2P) networks within the RFID data set to detect and remove inaccurate readings. The system works by breaking the readings into detection nodes, which are constantly sending and receiving messages. From these transmitted messages, false negatives and false positives are able to be detected and corrected resulting in a cleaner data set.

Ziekow and Ivantysynova have presented a method designed to correct RFID anomalies probabilistically by employing maximum likelihood operations (Ziekow & Ivantysynova, 2008). Their method utilises the position of a tag which may be determined by measuring properties associated with the Radio Frequency signal.

The Cost-Conscious cleaning method is a cleaning algorithm which utilises a Bayesian Network to judge the likelihood that read tags correctly depict reality when based upon the previously read tags (Gonzalez et al., 2007). The Cost-Conscious cleaning approach houses several different cleaning algorithms and chooses the least costly algorithm which would offer the highest precision in correcting the raw data. A similar approach has also been proposed that utilises a Bayesian Network to judge the existence of tags scanned (Floerkemeier, 2004). It lacks, however, the cost-saving analysis that would increase the speed of the clean.

Data Mining Techniques refer to the use of mining past data to detect inaccuracies and possible solutions to raw RFID readings. A study which has used data mining techniques extensively to correct the entire data set table is the Deferred Rule Based Approach proposed in (Rao et al., 2006). The architecture of the system is reliant on the user defining rules which are utilised to determine anomalies in the data set and, possibly, to correct them.

Probabilistic Inference refers to a process by which the in-coming data node will be evaluated. This is primarily based upon the weight of its likelihood and the weight of the remainder of the readings (Cocci et al., 2007; 2008). The cleaning algorithm utilises several techniques to correct that data such as Deduplication, Time conversion, Temporal Smoothing and Anomaly Filtering, and, additionally, uses a graph with probabilistic weights to produce further inferences on the data.

Probabilistic High Level Event Transformations refers to the process of observing the raw partial events of RFID data and transforming these into high level probable events. It has been primarily used in a program entitled Probabilistic Event EXtractor (PEEX) which has evolved from several publications. In its embryonic phase, Khoussainova, Balazinska and Suciu published a paper detailing the use of an algorithm called StreamClean which employ probabilistic inference to correct incoming data (Khoussainova et al., 2006).

A year after this article, the first papers for PEEX were published. This described the method which enabled high level event extraction based upon probabilistic observations (Khoussainova et al., 2007; Khoussainova, Balazinska & Suciu, 2008). The system architecture deciphers the raw RFID information searching for evidence which a high level event transpired. The system uses a Confidence Learner, History Lookup and Event Detector to enhance the reliability of the returned events. By transferring these low level readings into high level events, PEEX engages in cleaning as the process of probabilistically by categorising the results of these events, and in the process, caters for missed and inaccurate readings.

Currently, PEEX is being incorporated into a new a system named Cascadia where it will be utilised to help perform high level management of RFID tracking in a building environment (Khoussainova, Welbourne, Balazinska, Borriello, Cole, Letchner, Li, Ré, Suciu & Walke, 2008; Welbourne et al., 2008). Bayesian Networks have also been implemented in several studies to infer high level behaviour from the raw readings. The specific application was first demonstrated on a traveller moving through an urban environment (Patterson et al., 2003) and the second using RFID tags to track the activities of daily living (Philipose et al., 2004). In previous work, we have proposed the concept of using high level classifiers coupled with intelligent analysis to correct the various anomalies found in RFID data. First, we examined the potential of employing a simple algorithm that corrects a simple missed reading (Darcy et al., 2007). We then proposed the utilisation of highly intelligent analytical processes coupled with a Bayesian Network (Darcy et al., 2009b;c), Neural Network (Darcy, Stantic & Sattar, 2010a) and Non-Monotonic Reasoning (Darcy et al., 2009a; Darcy, Stantic & Sattar, 2010b) to correct missing RFID Data. Following this, we applied our Non-Monotonic Reasoning approach to both false-negative and false-positive data anomalies (Darcy, Stantic & Sattar, 2010d). We then also introduced a concept to extract high level events from low level readings using Non-Monotonic Reasoning (Darcy, Stantic & Sattar, 2010c). Finally, we proposed a methodology that considers and differentiates between a false-positive anomaly and breach in security using Non-Monotonic Reasoning (Darcy, Stantic, Mitrokotsa & Sattar, 2010).

6. Drawbacks and proposed solutions for current approaches

In this section, we highlight several drawbacks we have found associated with the various methodologies currently employed to correct RFID captured data. We also supply our suggested solutions to these problems where possible in an effort to encourage further interest in this field of research. Finally, we conclude with an overall analysis of these methodologies and their respective drawbacks.

6.1 Physical drawbacks and solutions

With regard to Physical Approaches, we have highlighted three main drawbacks and our suggested solutions to correct these issues where possible:

- **Problem:** The main problem that we foresee with the utilisation of Physical Approaches is that it usually only increases the likelihood that the missed objects will be found.
  Solution: We do not have a solution to the problem of physically correcting wrong or duplicate anomalies other than suggesting to utilise Middleware and/or Deferred solutions.
- **Problem:** Physical Approaches generates artificial duplicate anomalies in the event that all the tags attached are read.
  Solution: Specific software tailored to the application to automatically account for the artificially generated duplicate anomalies could be used for correction filtering at the edge.
- **Problem:** Physical Approaches suffer from additional cost to the user or more labour to purchase extra tags, equipment or time to move the objects.
  Solution: We do not believe there is a solution to this as Physical Approaches demand additional labour for the user to correct the mistakes as opposed to Middleware or Deferred Approaches.

6.2 Middleware drawbacks and solutions

We found three major drawbacks to the Middleware Approaches that prevent these from acquiring their maximum integrity. These issues include:

- **Problem:** Correcting incoming data at the edge of the RFID capture process will not provide the cleaning algorithm with adequate information needed to deal with highly

ambiguous and complex anomalies.
Solution: We believe that to correct this drawback, the user must employ a Deferred methodology in addition to the Middleware Approach to utilise all stored readings. This would result in more observational data eliminating highly ambiguous anomalies.

- **Problem:** When utilising probabilistic algorithms such as Bayesian Networks to correct anomalies, there is a risk of the methodology introducing artificially generated anomalies. This may occur in cases such as the training set not reflecting the reality of the scenarios or the system probabilistically choosing the incorrect action to take in a situation.
Solution: To correct this issue, the user may be able combine various probabilistic techniques together or to employ a deterministic approach in order to enhance the method of cleaning the database.

- **Problem:** RFID data streams that are captured by readers can be accumulated quickly resulting in data collisions. Simultaneous transmissions in RFID systems will also lead to collisions as the readers and tags typically operate on the same channel. There are three types of collisions possible to occur: Reader-Tag collision, Tag-Tag collision, and Reader-Reader collision.
Solution: It is crucial that the RFID system must employ *anti-collision* protocols in readers in order to enhance the integrity of the captured data. However, the step of choosing the right *anti-collision* protocol is also very important, since we cannot depend solely on the capability of anti-collision protocol itself, but also on the suitability of each selected technique for the specific scenario. The user may employ decision making techniques such as both the Novel Decision Tree and the Six Thinking Hats strategy for complex selective technique management to determine the optimal anti-collision protocol. The novelty of using complex selective technique management is that we will get the optimal outcome of *anti-collision* method for the specific scenario. This will, in turn, improve the quality of the data collection. It will also help over long period of use when these captured data are needed for transformation, aggregation, and event processing.

6.3 Deferred drawbacks and solutions

While reviewing the Deferred Approaches to correct RFID anomalies, we have discovered that there are certain shortcomings when attempting to clean captured observational data.

- **Problem:** Similar to the Middleware Approaches which utilise probabilistic calculations, a major problem in the Deferred Approaches is that due to the nature of probability, false positive and negatives may be unintentionally introduced during cleaning.
Solution: As stated previously, the inclusion of multiple probabilistic techniques or even deterministic approaches should increase the intelligence of the methodology to block artificial anomalies from being generated.

- **Problem:** Specifically with regard to the Data Mining technique, it relies on the order the rules appear as opposed to using any intelligence to decipher the correct course of action.
Solution: It is necessary to increase the intelligence of the order of the rule order by integrating high level probabilistic or deterministic priority systems.

- **Problem:** With regard to the Cost-Conscious Cleaning method, due to the fact that the method only utilises immediate previous readings and focuses on finding the least costly algorithm, accuracy may be lowered to ensure the most cost-effective action.
Solution: In the event that this algorithm is applied at a Deferred stage, it will not require

the data to be corrected as fast as possible. Therefore in this situation, the emphasis on cost-effectiveness is not relevant as is usually the case and other actions could be examined to derive the highest accuracy.

- **Problem:** As a general constraint of all Deferred Approaches, it is necessary to apply the correction algorithm at the end of the capture cycle when the data is stored in the Database. The main problem with this characteristic is that the methodologies will never be able to be applied as the data is being captured and, therefore, cannot correct in real-time.
  Solution: As most of the Deferred Approaches, especially the Data Mining and Highly Intelligent Classifier, requires certain observational data to correct anomalies, we propose the use of a buffering system that runs as the data is being captured and takes snapshots of the read data to correct any anomalies present. Unfortunately, due to the need that the methodology is run in real-time, it may not be able to include all the complexities of the current Deferred Approaches such as dynamic training of the classifiers.

6.4 Drawback analysis

In this research, we evaluated the current state-of-the-art approaches designed to correct the various anomalies and issues associated with RFID technology. From our findings, we have found that, while Physical Approaches do increase the chances of a tag being captured, it does generate duplicate anomalies and places cost in both time and labour onto the user that may not be beneficial. With regard to Middleware Approaches, we found that most anomalies are corrected through these techniques. However, due to the limited scope of information available, the more complex procedures such as dealing with highly ambiguous errors or transforming the raw observations into high-level events is not possible. In contrast, Deferred Approaches have an advantage to correct highly ambiguous anomalies and transform events. Its main issue, however, is not being available to process the observational information in real-time limiting its cleaning to a period after the records have been stored.

Overall, we have found from our research that a truly robust RFID system that eliminates all possible natural and artificial anomalies generated will require the integration of most approaches we have recognised. For example, various real-time anomalies are best filtered at the edge while increasingly ambiguous anomalies can only be corrected at a deferred stage of the capture cycle. Additionally, we found that there is a need to, not only employ probabilistic techniques, but also deterministic where possible as it theoretically should reduce the artificial anomalies produced. We, therefore, recommend the inclusion of all methods where possible, at least one of the Middleware and Deferred categories, and, where applicable, the inclusion of both deterministic and probabilistic techniques.

7. Conclusion

In this study, we have examined RFID technology and its current uses in various applications. We have also examined the three various issues among the integration of the systems including security, privacy and data abnormalities. Furthermore, we have examined the data abnormality issue to find that four problems exist including low-level nature, large intakes, data anomalies and complex spatial and temporal aspects. There have been various methodologies proposed in the past to address the various problems in the data abnormalities categorised into physical, middleware and deferred solutions. Unfortunately, due the various drawbacks such as application-specified solutions, lack of analytical information or reliance

on user-specified/probabilistic algorithms, current approaches do not provide the adequate support needed in RFID systems to be adopted in commercial sectors.
Specifically, we contributed the following to the field of RFID study:

- We provided a detailed survey of RFID technology including how it was developed, its various components and the advantages of integrating its technology into business operations.
- We highlighted the current usages of RFID categorising it into either "Integrated RFID Applications" and "Specific RFID Applications".
- We examined the various issues preventing the adoption of RFID technology including the concerns of security, privacy and characteristics. We also focused on the specific Anomalies generated by the capturing hardware including wrong, duplicate and missing errors.
- After examining the issues surrounding RFID, we investigated the state-of-the-art approaches currently employed for correction. We categorised these methodologies into Physical, Middleware or Deferred Approaches.
- Finally, we explored the drawbacks found in currently employed Approaches and suggested several solutions in the hope of generating interest in this field of study.

With regard to future work, we specifically would like to extend our previous studies discussed in Section 5.3 by allowing it to function in real-time. We would do this through the creation of a buffer system discussed in Section 6.3 by taking snapshots of incoming data and correcting anomalies where found. We also firmly believe that this sincerely is the next step of evolution of our approach to allow it to be employed as the observational records are read into the Middleware.

8. References

Bai, Y., Wang, F. & Liu, P. (2006). Efficiently Filtering RFID Data Streams, *CleanDB*, pp. 50–57.

Brusey, J., Floerkemeier, C., Harrison, M. & Fletcher, M. (2003). Reasoning About Uncertainty in Location Identification with RFID, Workshop on Reasoning with Uncertainty in Robotics at IJCAI.

Carbunar, B., Ramanathan, M. K., Koyuturk, M., Hoffmann, C. & Grama, A. (2005). Redundant Reader Elimination in RFID Systems, *SECON*.

Chawathe, S. S., Krishnamurthy, V., Ramachandran, S. & Sarma, S. E. (2004). Managing RFID Data, *VLDB*, pp. 1189–1195.

Chen, W. T. (2006). An Efficient Anti-Collision Method for Tag Identification in a RFID System, *IEICE Transactions* 89-B(12): 3386–3392.

Cocci, R., Diao, Y. & Shenoy, P. (2007). SPIRE: Scalable Processing of RFID Event Streams, *5th RFID Academic Convocation*.

Cocci, R., Tran, T., Diao, Y. & Shenoy, P. J. (2008). Efficient Data Interpretation and Compression over RFID Streams, *ICDE*, IEEE, pp. 1445–1447.

Collins, J. (2004). Boeing Outlines Tagging Timetable, RFID Journal. Available from: <`http://www.rfidjournal.com/article/view/985/1/1`>.

Collins, J. (2005). DOD Tries Tags That Phone Home, RFID Journal. Available from: <`http://www.rfidjournal.com/article/articleview/1458/1/1/`>.

Darcy, P., Stantic, B. & Derakhshan, R. (2007). Correcting Stored RFID Data with Non-Monotonic Reasoning, *Principles and Applications in Information Systems and Technology (PAIST)* 1(1): 65–77.

Darcy, P., Stantic, B., Mitrokotsa, A. & Sattar, A. (2010). Detecting Intrusions within RFID Systems through Non-Monotonic Reasoning Cleaning, *Intelligent Sensors, Sensor Networks and Information Processing (ISSNIP 2010)*, pp. 257–262.

Darcy, P., Stantic, B. & Sattar, A. (2009a). A Fusion of Data Analysis and Non-Monotonic Reasoning to Restore Missed RFID Readings, *Intelligent Sensors, Sensor Networks and Information Processing (ISSNIP 2009)*, pp. 313–318.

Darcy, P., Stantic, B. & Sattar, A. (2009b). Augmenting a Deferred Bayesian Network with a Genetic Algorithm to Correct Missed RFID Readings, *Malaysian Joint Conference on Artificial Intelligence (MJCAI 2009)*, pp. 106–115.

Darcy, P., Stantic, B. & Sattar, A. (2009c). Improving the Quality of RFID Data by Utilising a Bayesian Network Cleaning Method, *Proceedings of the IASTED International Conference Artificial Intelligence and Applications (AIA 2009)*, pp. 94–99.

Darcy, P., Stantic, B. & Sattar, A. (2010a). Applying a Neural Network to Recover Missed RFID Readings, *Australasian Computer Science Conference (ACSC 2010)*, pp. 133–142.

Darcy, P., Stantic, B. & Sattar, A. (2010b). Correcting Missing Data Anomalies with Clausal Defeasible Logic, *Advances in Databases and Information Systems (ADBIS 2010)*, pp. 149–163.

Darcy, P., Stantic, B. & Sattar, A. (2010c). Intelligent High-Level RFID Event Transformation Utilising Non-Monotonic Reasoning, *Wireless Communications Networking and Mobile Computing (WiCOM), 2010 6th International Conference on*, pp. 1–4.

Darcy, P., Stantic, B. & Sattar, A. (2010d). X-CleLo: Intelligent Deterministic RFID Data Transformer, *International Workshop on RFID Technology*, pp. 59–68.

Decker, C., Kubach, U. & Beigl, M. (2003). Revealing the Retail Black Box by Interaction Sensing, *ICDCS Workshops*, pp. 328–333.

Derakhshan, R., Orlowska, M. E. & Li, X. (2007). RFID Data Management: Challenges and Opportunities, *RFID 2007*, pp. 175 – 182.

Deshpande, A., Guestrin, C., Madden, S., Hellerstein, J. M. & Hong, W. (2004). Model-Driven Data Acquisition in Sensor Networks, *VLDB*, pp. 588–599.

Deshplande, A., Guestrin, C., Madden, S., Hellerstein, J. & Hong, W. (2005). Model-based Approximate Querying in Sensor Networks, *VLDB*, pp. 417–443.

Ding, J. & Liu, F. (2009). Novel Tag Anti-Collision Algorithm with Adaptive Grouping, *Wireless Sensor Network (WSN)* 1(5): 475–481.

Embry, W. (2005). Are you ready for RFID? The promise and challenges of implementing Radio Frequency Identification (RFID) systems across the extended supply chain, *Technical report*, SAS.

Emerson & Cuming Microwave Products (2008). Emerson & Cuming Microwave Products – ECCOPAD [online], Emerson & Cuming Microwave Products. Available from: <http://www.eccosorb.com/europe/english/page/66/eccopad>.

Engels, D. W. (2005). On Ghost Reads in RFID Systems, *Technical Report AUTOIDLABS-WP-SWNET-010*, Auto-ID Labs.

EPCGlobal (2005). EPC Radio-Frequency Identity Protocols Class-1 Generation-2 UHF RFID Protocol for Communications at 860MHz-960MHz Version 1.0.9. Available from: <http://www.epcglobalinc.org/standards/>.

EPCGlobal (2006). EPCGlobal Tag Data Standards Version 1.3: Ratified Specification. Available from: <http://www.epcglobalinc.org/standards/tds/>.

EPCGlobal (2008). EPCGlobal Tag Data Standards Version 1.4: Ratified Specification. Available from: <http://www.epcglobalinc.org/standards/tds/>.

Ferguson, R. B. (2005). Logan airport to demonstrate baggage, passenger rfid tracking, eWeek. Available from: <http://www.eweek.com/c/a/Mobile-and-Wireless/Logan-Airport-to-Demonstrate-Baggage-Passenger-RFID-Tracking/>.

Floerkemeier, C. (2004). A Probabilistic Approach to Address Uncertainty in RFID, *Auto-ID Labs Research Workshop*.

Floerkemeier, C. (2007). Bayesian Transmission Strategy for Framed ALOHA Based RFID Protocols, *RFID, 2007. IEEE International Conference on RFID Gaylord Texan Resort*, Grapevine, TX, USA, pp. 228–235.

Floerkemeier, C. & Lampe, M. (2004). Issues with RFID Usage in Ubiquitous Computing Applications, *in* A. Ferscha & F. Mattern (eds), *Pervasive Computing: Second International Conference, PERVASIVE 2004*, pp. 188–193.

Floerkemeier, C., Lampe, M. & Schoch, T. (2003). The Smart Box Concept for Ubiquitous Computing Environments, *Proceedings of sOc'2003 (Smart Objects Conference)*, pp. 118–121.

Gonzalez, H., Han, J. & Shen, X. (2007). Cost-Conscious Cleaning of Massive RFID Data Sets, *ICDE*, pp. 1268–1272.

Harrop, P. (2005). RFID in the Postal Service, MoreRFID. Available from: <http://www.infowars.com/articles/big_brother/rfid_trackers/rfid_in_the_postal_service.html>.

Jeffery, S. R., Franklin, M. J. & Garofalakis, M. N. (2008). An Adaptive RFID Middleware for Supporting Mtaphysical Data Independence, *VLDB Journal* 17(2): 265–289.

Jeffery, S. R., Garofalakis, M. N. & Franklin, M. J. (2006). Adaptive Cleaning for RFID Data Streams, *VLDB*, pp. 163–174.

Juels, A. (2006). RFID Security and Privacy: a Research Survey, *Selected Areas in Communications, IEEE Journal on* 24(2): 381–394.

Karygiannis, T., Eydt, B., Barber, G., Bunn, L. & Phillips, T. (2007). Guidelines for Securing Radio Frequency Identification (RFID) Systems. National Institute of Standards and Technology.

Khoussainova, N., Balazinska, M. & Suciu, D. (2006). Towards Correcting Input Data Errors Probabilistically using Integrity Constraints, *MobiDE*, pp. 43–50.

Khoussainova, N., Balazinska, M. & Suciu, D. (2007). Probabilistic RFID Data Management, *UW CSE Technical Report UW-CSE-07-03-01* .

Khoussainova, N., Balazinska, M. & Suciu, D. (2008). PEEX: Extracting Probabilistic Events from RFID Data, *International Conference on Data Engineering (ICDE)*, pp. 1480–1482.

Khoussainova, N., Welbourne, E., Balazinska, M., Borriello, G., Cole, G., Letchner, J., Li, Y., Ré, C., Suciu, D. & Walke, J. (2008). A Demonstration of Cascadia through a Digital Diary Application, *SIGMOD '08: Proceedings of the 2008 ACM SIGMOD International Conference on Management of Data*, pp. 1319–1322.

Konidala, D. M., Kim, Z. & Kim, K. (2007). A simple and cost-effective RFID tag-reader mutual authentication scheme, *International Conference on RFID Security (RFIDSec)'07*, pp. 141–152.

Lampe, M. & Floerkemeier, C. (2004). The Smart Box Application Model, *In: Alois Ferscha, Horst Hoertner, Gabriele Kotsis (Eds.): Advances in Pervasive Computing, Austrian Computer Society (OCG)*, pp. 351–356.

Landt, J. (2001). *Shrouds of Time The history of RFID*, The Association for Automatic Identification and Data Capture Technologies.

Landt, J. (2005). The history of RFID, *Potentials, IEEE* 24(4): 8–11.

Langheinrich, M. (2009). A Survey of RFID Privacy Approaches, *Personal Ubiquitous Comput.* 13: 413–421.

Lee, J. G., Hwang, S. J. & Kim, S. W. (2008). Performance Study of Anti-collision Algorithms for EPC-C1 Gen2 RFID Protocol, *Information Networking. Towards Ubiquitous Networking and Services*, Vol. 5200/2008, Springer Berlin/Heidelberg, Estoril, Portugal, pp. 523–532.

Lee, S. R. & Lee, C. W. (2006). An Enhanced Dynamic Framed Slotted ALOHA Anti-collision Algorithm , *Emerging Directions in Embedded and Ubiquitous Computing*, Springer Berlin/Heidelberg, Seoul, Korea, pp. 403–412.

Mirowski, L. & Hartnett, J. (2007). Deckard: A System to Detect Change of RFID Tag Ownership, *International Journal of Computer Science and Network Security* 7(7): 89–98.

Mitrokotsa, A., Rieback, S. & Tanenbaum, A. (2010). Classifying RFID Attacks and Defenses, *Special Issue on Advances in RFID Technology, Information Systems Frontiers* pp. 491–505.

Myung, J. & Lee, W. (2006a). Adaptive binary splitting: a RFID tag collision arbitration protocol for tag identification, *Mob. Netw. Appl.* 11(5): 711–722.

Myung, J. & Lee, W. (2006b). Adaptive splitting protocols for RFID tag collision arbitration, *MobiHoc '06: Proceedings of the 7th ACM international symposium on Mobile ad hoc networking and computing*, ACM, New York, NY, USA, pp. 202–213.

Patterson, D. J., Liao, L., Fox, D. & Kautz, H. A. (2003). Inferring High-Level Behavior from Low-Level Sensors, *Ubicomp*, pp. 73–89.

Peng, X., Ji, Z., Luo, Z., Wong, E. C. & Tan, C. J. (2008). A P2P Collaborative RFID Data Cleaning Model, *The 3rd International Conference on Grid and Pervasive Computing GPC 2008*, pp. 304–309.

Philipose, M., Fishkin, K. P., Perkawitz, M., Patterson, D. J., Fox, D., Kautz, H. & Hahnel, D. (2004). Inferring Activities from Interactions with Objects, *Pervasive Computing* pp. 10–17.

Polniak, S. (2007). *The RFID Case Study Book RFID application stories from around the globe*, Abhisam Software.

Potdar, V., Hayati, P. & Chang, E. (2007). Improving RFID Read Rate Reliability by a Systematic Error Detection Approach, *RFID Eurasia, 2007 1st Annual*, pp. 1–5.

Preradovic, S., Balbin, I., Karmakar, N. & Swiegers, G. (2008). A Novel Chipless RFID System Based on Planar Multiresonators for Barcode Replacement, *RFID, 2008 IEEE International Conference on*, pp. 289–296.

Preradovic, S. & Karmakar, N. (2009). Design of Short Range Chipless RFID Reader Prototype, *Intelligent Sensors, Sensor Networks and Information Processing (ISSNIP 2009)*, pp. 307–312.

Pupunwiwat, P. & Stantic, B. (2009). Unified Q-ary Tree for RFID Tag Anti-Collision Resolution, *in* A. Bouguettaya & X. Lin (eds), *Twentieth Australasian Database Conference (ADC 2009)*, Vol. 92 of *CRPIT*, ACS, Wellington, New Zealand, pp. 47–56.

Pupunwiwat, P. & Stantic, B. (2010a). A RFID Explicit Tag Estimation Scheme for Dynamic Framed-Slot ALOHA Anti-Collision, *Wireless Communications, Networking and Mobile Computing (WiCOM 2010)*, IEEE, Chengdu, China, pp. 1–4.

Pupunwiwat, P. & Stantic, B. (2010b). Dynamic Framed-Slot ALOHA Anti-Collision using Precise Tag Estimation Scheme, *in* H. T. Shen & A. Bouguettaya (eds), *Twenty-First Australasian Database Conference (ADC 2010)*, Vol. 104 of *CRPIT*, ACS, Brisbane, Australia, pp. 19–28.

Pupunwiwat, P. & Stantic, B. (2010c). Joined Q-ary Tree Anti-Collision for Massive Tag Movement Distribution, *in* B. Mans & M. Reynolds (eds), *Thirty-Third Australasian Computer Science Conference (ACSC 2010)*, Vol. 102 of *CRPIT*, ACS, Brisbane, Australia, pp. 99–108.

Pupunwiwat, P. & Stantic, B. (2010d). Resolving RFID Data Stream Collisions using Set-Based Approach, *The Sixth International Conference on Intelligent Sensors, Sensor Networks and Information Processing (ISSNIP 2010)*, IEEE, Brisbane, Australia, pp. 61–66.

Rahmati, A., Zhong, L., Hiltunen, M. & Jana, R. (2007). Reliability Techniques for RFID-Based Object Tracking Applications, *DSN '07: Proceedings of the 37th Annual IEEE/IFIP International Conference on Dependable Systems and Networks*, IEEE Computer Society, Washington, DC, USA, pp. 113–118.

Rao, J., Doraiswamy, S., Thakkar, H. & Colby, L. S. (2006). A Deferred Cleansing Method for RFID Data Analytics, *VLDB*, pp. 175–186.

Raskino, M., Fenn, J. & Lenden, A. (2005). Extracting Value From the Massively Connected World of 2015, *Technical Report G00125949*, Gartner Research.

Ryu, J., Lee, H., Seok, Y., Kwon, T. & Choi, Y. (2007). A Hybrid Query Tree Protocol for Tag Collision Arbitration in RFID systems, *MobiHoc '06: Proceedings of the 7th ACM international symposium on Mobile ad hoc networking and computing*, IEEE Computer Society, Glasgow, UK, pp. 5981–5986.

Schoute, F. C. (1983). Dynamic Frame Length ALOHA, *IEEE Transactions on Communications* 31(4): 565–568.

Schuman, E. (2005). Will Users Get Buried Under RFID Data?, Enterprise Technology News and Reviews. Available from: <http://www.eweek.com/c/a/Enterprise-Applications/>.

Shih, D.-H., Sun, P.-L., Yen, D. C. & Huang, S.-M. (2006). Taxonomy and Survey of RFID Anti-Collision Protocols, *Computer Communications* 29(11): 2150–2166.

Stockman, H. (1948). Communication by Means of Reflected Power, *Insistute of Radio Engineers (IRE)*, pp. 1196–1204.

Swedberg, C. (2005). Hospital Uses RFID for Surgical Patients, RFID Journal. Available from: <http://www.rfidjournal.com/article/articleview/1714/1/1/>.

Tan, C. C., Sheng, B. & Li, Q. (2008). How to monitor for missing RFID tags, *IEEE ICDCS*, Beijing, China, pp. –.

TecO & SAP/CEC (2003). Projects - SmartShelf, TecO. Available from: <http://www.teco.edu/projects/smartshelf//>.

Thamilarasu, G. & Sridhar, R. (2008). Intrusion Detection in RFID Systems, *Military Communications Conference, 2008. MILCOM 2008. IEEE*, pp. 1 –7.

Vogt, H. (2002). Efficient Object Identification with Passive RFID Tags, *Pervasive '02: Proceedings of the First International Conference on Pervasive Computing*, pp. 98–113.

Wan, D. (1999). Magic Medicine Cabinet: A Situated Portal for Consumer Healthcare, *Handheld and Ubiquitous Computing*, pp. 352–355.

Wang, F. & Liu, P. (2005). Temporal Management of RFID Data, *VLDB*, pp. 1128–1139.

Wang, F., Liu, S. & Liu, P. (2010). A temporal RFID data model for querying physical objects, *Pervasive and Mobile Computing* 6(3): 382–397.

Want, R., Fishkin, K. P., Gujar, A. & Harrison, B. L. (1999). Bridging Physical and Virtual Worlds with Electronic Tags, *Proceedings of the SIGCHI conference on Human factors in computing systems: the CHI is the limit*, pp. 370–377.

Ward, M., van Kranenburg, R. & Backhouse, G. (2006). *RFID: Frequency, standards, adoption and innovation*, JISC Technology and Standards Watch.

Welbourne, E., Khoussainova, N., Letchner, J., Li, Y., Balazinska, M., Borriello, G. & Suciu, D. (2008). Cascadia: A System for Specifying, Detecting, and Managing RFID Events, *Mobile systems, applications, and services (MobiSys)*, pp. 281–294.

Ziekow, H. & Ivantysynova, L. (2008). A Probabilistic Approach for Cleaning RFID Data, *RFDM'08 Workshop in conjunction with ICDE 2008*, pp. 106–107.

2

RFID Components, Applications and System Integration with Healthcare Perspective

Kamran Ahsan
Staffordshire University
UK

1. Introduction

RFID (radio frequency identification) technology has already proved its use in various areas such as security, library, airline, military, animal forms, sports and other areas. RFID is being used for various applications in many industries. For example, equipment tracking, access controls including personal and vehicle, logistic, baggage, items security in departmental store. The main advantages RFID provide is resource optimization, quality customers' care, enhanced accuracy, efficient business processes, and effective business and healthcare processes. RFID can help is recognizing contextual knowledge and can help to improve objects predictability for certain processes. However, it is necessary to study RFID components for using these in healthcare environment. RFID main components are antennas, tags and readers. The investigation of these components provides an understanding of its use in healthcare settings and integration in healthcare processes.
This chapter studies the RFID components such as Antenna and reader. This chapter discusses the RFID active and passive tags, and compare these tags including advantages and disadvantages of RFID system. In this chapter, RFID applications are explored and technical model is analyzed. It also considers the healthcare perspectives and RFID use within healthcare settings. This study constructs a model for connected RFID applications which provides quick support for various healthcare functions and enhances flexibility for various systems' components integration.

2. Motivation of RFID technology

Existing research suggests that healthcare organisations are adopting information technology, specifically mobile technology throughout the world including the USA, Europe and UK (Bharadwaj et al., 2001). In the UK, the NHS (NHS-UK, 2009) is keen to adapt mobile technology for better information handling and this argument is supported in this chapter. However, real-time techniques and contextual knowledge management concepts for instant care is somehow neglected (Watson, 2006). Healthcare processes are volatile and the context of information changes rapidly. New technology has not considered information within their context. The context of information is more complex in healthcare in comparison to other industries. Although businesses have already started to develop and implement mobile technology for handling contextual information to improve processes but the same approaches cannot be adopted in the healthcare industry due to dominant knowledge use rather than just information and substantial human involvement (Connecting for health, 2009). However, the proven technology in business scenarios such as RFID can be adopted for a healthcare situation with the appropriate modelling of its use. Managing context for any information is a difficult task but information systems play an important role into it but contextual knowledge is even more difficult and need location, time and duration for information for providing context to any knowledge (Bharadwaj et al., 2001).

If knowledge gets support with context of objects' location, duration and time then this contextual knowledge can improve various situations for resource optimization and instant better actions. RFID technology use is critical to get this knowledge and providing context to it. RFID can also support tacit knowledge on a real-time basis in healthcare situations such as patients moving between locations to get medical treatment and a change in their medical condition at the same time. The utilization of tacit knowledge is crucial but it needs context environmental knowledge for instance actions. One of the properties of RFID is to provide instant location information of any object associated to it and this can play a vital role for tacit knowledge support and managing other environmental knowledge. Advanced use of RFID technology can integrate patients' flow processes appropriately and support patients' treatment processes by deterministic patients' movement knowledge (location and time etc.) within hospital settings (Connecting for health, 2010).

In a healthcare situation the patients' movement processes are subject to change due to various reasons including a change in the patients' medical condition, due to the unavailability of a particular resource at any given time and the unpredictable duration of any medical procedure (DH-UK, 2009). When processes are executed according to a plan and schedule then it consumes healthcare resources in a predicted way, if processes change due to any of the reasons described above then time and resources may be misused or processes become unpredictable. These situations consume resources unnecessarily and the instability of one process at one location may affect other processes at another location. So, the use of RFID technology is crucial for determining situations through getting time and location of an object within healthcare settings. Use of RFID technology is important for better process management including improved decision-making.

3. RFID utilisation

RFID works for identification of items/objects (Bohn, 2008). Sometime it only identifies item category or type but it is capable of identify items/objects uniquely. RFID also enables data storage for remote items/objects through remotely access items information (Schwieren1 & Vossen, 2009). RFID system consists of RFID tags, RF Antennas, RFID readers and back-end database for storing unique item's ID. In RFID systems, RFID tags use as unique identifier, these tags associate with any items, when system reads these unique tags then information associated with that tags can be retrieved. Antennas are first point of contact for tags reading. Reader can only work with software resides in reader's ROM (Glover & Bhatt, 2006). RFID system is based upon tags and reader's communication and range of communication/reading depends on operating frequency. When antennas deduct tags then an application which is part of reader manipulates tags' information in readable format for the end user. There is a great amount of research being conducted to improve the efficiency of RFID systems, increasing the accuracy of RFID reader and the feasibility of RFID tags. Although RFID accuracy needs more enhancement and efficiency yet to be increase but still RFID system is used in many applications (Bohn, 2008). There are a variety of tags, readers and antennas types are available. Before implementing RFID system, selection among these types needs understanding of these types in relation to their feasibility, capabilities and reliability. It is also necessary to understand combinational use of these types for implementing a single feasible RFID system.

4. Research approach

Qualitative research methodology is followed for observing the patients' flow situation within hospital settings. It includes observation and open interviews. This study tries to find out the pattern within hospital condition, knowledge elements for healthcare processes and priority of each knowledge element for knowledge factor integration with the help of location deduction technology (RFID). Some individual scenarios are considered within patients' movement processes and understanding is build for integration of RFID integration within these processes. In this respect, qualitative methodology is sufficient for including each knowledge element and device a way of handling these elements through location deduction technology. This chapter explores RFID technology with its kinds, types and capabilities. It is conferred that how RFID technology can be generalised through generalise technical model. It is discussed that how component layering approach can be feasible for integrating various healthcare management disciples for providing improved management. Healthcare knowledge factors are considered for supporting knowledge elements through RFID technology to improve healthcare situation.

5. RFID evaluation

RFID technology continues to evolve in past years in terms of various shapes of tags for increase its feasibility of its use, fast reading rate of reader and range of antennas etc. The use of RFID also evolves due to enhancement in its components. As the accuracy increases, the use of technology also increases such as baggage handling, goods delivery tracking and courier services. RFID system enhancement also evolves automation applications development e.g. automatic toll payments, automatic equipment tracking and document management etc. (Garfinkel & Rosenberg, 2005). In this connection, the evolution process of RFID with respect to past few decades can be seen in figure 1.

6. How RFID system works

The basic unit of RFID system is tags and tags have its own unique identification number system by which it recognizes uniquely. These unique identification numbers save in tags' internal memory and it is not changeable (read-only). However, tags can have other memory which can be either read-only or rewrite able (Application Notes CAENRFID, 2008). Tag memory may also contain other read-only information about that tag such manufactured date. RFID reader generates magnetic fields through antennas for getting acknowledgement from tags (Garfinkel & Rosenberg, 2005). The reader generates query (trigger) through electromagnetic high-frequency signals (this frequency could be up to 50 times/second) to establish communication for tags (Srivastava, 2005). This signal field might get large number of tags data which is a significant problem for handling bulk of data together. However, this problem can be overcome through filtering these data. Actually software performs this filtering and information system is used to supply this data to data

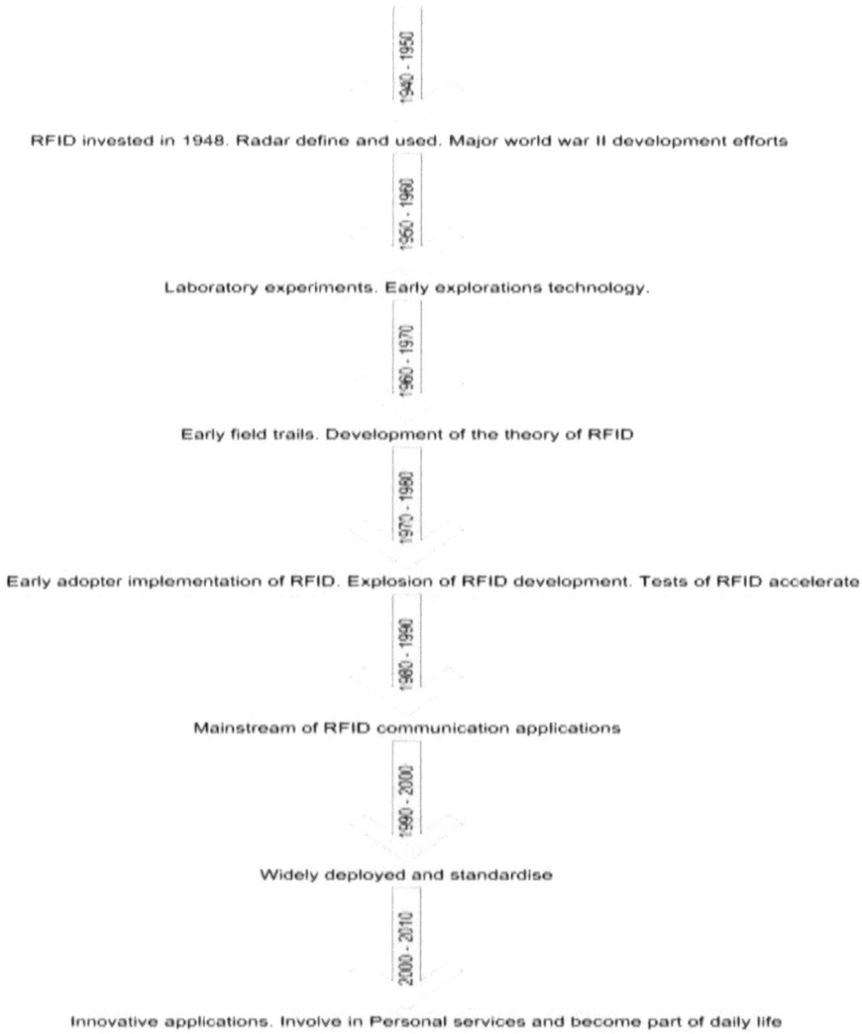

Fig. 1. RFID evolution: over past the few decades (Srivastava, 2005)

repository or use any other software procedures to control data according to the need and system capability (Srivastava, 2005; Application Notes CAENRFID, 2008). This piece of software works as a middle layer between user application and reader because the reader normally does not have the capability to handle bulk data at once; it has the job to supply reading data to user application for further process (Frank *et al.*, 2006). This buffering capability may supply data from reader to information system interface (user interface) directly or may provide and use some routine to save into database for later exploit, it is depend on user requirement.

Reader and tags communication can be maintained through several protocols. When the reader is switched on then these protocols start the identification process for reading the tags, these important protocols are ISO 15693, ISO 18000-3, ISO 18000-6 and EPC. ISO 15693 and ISO 18000-3 protocols are used for high frequency (HF) and, ISO 18000-6 and EPC protocols are used for ultra high frequency (UHF). Frequency bands have been defined for these protocols and they work within specified range such as HF has 13.56 MHz and UHF between 860 - 915 MHz (Application Notes CAENRFID, 2008). Reader modulates tags responses within frequency field (Parks *et al.*, 2009).

The reader handles multiple tags reading at once through signal collision detection technique (Srivastava, 2005). This signal collision detection technique uses anti-collision algorithm, the use of this algorithm enables multiple tag handling. However, multiple tags handling depend on frequency range and protocol use in conjunction with tag type which can enable up to 200 tags reading at single time. Reader protocol is not only use for reading the tag but also perform writing on to tags (Application Notes CAENRFID, 2008).

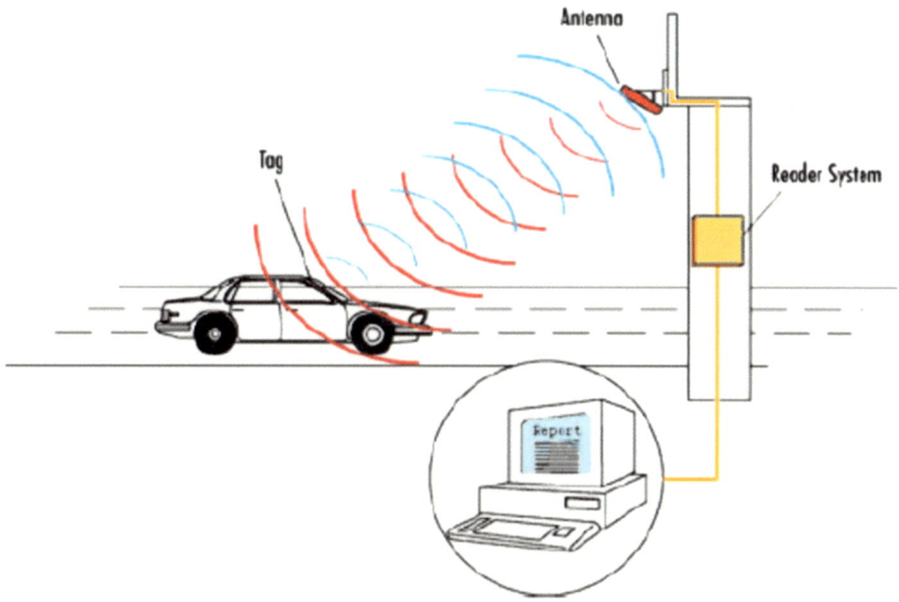

Fig. 2. A typical RFID system (Application Notes CAENRFID, 2008)

The use of the reader within RFIFD system can be seen in figure 2. This figure also define the overall cycle of tag reading by reader through antenna and transforming data into communicate able form to user applications.

7. How RFID system works

RFID system deducts tags within antennas' range and performs various operations onto each tag. The RFID system can only work effectively if all RFID components logically connect together and these components need to be compatible with each other. Thats' why understanding of these separate components is necessary. Implementation of complete RFID solution is only possible through integration of these components which needs understanding of compatibility for each component, realisation of each components compatibility needs property study for these components (Sandip, 2005). These components are gathered and defined as under. Also integration of these components can be understood with figure 3.

- Tag has unique ID and use for unique identification; tags are attached with objects in RFID solutions.
- Antenna use for reading tags; antenna has its own magnetic field and antenna can only read tags within these magnetic fields.
- Reader works for handling antenna signals and manipulate tags' information.
- Communication infrastructure use for reader to communicate with IT infrastructure and work as middle layer between application software and reader.
- Application software is a computer base software which enable user to see RFID information, this can be database, application routines or user interface.

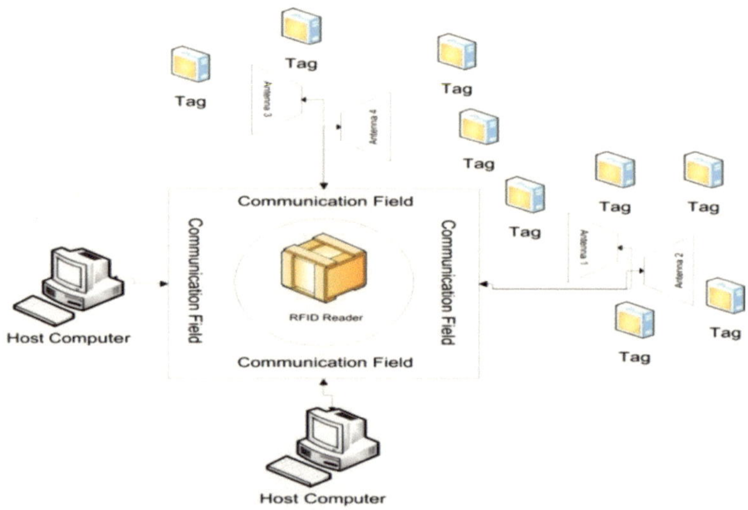

Fig. 3. Components of an RFID system

8. RFID tags

RFID tag has memory in the form of a microchip which store unique code for tag's identification, this unique identification called tag's ID (Application Notes CAENRFID, 2008). The microchip is a small silicon chip with embedded circuit. Numbering technique is used for providing unique identification (Garfinkel & Rosenberg, 2005). This microchip could have read-only or writeable characteristics depending on tag type and its application within RFID solution. These characteristics depend on the microchip circuitry which has form and initialize during tag manufacturing (Miller & Bureau, 2009). Some tags (read-only) re-programming is possible but need separate electronic equipment for re-programming read-only tag's memory. Writable tags also know as re-write tags do not need any separate equipment and reader can write data on it, depend on the protocol support, if reader have writing command capability and tags are in range. Tags selection is very important for feasible use in RFID solution. This selection is dependent on the tag size, shape and material. Tags can be integrated in varity of material depending on the need of the environment. The tag is embedded in plastic label in form of a microchip, stick able material for documents handling, plastic material with use of pin for use in cloths material are the good examples to be consider (Frank *et al.*, 2006). Various forms of tags with respect to its sizes and shapes can understand with figure 4.

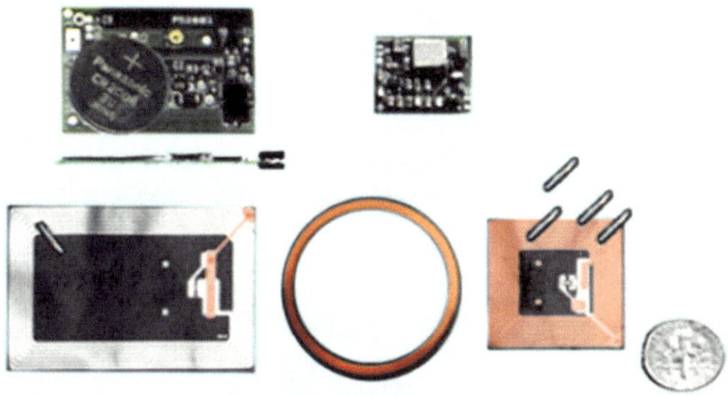

Fig. 4. Varity of RFID tags (various shape & sizes) (Frank *et al.*, 2006)

Classification of RFID tags is also possible with respect to their capabilities such as read-only, re-write and further data recoding. Further data recording examples are temperature, motion and pressure etc. (Narayanan *et al.*, 2005). Compiled tags classification into five classes previously gathered by Narayanan *et al.* (2005) is shown in figure 5.

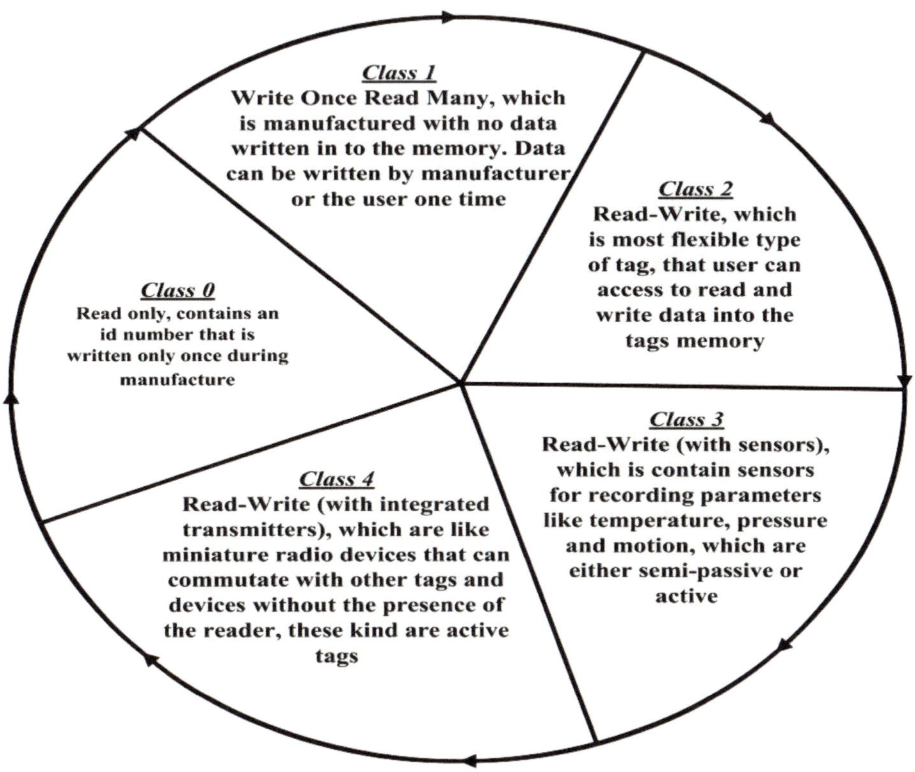

Fig. 5. RFID tags classifications (Narayanan et al., 2005)

Active, semi-active and passive are the three main tags types. Tags made up with few characteristics which may vary slightly depending on type of tag, due to which their use can be change in RFID solution (Zeisel & Sabella, 2006). So, selection of tags depends on the functional need of RFID application. The main difference is between active and passive tags because semi-active tags have mix of both tag's characteristics (Application Notes CAENRFID, 2008). These types differentiate upon memory, range, security, types of data it can record, frequency and other characteristics. The combinations of these characteristics effects tags' performance and change its support and usefulness for RFID system (Intermec, 2009). The main tag types (active and passive tags) are compared in following figure 6.

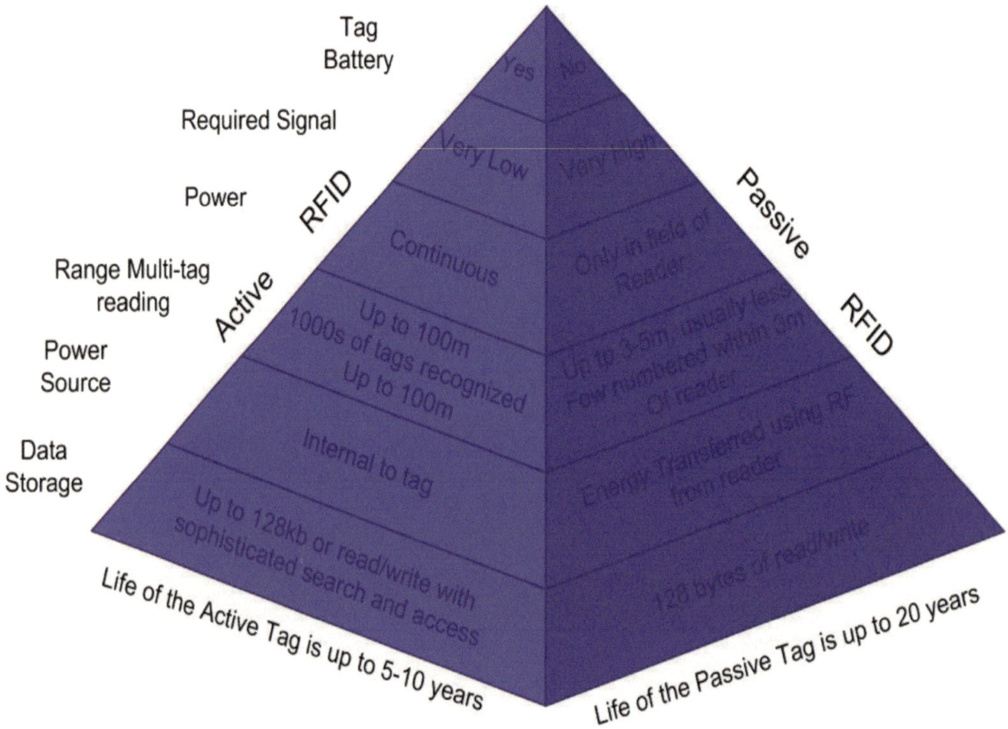

Fig. 6. RFID active and passive tags comparison

8.1 Tags physical features

The tags have various physical features such as shape, size and weight. Consideration of these features depends on environment tag being used. Classified tag's physical features are as under.

- Smart labels can embed in layers type materials such as papers.
- Small tags can embed objects other then flat panel such as clothes and keys.
- Plastic disks can use for attaching with durable objects and use in tough environments such as pallets tagging use in open air.

8.2 Tags capabilities

The tags can also be differentiated with respect to tags capabilities and performance (Schwieren1 & Vossen, 2009; Garfinkel & Rosenberg, 2005). Following is the list for tags capabilities.

- Anti-collision capability of a tag, tags having anti-collision can enable reader to recognize its beginning and ending which help reader to read all tags in its range.
- How tags get its power source such as active has its own battery and passive get power from reader through its magnetic field.
- Conditions of tag environment such as use in water.
- Tags data writing capabilities such as write one time or many times onto tag memory.
- Coupling mechanism tag use such as magnetic, inductive, capacitive and backscatter. Coupling mechanism determines tags information and power sharing methods.
- If tag can work for more than one protocol which enable tags to work with different types of readers.
- Tags with encrypted data handling feature.
- Either tag has two way communication (full duplex) or one way communication (half duplex).

8.3 Tags standards

Understanding of tags standards is necessary for working with various systems, protocols and procedures. It is dependent on organisational policies and scope of RFID system. Tags standards enable interoperability capability to RFID solutions (Sandip, 2005). For example, if tags have standardization and its uniqueness can be identified across different systems then it enhances the use of standard tags (Schwieren1 & Vossen, 2009). The spectrum of tags can be single situation such as tags use in single warehouse, multiple spectrums such as same tags use in logistic and supply chain and need recognition across different organisations and various systems (Shepard, 2005). The selection of tags standards within RFID solutions depend on these spectrum. Following three standards are gathered by (Shepard, 2005).
ISO/IEC 18000 tags: This standard works for various frequency ranges including long range (UHF), high frequency (HF), low frequency (LF), and microwave. This standard supports various principle and tags architectures. The range of tag identification includes 18000-(1 to 7).
ISO 15693: In this standard tag IDs are not as unique as ISO 18000. Although vendors try to build unique tags with certain specification and coding but it is not globally unique. These standard tags most often use in smart cards for contact-less mechanism. However, it is also use in other application but in local scenario (not global) e.g. supply chain and asset tracking etc.
EPC tags: It is the standard for maintaining the uniqueness under certain management bodies. It carries out tags uniqueness with all the vendors associated with one management entity. Management entities carry their own EPC number technique and own the certain object class.

8.4 Tags states

Tags process recognize with its state within RFID working environment. Tags cannot have multiple states simultaneously. The set of tag states depend on the type of tag. However, these states generally include open state, reply state, ready state, acknowledge state, arbitrate state, killed state and secured state (Shepard, 2005).

8.5 Tags frequencies and range

RFID tags capability and working feasibility change according to its frequency and range. Tags prices and its use also vary in relation with tags frequency and range. Various frequencies and its range (working distance) can be seen in following figure 7.

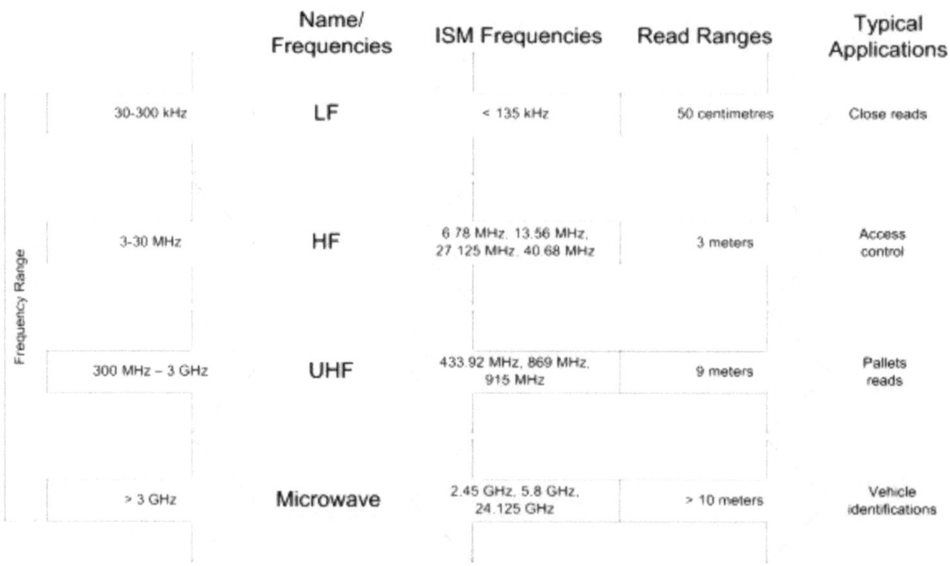

Fig. 7. RFID frequencies and ranges

The performance, range and interference feasibility depend on the frequency at which tags' operate (Zeisel & Sabella, 2006). Different tags standard uses different frequency bands in which ISO and EPCglobal standard are major organisations working for UHF bands for developing international standards (Narayanan *et al.*, 2005). However, full compatibility is still not achieved that's why most of the organisation obligated to use International Telecommunication Union principles (DHS, 2006). These principles include following frequency bands.

- **High frequency** can work up to one meter. It can embed with thin objects such as papers, that's why it is mostly use in sales points and for access controls. 13.56MHz is the frequency at which it work and it is less expensive to implement (Srivastava, 2005; Application Notes CAENRFID, 2008).
- **Low frequency** fulfils short range applications' needs. It is not effective for metal or wet surfaces and only works half of the high frequency range (maximum half a meter) (Frank et al., 2006). Low frequency works on 125 KHz (Application Notes CAENRFID, 2008).
- **Ultra high frequency** has better read rate and large number of UHF tags can be recognize at one time. It has also good better read range and three times with high frequency, it is capable to read tags up to three meters. However, range can be reduced in wet environment. It works between 860-930 MHz frequencies (Srivastava, 2005).
- **Microwave** has less read range and it works within one meter. But it has rate of reading is faster than UHF with very little affect on wet and metal surfaces. It works on Giga Hertz frequency and faster than LF, HF and UHF, that's why it can work better for vehicle access application (Application Notes CAENRFID, 2008).

8.6 Tags fields

Active tags have its own power but passive tags get the power from antenna based on readers' signal to antenna (Application Notes CAENRFID, 2008). Passive tags response or communication signal is based on the power it gets from antenna. Following two methods passive tags use for getting power from reader.

Far field uses coupling methods with the electric signals within field of antenna as shown in figure 8. These tags embed their signal in reverse order with antenna signal using some standard format so that reader can recognize the tag signals (Frank *et al.*, 2006).

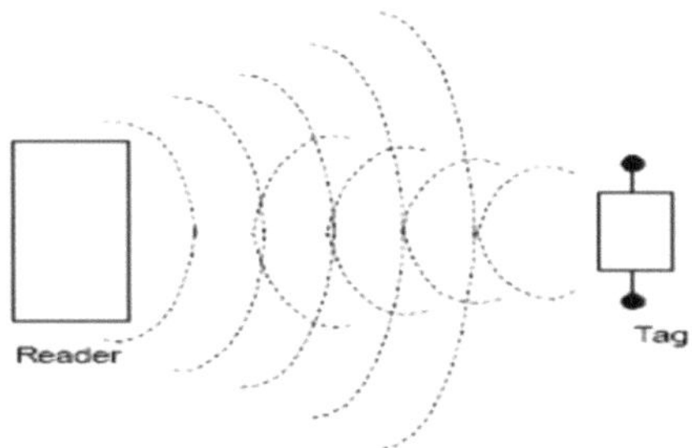

Fig. 8. RFID far field methodology (Application Notes CAENRFID, 2008)

Near field uses inductive coupling within magnetic field of an antenna as shown in figure 9 (Application Notes CAENRFID, 2008).

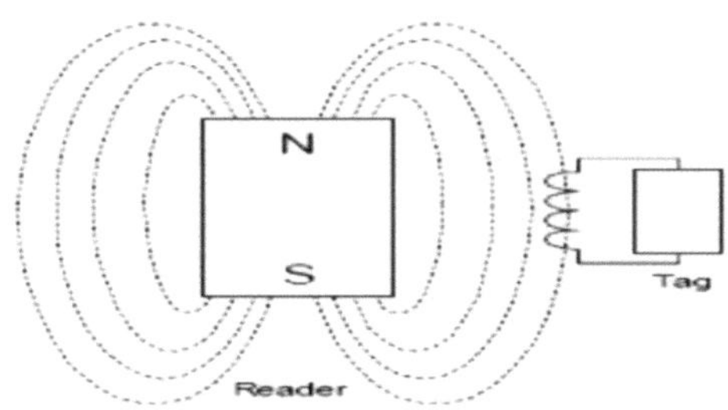

Fig. 9. RFID near field methodology (Application Notes CAENRFID, 2008)

These methods are use in different kind of applications and system is based on different circuitry (Meiller & Bureau, 2009). Far field is appropriate for microwave and UHF because it can work in longer range and near field is suitable for LF and HF because it can only work within shorter range (Meiller & Bureau, 2009; Parks *et al.*, 2009).

9. RFID antennas

RFID antenna is the middle-ware technology or component, it work between reader and tag and provide energy to tags in some cases (passive tags). It performs tags data collection. It shapes can be altered depend on the application and feasibility of use but shapes varies the range of antenna.

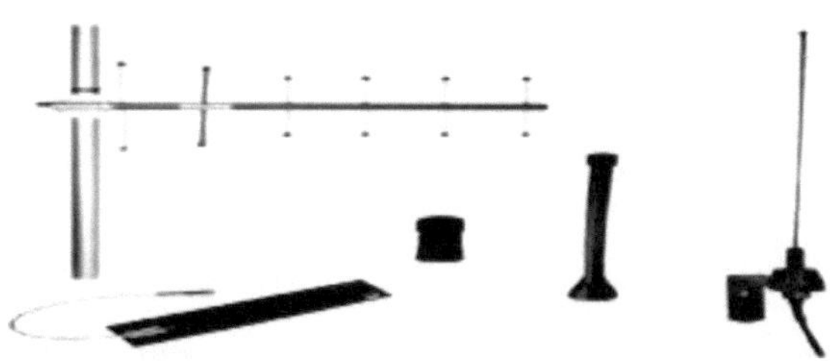

Fig. 10. RFID antennas types (Intermer, 2009)

Antenna has various shapes and some of them can be seen in figure 10. Antennas can be differentiated with various properties such as direction of signals (tags reading direction) and polarities. Stick antennas, gate antennas, patch antennas, circular polarized, di-pole or multi-pole antennas, linear polarized, beam-forming or phased-array element antennas, Omni directional antennas and adaptive antennas are the types of antenna commonly use in various applications (Zeisel & Sabella, 2006).

10. RFID readers

RFID reader is a external powered equipment used in RFID system for producing and accepting radio signals (GAO, 2005). A single reader can operate on multiple frequencies and this functionality depends on the vendor (Application Notes CAENRFID, 2008; Frank *et al.*, 2006), it can have anti-collision algorithm/procedures for deducting multiple tags at one time. RFID reader works as middle-ware between tag and user application. Reader is the central part of the RFID system and communicates with tags and computer program, it supply tags information to a computer program after reading each tags unique ID. It can also perform writing onto tag, if the tag is supported. Although the reader can have multiple frequency capability but it works on a single frequency at a time. The reader can communicate with the computer program and need either wired or wireless connection with the computer. This reader can use a wire connection with any of the following: USB, RS-232, and RS485. Otherwise, the reader can connect with the computer through Wi-Fi (known as network reader) (Sandip, 2005; Zeisel & Sabella, 2006). The reader provides various management techniques and functionality to computer programs (Zeisel & Sabella, 2006) through various built-in functions/components, these components can be understood with following figure 11.

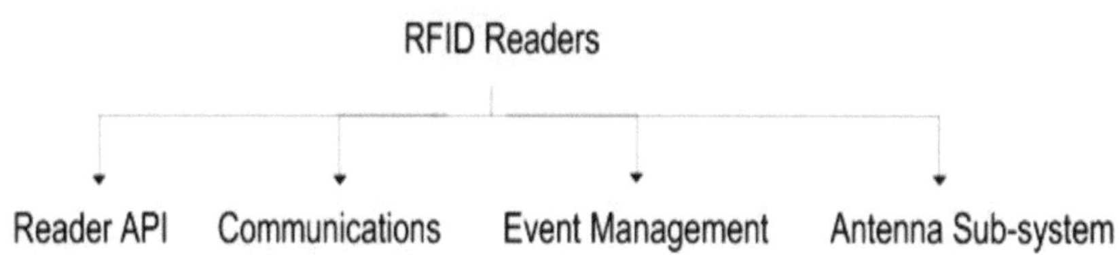

Fig. 11. RFID reader logical components

10.1 Reader protocols

Although vendors are trying to implement reader with common protocols but the standardization of RFID readers' protocol is not achieved yet which is why readers are not interoperable (Glover & Bhatt, 2006). An organisation cannot replace a reader easily after RFID solution implementation. However, there are some common capabilities RFID readers provide. Command, sensor, observation, alert, transport, host and trigger are the most common capabilities provide by RFID readers.

Synchronous and asynchronous are two types of communications used by readers (Shepard, 2005). In synchronous readers' communication with host, the host requests the update with the reader (Garfinkel & Rosenberg, 2005). In response to that, the reader sends the list of updates to the host. In case of asynchronous communication, the reader sends notification to the host about its observation. This notification can be sent to host upon request or immediately after new observations, it is dependent on the requirement and trigger mechanism of RFID system (Shepard, 2005). Both types of communication can be understood from the figure above 12.

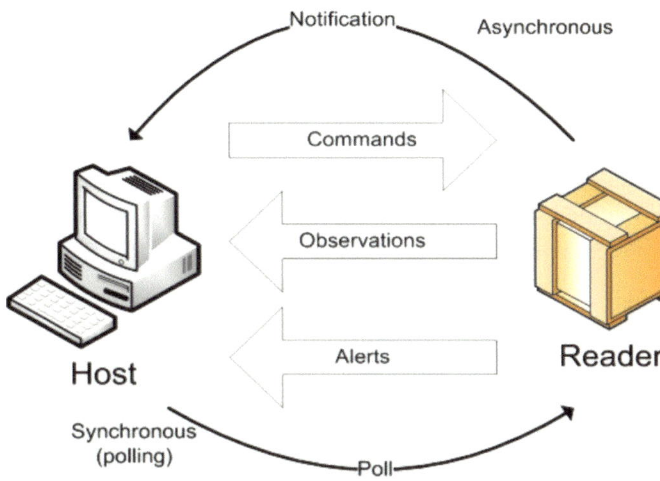

Fig. 12. Information flow and a/synchronous communications (Shepard, 2005)

In both of these communication methods the information flow has three types which include; observation, host pass commands to reader and reader pass alerts to host (Shepard, 2005).
EPCglobal is the most common and most accepted protocol. EPCglobal provides three layers for communications; these layers are message, transport and reader (Zeisel & Sabella, 2006). The messaging layer use transport layer to pass messages according to the format defined by the reader layer (Garfinkel & Rosenberg, 2005). Connection commands, host commands, security and reader notifications are the most common command deal by message layer. Reader layer identifies the format of the message transport between host and reader. The transport layer is responsible for network support and establishes communication between reader hardware and computer operating system (Zeisel & Sabella, 2006).

10.2 Reader interfaces

RFID reader communicates with the computer program by using the reader's protocol as described in the previous section. The reader should be capable to handle various types of commands which include management of events, communicate with applications and adapter. These also provide various kinds of interface with the reader. Figure 13 shows the three kinds of interfaces most commonly any reader provide.
The reader provides a command set for communicating with user interface of computer programs. These command set understands the reader properties and provides functionality for using a particular reader (GAO, 2005). These command sets are known as application program interface (APIs) (Frank *et al.*, 2006). If organisation builds their application program based on a specific reader then this computer program needs to use APIs provided by particular reader (Application Notes CAENRFID, 2008). Customize application might not be compatible with other reader but in this case a vendor upgrades their readers hardware, organisation might be able to use those readers. Vendors most often provide the compatibility of previous APIs in the case of an upgrade (Shepard, 2005; Frank *et al.*, 2006).

In that case, organisations can upgrade their hardware using the same application but organisations must refer to vendors' device specifications before any upgrade. Some vendors also provide application compatibility with a range of their hardware through consistent APIs set (Zeisel & Sabella, 2006). This mechanism provides adaptability of various readers with same application for some extent. Reader interface within RFID reader provides the filtering for the reader data (raw data) before sending to application program. Reader provides the raw data to reader interface, this data could be bulk (depend on the environment), reader interface need to find relevant data within bulk data provided by reader. This functionality reduces the overhead of program interface or application program and as well as provide low traffic for communication between computer software and reader due to sending only relevant data.

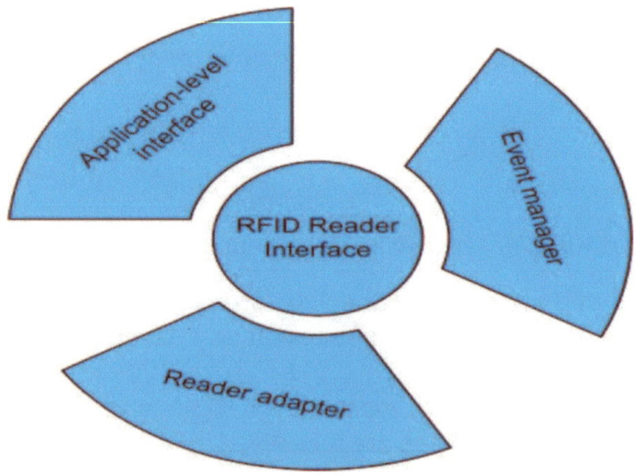

Fig. 13. Reader interface (Frank *et al.*, 2006)

11. Advantages and disadvantages of RFID systems

The use of RFID solutions have been recognize by many industry. However, the appropriate level of RFID components combination and selection of these components according the suitability of organisational situation and environment can make it beneficial. Otherwise, RFID system with the inappropriate combination and selection of RFID components may generates error or does not work effectively which could be increase organisational operational cost and may affect customers' good will. The list of advantages and disadvantages can be seen in table 1 (Meiller & Bureau, 2009).

Advantage	Disadvantage
High speed	Interference
Multipurpose and many format	High cost
Reduce man-power	Some materials may create signal problem
High accuracy	Overloaded reading (fail to read)
Complex duplication	
Multiple reading (tags)	

Table 1. Advantages and disadvantages of RFID system

12. RFID general technical model

So far it has been studied that RFID system varies with respect to various features. These features include physical features, components, standards, capabilities, frequencies, states, ranges, protocol, interfaces and readers. Due to variable RFID features and compatibility issues, it is very difficult to develop integrated RFID solution (Glover & Bhatt, 2006; Application Notes CAENRFID, 2008). If organisation tries to build RFID solution with future compatible hardware then it makes RFID components' selection, implementation and integration even more complex. However, RFID regulatory bodies try to provide safe and less conflict (radio and other frequency using equipment) RFID standards and vendors try to provide interoperable equipments. But true interoperability is not possible until globally accepted standard not developed and manufacture adapt single standard or at least limited standards. In this context, two main organisations are doing efforts for providing globally accepted standards (Application Notes CAENRFID, 2008). These organisations (EPCglobal and ISO) are trying to develop unique standard for RFID tags so that tags can be used in wide spectrum throughout the world including supply chain and transportation. However, there still no standard is available for compete RFID system or solution. In this connection, it is necessary to understand RFID tags, air interface, reader, reader's programs including data protocol processor and physical interrogator, needs of application programs, and application commands and responses in integrated way. For this purpose, a generalise model for RFID system is provided for better understanding (see figure 14).

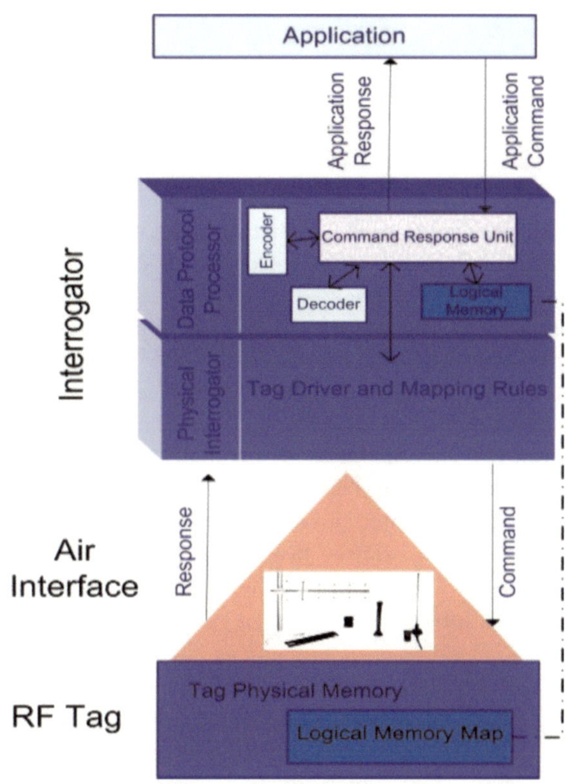

Fig. 14. RFID general technical model

13. RFID applications

Before understanding the use of RFID system for contextual knowledge management, it is appropriate to study the use of RFID applications for other scenario or situations with respect to short and long range RFID applications. For this purpose a generalised RFID application model has been presented as shown in figure 15.

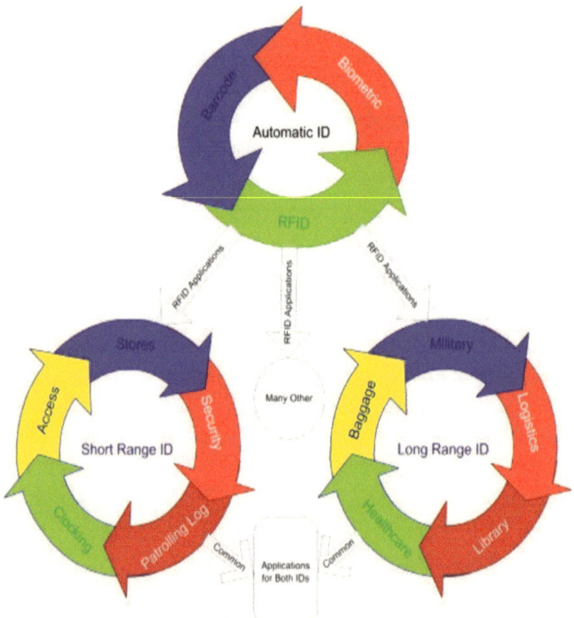

Fig. 15. General RFID applications according to its capabilities

From a RFID application perspective, RFID has two categories: short range and long range. In short range application tag need to show to reader. Application works perfectly in short range if object or tag access the reader one by one. For example, in access control, only employee or specified person can access the secure building and building has several divisions each division has particular set of authorized people, who can access certain divisions. Person need to show the tag near the reader before every secure door for enabling the system to decide whether door should open for that person or not. In long range applications, tags do not need reader near, as compare to short range. For example, tags place on every book in the library and user can easily access the exact book shelf for the required book. This enable automated inventory control. Readers can read the multiple items simultaneously from the distance. Reading distance can vary depending on the frequency and type of reader. This section further discusses the most common use within short and long range applications.

13.1 Security and control applications

RFID system can be used for control access and security; it is also useful for audit purposes. These applications are not only granted permission to access a particular secure zone but also record who is accessing, from which location/areas, at what time and for how long duration. These types of RFID systems can maintain building and departmental security including privacy. This type of RFID solution is also workable for equipments and objects controls (Shepard, 2005).

13.2 Patrolling log applications

In this application, security firms use RFID system to control their security guards and use RFID data for various purposes. These purposes include performance checks, data can be used in reference to unexpected event and audits for a secure area etc. The main difference in this type of application with other applications is that the reader is variable and the tag is fixed, i.e. the reader go near to the tag rather tag swap near to the reader. In security patrolling several numbers of tags can be fixed throughout the building and the security guard needs to swap the reader with each tag (checkpoint) in a sequential order within an allocated time and repeat this process throughout his/her shift. The reader records each swap which can be transferable to a computer program later on for audit or other purposes.

13.3 Baggage applications

Baggage handling and package delivery is a complex task and needs a large number of human involvement which is an expensive resource. Humans do various operations from receiving packages, sorting, assembling and distribution. Due to human involvement the error rate can be high. While the use of RFID tagging system not only reduces human involvement but also automates the process for certain extend which enable fast packages delivery. RFID solutions for baggage and packaging firm including airline industry are better for their effective operation and it reduce the complexity for overall system.

13.4 Toll road applications

RFID can provide the automated toll collection and maintain the traffic flow without stopping vehicles for payment. In these type applications, vehicles either pre-pay their toll yearly quarterly or monthly or other kind of scheme can be applicable such as pay-as-you go. In any case, the reader can recognize and record the vehicle entry at each toll which can be calculated later on by an application program. These applications not only help in toll collection and maintain traffic flow but also provide statistical data for any road which can be utilized for analysis and improvements (Shepard, 2005).

14. Healthcare RFID applications

RFID usefulness already proved its importance in security control, patrolling logs, baggage and packaging, and toll collections. Similarly, RFID can improve healthcare processes in various ways such as equipment handling, drugs transportation, blood samples administration, patients' notes management and others. RFID provide timely information about the objects associated with RFID solution, this feature enables better and up-to-date information about the processes link with these objects. The better information about objects around healthcare processes can further reduce the number of errors.

Information systems have many limitations, including failing to successfully update information automatically in relation to the location of an object, while RFID system can overcome this limitation through RFID advanced features. These RFID feature can save precious healthcare resources which can be utilized for patients' care such as a reduction in consultants' time for managing patients can enable consultant to give more time for patients' direct care. Healthcare processes involved with both processes patients' care and non-patients' care. Patients' care includes; fever test, blood pressure and scan etc. which is direct care. Non-patients' care includes notes management, movement of patients from one place to another, equipment handling from one ward to another ward, sending blood samples to pathology, preparing management reports and managing beds etc. These non-patients' care processes are known as indirect care. If indirect care processes improve within healthcare settings then it saves healthcare practitioners' time which can be utilized for direct care. RFID systems can also provide the automation to some extent, this not only provide better planning but also enables real-time management. Most of RFID solutions focus on one issue such as management of objects; this management can be implemented for various things such as documents and equipments. If RFID solution only consider better management for healthcare then it might not be effective in various situations such as RFID solution for notes scanning by using low or high frequency tags. If RFID solution for notes scanning are implemented through low or high frequency then the RFID system may need physical scanning for each move like sales point scanning. In this case, healthcare practitioner intervention is required for each transaction. For example, if notes have to move from ward to theatre, it needs scanning at the ward for a system update and the theatre might need to scan again to enable the system to update the new location of the notes. When time is scarce this process may not occur effectively and the scanning chain may be broken at any time. In most of the cases, healthcare processes need automation to minimize the overhead from healthcare practitioners.

Implementation of RFID system for one modality or discipline can solve issues related to one modality or discipline. In case various RFID systems need implementation for different disciplines within healthcare, then these should be synergetic with each other. Otherwise it can increase the overhead for using more than one interface and systems for dealing with various objects and disciplines. RFID solutions can help direct care staff/practitioners to improve care but also effective for other staff such as radiologist and porters to play better role for overall process improvement. The use of RFID systems can be different for various stakeholders and each stakeholder can see different sets of information but it needs integration for consistent information.

The complexity of healthcare processes need appropriate level of RFID integration for several management disciplines. For example, better managing of notes can enable ward staff to handle patients in timely manner but if the equipment is not available staff still cannot execute patients' treatment procedures until right equipment is available. So, any procedure needs all process elements to handle patients' treatment efficiently. RFID can maintain elements availability for healthcare processes. However, it needs the study for RFID solutions in the context of process automation rather another overhead for healthcare practitioners such as automatic scanning when notes are presented in any unit within hospital settings rather than physical scanning of each set of notes. It is also necessary to investigate the integration level required for healthcare processes and classify each management disciplines for integration RFID application. In this connection, a model for possible management discipline within healthcare setting need RFID application in integrated way is presented (see figure 16).

It is very complex to provide a generic system for RFID application for every hospital setting due to a variation in processes. However, it appropriate to provide a generic model for adapting RFID technology solution in hospital setting which enable healthcare management to visualize healthcare processes in holistic way. Figure 16 is not the complete list for every healthcare management discipline and RFID technology can be used for other purposes within healthcare environment. However, this list provides the most urgent, appropriate and related management discipline which need integration through RFID application for overall processes automation. This section also elaborates few scenarios for further understanding of these separate but synergetic management disciplines. It is important to describe few scenarios in relation with RFID usability, it helps to understand generalized RFID application model for healthcare and it capability to improve healthcare processes within specific scenarios.

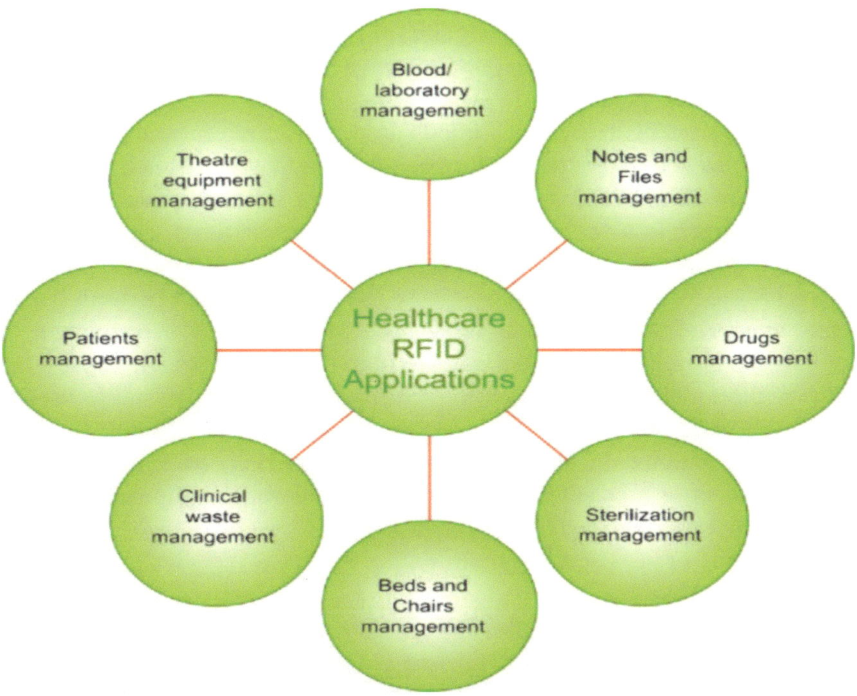

Fig. 16. Healthcare RFID applications

14.1 Theatre inventory management

Theatre inventory management is a separate healthcare management discipline which facilitates surgery and support theatres' other processes. Theatre equipment management is necessary because without equipment availability most surgeries are not possible such as hip replacement surgery needing the exact size and shape bowl for patient's surgery. If the exact surgery equipment is not available at the time of surgery then the surgery has to be cancelled. The theatre inventory management is complex due to various conditions. For examples, for some types of equipment the expiry date is very short and the theatre cannot keep this type of equipment for long periods of time that is why management like to keep inventory at low level for these types of equipments in order to avoid loss. Surgeons want to use new equipments in some cases depending on the patients' age and condition. Maintaining the level of inventory for these kind of equipments are very difficult. RFID can provide automation for the inventory update and support management to maintain their critical inventory effectively within low budget and meet the theatre requirement at the same time.

14.2 Equipment sterilization management

Sterilisation equipment control and maintenance is very complex due to a variety of functions. Sterilisation unit works in conjunction with theatre. Sterilisation unit deals with two type of equipment: consumable and reusable. Both types of equipment management are difficult. Consumable items either goes into recycle after use or it remains inside patients' body such as screws use for fix patients' bone. Although consumable items' management is easier then reusable but sterilisation unit still needs to maintain record for all equipment including consumable items for audit purposes, analysis and research. In case, patients get infection after surgery then the sterilisation unit needs to record those cases in association of equipment used for those patients.

Reusable items management is more difficult due to various functions such as theatre supply, theatre collection, recorded use for each equipment, synchronisation with surgery scheduling, equipment orders with suppliers, equipment washing, packaging, sterilisation and others. Equipment is used in the form of a set of items in surgeries. Every surgery requirement is different and sterilisation needs to fulfil the theatre requirement according to the theatre schedule for every surgery otherwise the surgery cannot perform better. If any patient gets infected then sterilisation needs to back track all the equipment used for particular surgery including individual items within equipment set which may be part of other equipment set after use. In this complex scenario, it is very difficult to keep accurate record. Accuracy of record even more necessary in case of infected equipment handling. RFID can provide automation, minimize the errors and keep track individual items; it allows quick and more efficient information for sterilisation equipment which further enables better equipment management.

14.3 Notes management

Patients' notes are important for patients' treatment. Without patients' history and notes doctor cannot see patient. Notes are necessary for in-patients and out-patients. Out-patients such as clinic appointment, if doctor don't have the notes at the time of patients' clinical appointment then appointment needs to be cancel. In this case, hospital resource and time cannot be utilized at optimum level. Staff need in-patients notes on a regular basis in comparison to out-patients notes, due to the frequency of treatment in relation to the following; wards, scan and theatre units. In this case there is a need to see patients' notes at the time of treatment or before treatment. So, it is critical for healthcare practitioners and clerical staff to make patients' notes available at the location and time of treatment. When patients have multiple modalities and multiple staff need to update notes then notes management is even more complex because notes have to move from one place to another very frequently. During transfer process notes can go missing and as well as notes can delay which cause delay or cancellation (depend on medical condition) in treatment. RFID can improve notes management and automate records for notes location which can help staff to visualize the status of the notes. This enables staff to collect notes quickly if necessary without patients' treatment cancellation. It makes the archiving process easy and secure. RFID solution also stops unauthorized access which certainly improves the quality of care.

15. RFID connecting model

It has been investigated in section 9 and 10 that as technology evolves each time, tags and hardware increase their performance for better RFID use. Although it is recommended in figure 14, that the vendor can minimize the complexity at the technological level with consistent technological upgrades. However, there is no single standardization is available at technical level and it is very difficult to achieve standardization at technical level too. Due to lack of standardization it is difficult to rely on one technological solution. In that case, future technological upgrade may affect application (see section 13 & 14) usability and application may not compatible with new technological upgrades. However, adoption of new advancement in technology is also good for better performance. So, it is better to adapt middle level approach, in which RFID solutions should not stop adaption of new technological advancement and also does not affect application interface. Especially in the case of healthcare application interfaces because healthcare applications their interfaces and integration are really complex. Moreover healthcare applications are significantly big and need major investment. However, it improves overall organisational performance with resource optimization significantly.
This research uses RFID for context management and support practitioners knowledge in real-time environment. Practitioners need constant support with appropriate level of knowledge management interface. Section 14 discusses the various application need to use RFID hardware for constant update of equipment, notes and other stuff within healthcare for better overall healthcare management which is necessary for patient processes. In this connection, a RFID connecting model for healthcare applications is developed, it supports RFID application interface should not affect if RFID solution adapts RFID technology change or upgrade. Figure 17 shows this model, it provide the flexibility to RFID applications to adopt future technology advancements without changing frontend.

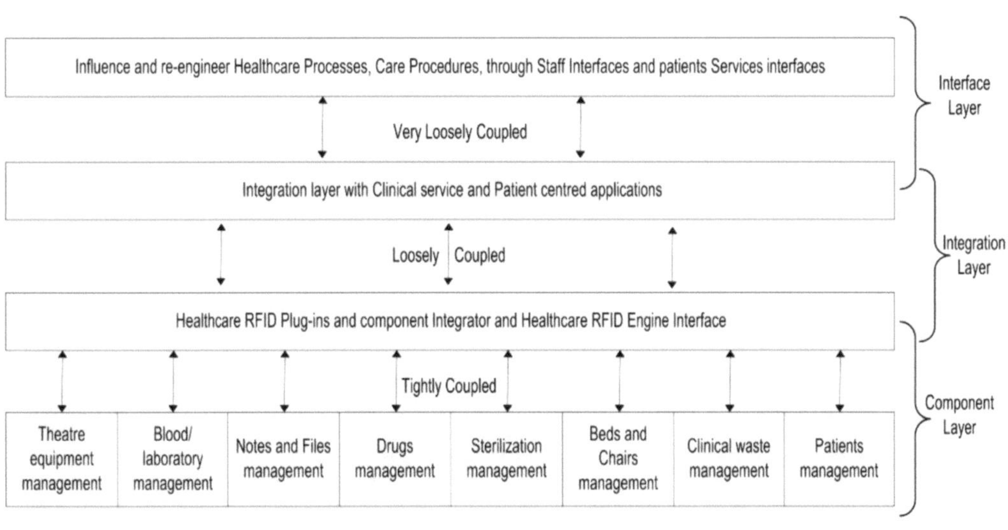

Fig. 17. Hospital RFID application model

It is feasible for staff and healthcare processes to work through the same interface layer. The interface layer should not need changing due to the integration layer which is based on patient centred application and healthcare services, and use RFID engine. The foundation of engine interface is based on RFID plug-ins and component integrator. In component layer each management scheme utilizes various types of tags, readers and hardware. Each component such as drugs management, theatre equipment management can use the same or different implementation logic. However, it provides feasibility and flexibility for interaction with healthcare interface through variable set of plug-ins and component integrator (technical procedures). This model further provide the feasibility to integrate all management schemes appropriately for better patients process management which can minimize the error and improvement the performance with resource optimization.

16. Conclusion

This chapter considered the RFID components with its potential alternatives and possible healthcare applications. The present research defines and analyses the most important RFID components (tag and reader) with its' alternatives and its use in various situations. It is conceived that RFID is very important for resource optimisation, increasing efficiency within organisational processes, providing enhanced service, and making organisational staff overall experience better. The research observed various cases in healthcare settings and analyses the complexity of healthcare processes. However, it is pragmatic to put RFID for healthcare objects' (e.g. notes, equipments and drugs etc.) tracking for improved healthcare service with optimised use of resources.

The first part of this chapter has explained and described the RFID technology and its components, and the second part has discussed the main considerations of RFID technology in terms of advantages and study model. The last part explores RFID technology applications. This chapter considers RFID technology as a means to provide new capabilities and efficient methods for several applications. For example, in healthcare, access control, analyzing inventory information, and business processes. RFID technology needs to develop its capability to be used with computing devices. This allows businesses to get real potential benefits of RFID technology. This study facilitates adoption of location deduction technology (RFID) in a healthcare environment and shows the importance of the technology in a real scenario and application in connection with resource optimization and improving effectiveness. However, there is no doubt in the future that many companies and organisations will benefit from RFID technology especially healthcare.

17. Acknowledgment

We would like to thank the hospital management and NHS Trust chair for allowing us access to the hospital for our research. We are grateful to all the hospital staff: managers, surgeons, doctors, IT managers, IT developers, nurses and ward staff for their support and time in providing us with information about patients' movements for medical treatment within the hospital. The resulting knowledge and analysis has provided a useful foundation for our research in exploiting the RFID usability for healthcare.

18. References

Application Notes CAENRFID, (2008), Introduction to RFID Technology, *CAENRFID: The Art of Identification*

Bharadwaj, V., Raman, R., Reddy, R. & Reddy, S., (2001), Empowering mobile healthcare providers via a patient benefits authorization service, WET ICE 2001. *Proceedings. Tenth IEEE International Workshops on Enabling Technologies: Infrastructure for Collaborative Enterprises*, IEEE.
Bohn, J., (2008), Prototypical implementation of location-aware services based on a middleware architecture for super-distributed RFID tag infrastructures, *Personal Ubiquitous Computing*, ACM, 12 (2):155-166.
Connecting for health, http://www.connectingforhealth.nhs.uk/systemsandservices /nhsmail/using [access: 11th October, 2009].
Connecting for health, http://www.connectingforhealth.nhs.uk/ [access: 18th August, 2010].
DHS (Department of Homeland Security) , (2006), Enhanced Security Controls needed for US-Visit's System using RFID Technology, *U.S. Department of Homeland Security (Office of Inspector General)*, available at: www.dhs.gov/xoig/assets/mgmtrpts /OIG_06-39_Jun06.pdf, OIG-06-39.
DH-UK, http://www.dh.gov.uk/en/index.htm [access: 30th September, 2009].
Frank, T., Brad, H., Anand, M., Hersh, B., Anita, C. & John, K., (2006), *RFID Security*, ISBN: 1-59749-047-4.
GAO (Government Accountability Office), (2005), Information Security: Radio Frequency Identification Technology in the Federal Government, *Report to Congressional Requesters*, US. Government Accountability Office, available at: www.gao.gov/new.items/d05551.pdf, GAO-05-551.
Garfinkel, S. & Rosenberg, B., (2005), *RFID Application, Security, and Privacy*, ISBN: 0-321-29096-8.
Glover, B. & Bhatt, H., (2006) RFID Essentials, *O'Reilly Media, Inc, Sebastopol*, ISBN 0-596-00944-5.
Intermec, (2009), ABCs of RFID: Understanding and using radio frequency identification, White Paper, *Intermec Technologies Corporation*, available at: http://epsfiles.intermec. com/eps_files/eps_wp/ABCsofRFID_wp_web.pdf [access: 3rd January, 2010].
Meiller, Y. & Bureau, S. (2009), Logistics Projects: How to Assess the Right System? The Case of RFID Solutions in Healthcare, *Americas Conference on Information Systems (AMCIS) 2009 Proceedings*, Association for Information Systems.
Narayanan, A., Singh, S. & Somasekharan, M., (2005), Implementing RFID in Library: Methodologies, Advantages and Disadvantages, *Scientific Information Resource Division, IGCAR*, Kalpakkam, Government of India, available at: http://library.igcar.gov. in/readit-2005/conpro/lgw/s5-8.pdf [access: 15th February, 2010].
NHS-UK, http://www.nhs.uk/Pages/HomePage.as px [access: 11th October, 2009].
Parks, R., Yao, W. & Chu, C. H., (2009), RFID Privacy Concerns: A Conceptual Analysis in the Healthcare Sector, *Americas Conference on Information Systems (AMCIS) 2009 Proceedings*, Association for Information Systems.
Sandip, L., (2005), *RFID Sourcebook*, IBM Press, ISBN: 0-13-185137-3.
Schwieren1, J. & Vossen, G., (2009), A Design and Development Methodology for Mobile RFID Applications based on the ID-Services Middleware Architecture, *Tenth*

International Conference on Mobile Data Management: Systems, Service and Middleware, IEEE Computer Society.

Shepard, S., (2005), *RFID Radio Frequency Identification*, McGraw-Hill, ISBN:0-07-144299-5.

Srivastava, L., (2005), RFID: Technology, Applications and Policy Implications, Spectrum Management Workshop, *International Telecommunication Union*, available at: http://www.itu.int/osg/spu/presentations/2005/srivastavaRFID2005.pdf.

Watson, M., (2006), Mobile healthcare applications: a study of access control, Proceedings of the 2006 International Conference on Privacy, Security and Trust: Bridge the Gap Between PST Technologies and Business Services, *ACM*, article no. 77, DOI: http://doi.acm.org/10.1145/1501434.1501524.

Zeisel, E. & Sabella, R. (2006), *RFID+ Exam Cram*, Pearson, Series 2, ISBN: 0-7897-3504-0.

3

Development of a Neonatal Interactive Simulator by Using an RFID Module for Healthcare Professionals Training

Loreana Arrighi, Jenny Cifuentes, Daniel Fonseca,
Luis Méndez, Flavio Prieto and Jhon J. Ramírez
Universidad Nacional de Colombia
Colombia

1. Introduction

This chapter of the book presents the experience and achievements attained in a project carried out by the National University of Colombia which is intended to design and implement tools for training students in medical and nursing techniques applied on neonatal patients. The main result to be shown in this chapter is a virtual and physical tool - based on RFID technologies - that simulates pathologies in neonates in order to teach students the correct use of medications by means of umbilical vein catheterization based on the medical interpretation of the patient's symptoms. In addition, professor's and student's testimonies are shown referencing their experience with the tool in the generation of different medical scenarios of diagnosis and in the application of dosification techniques.

This chapter is organized as follows: the project justification is presented in Section 2 along with other projects already carried out in this line of research; in Section 3, the design and the implementation is presented; next, in Section 4, the results are exposed and finally the conclusions and recommendations are stated by the authors.

2. Justification and background

2.1 Justification and problem definition

Around 100 million babies are born every year worldwide and approximately 10% of them need some assistance to start breathing; 1% of the total requires intensive resuscitation efforts such as endotracheal intubation and thoracic massages (Murphy & Halamek, 2005). In neonatology, undesirable events that emerge from medical practice can have a negative impact on the neonate's formation and growth. Hence, medical and nursery personnel training and learning processes with real patients carried out before become a decisive factor when saving lives and guaranteeing adequate prognosis.

The traditional learning method has two stages: the theoretical knowledge and clinic experience. The limitations of those stages are illustrated in Table 1: the class environment is characterized by being extremely theoretical and by the lack of realism and the clinical setting is where at some point apprentices refine his or her abilities with live patients but associated with a high risk for their health. In addition, clinics are compelled to ensure optimal treatment for their patients from the very first moment they are admitted (Hayes, 1994; Lynöe et al, 1998).

Class Environment
It is characterized by being passive in its learning opportunities
It is focused on teaching instead of learning
Lack of realistic signals, distractions or pressure
Incapable of preparing the apprentice adequately for a real environment
Clinical Environment
Exposes patients to some degree of risk
Learning opportunities are random
Learning is limited by the swiftness of the moment, pressure and high inherent cost

Table 1. Limitations of traditional methods (Halamek, 2008)

Tools that have led to a new way to teach and learn based on Medical Simulation (Murphy & Halamek, 2005); Ostergaard et al, 2004 ; Ziv et al, 2006) by using computational tools and mannequins are being used to avoid experimenting with real patients and overcome the limitations of conventional medical training

Simulation with training equipment allows saving lives and improving quality of life since medicine students can acquire skills and key competences such as the appropriation of new knowledge, making fast and safe decisions, and the acquisition of clinical experience in environments similar to those that take place in real emergencies.

Nevertheless, one of the greatest challenges of the simulation and the use of mannequins is that the condition of a real patient changes throughout time depending on the quality and swiftness of the diagnostic and the treatment; in contrast mannequins are stable and the pathology evolution is left to the imagination of the doctor or nurse because the symptoms are difficult to simulate in the dummy.

Even though the quality of simulators that can be acquired in the market is excellent, there are some disadvantages such as their high costs and that the controllers which allow practicing the development of pathologies cannot be used because they differ from the Colombian health sector conditions. The medicine faculty of the National University of Colombia has developed its own philosophies, methodologies and technical approaches to diagnose and to follow schemes under adverse conditions like those found in healthcare centers in any region in Colombia. Nevertheless there is an important barrier to teach and learn because commercial simulators do not allow the presentation of these philosophies, methods and techniques developed in this institution. (Currea, 2004)

On top of that, many of the commercial simulators have limited communication infrastructures among the different elements of such simulators; such is the case of wired connections to exchange data between the controllers and mannequins that can be replaced by radio frequency technologies and radio identification (RFID).

Due to the importance of the topic and the mentioned limitations, a variety of tools for medical simulation have been developed in the present project by members of the Master in Industrial Automation of the National University of Colombia using dynamic models that allow the generation of diverse biomedical signals of a neonate in order to work with a more real perception. In addition, a virtual and a physical tool for the simulation of neonatal pathologies has been created based on RFID technology in order to teach students the correct way to administrate medications through the umbilical cord based on the medical interpretation of the patient's symptoms which are recreated by virtual reality using animated graphics.

2.2 The medical simulation: context and background

A simulator is an artificial representation of the real world giving the enough fidelity to achieve a specific goal in the learning process (Halamek et al, 2000 ; Ostergaard et al, 2004; Rall & Dieckmann, 2005) .

Medical Simulation is also defined as the imitation of a real thing, situation or medical process for the practicing of skills and resolution problems (Halamek, 2008). It is a recent method for learning among healthcare areas, and it reduces the gap between cognitive skills and clinical experience.

In general, medical simulation has been structured into 5 categories; see Table 2, according to the method proposed by David Gaba (Small et al, 1999) : verbal, standardized patients, body parts trainers, computerized patients and electronic patients.

Category	Characteristics
Verbal Simulation	It is based on knowledge communication by using role plays.
Standardized Patients	Actors that perform and evaluation, for instance, on the way to obtain clinical data, the necessary skills to carry out physical checkups as well as communication and professionalism.
body parts trainers	Anatomical models of body parts showing a normal state or representing any illness or problem.
Computerized patients	Interactive patients that can be either software-based or part of an internet-based world.
Electronic Patients	These are software applications that operate over a virtual reality or a mannequin and the clinical environment mimicked is integral.

Table 2. Schemes of medical simulation

The main advantages of simulators are (Halamek, 2008 ; Murphy & Halamek, 2005 ; Ziv et al, 2006) :

- It does not generate any risk to the patients due to it reduces the error probability or undesirable events in human beings.
- It allows practice without interferences and interruptions.
- It facilitates feedback from both the professor and training environment to the student.
- Simulations can be organized in convenient moments for both trainees and trainers.
- It can be scalable in intensity in order to know the needs of apprentices in all levels of experience.
- It allows the practice of unusual routines and situations.
- It promotes the integration of cognitive, technical and behavioral skills.
- It facilitates the training of students into multidisciplinary teams.
- It promotes the use of multiple learning strategies.
- It facilitates an objective evaluation for each student.

Simulation has been formally used in medical training in the last decades. Nevertheless, representation of signs and symptoms referenced in the literature or in the theater can be actually considered as the predecessors of non-technical simulation. Application of this tool was delayed because of high costs and lack of rigorous testing which generated skepticism as well as resistance to change (Ziv et al, 2006) . Some of the most relevant predecessors of simulators for medical training are presented in the following sections.

2.2.1 Computerized simulators

Computerized simulation in the medical area started in 1960 with a graphic communication system (Khalifa et al, 2006). Computers facilitated the mathematical description of the human physiology and pharmacology as well as the worldwide communication and the design of virtual worlds (Smith & Daniel, 2000). This resulted in the development of a virtual reality prototype for medical training in which the user was represented by an avatar which was capable of handling its virtual instruments and carrying out medical procedures. This platform allowed several users and multiple modules of simulation that allowed the creation of a shared virtual environment (Stanfield et al, 1998); in this latter aspect, N T and Smith and the colleges from California University used their experience in cardiovascular physiology and anesthetics to develop "Sleeper" which was the precursor of the current BodySim® designed for practicing resuscitation (Cooper & Taqueti, 2004). Years later, MicroSim® would be released to the market; a CD-ROM of Laerdal intended to provide structured training in medical emergencies (Perkins, 2007).

Currently, all the branches of surgery including general surgery (McCloy & Stone, 2001), urology (Hoznek et al, 2003), neurosurgery (Spicer & Apuzzo, 2003), gynecology (Letterie, 2003), and orthopedic surgery (Tsai et al, 2001) have made use of virtual reality in one way or another. In addition, anesthesiology and other medicine subspecialties oriented to procedures such as gastroenterology, lung science and cardiology that have been included in the area of virtual reality (Gillies & Williams, 1987).

2.2.2 Physical simulators

Mannequins to teach obstetric skills and reduce high mortality in infants were patented in 1960 (Buck, 1991). In particular, Resusci Annie®, Laerdal's emblematic product; is one of the first landmarks in the history of medical simulation because even when it was initially designed for mouth to mouth respiration, it subsequently evolved by integrating a spring in its chest to allow cardiopulmonary resuscitation.

The first patient simulator at human scale was called Sim 1® and it was built by the University of California. Some features of this simulator include pupils that can dilate, jaw that can open, eyes that can blink, respiratory movements and heart beat synchronized with temporal and carotid pulse (Cooper & Taqueti, 2004).

Gaba built the Comprehensive Anesthesia Simulation Environment (CASE) prototype in 1986 en Stanford. Similar to other innovations, its high cost limited the acquisition of the mannequins to a reduced quantity in medical centers. Several European centers developed their own computerized mannequins for simulation. ACCESS®, Sophus® and Leiden® are three examples of inexpensive simulators developed worldwide (Chopra et al, 1994). After, the KISMET® simulator (1993) introduced distant-surgery, which initially had low realism in quirurgic simulations but was quickly improved parallel to the progress in technical elements and computer power. The partial mannequin Simulator-K was developed to assess cardiac abilities (1990) (Takashina et al, 1990).

At the same time, UltraSim® reproduced the relevant abdominal pathology in obstetrics and gynecology; then, the ophthalmic training system evolved into virtual reality with EYESI® produced by VRMagic; this one was initially designed as a simulator of vitreoretinal surgery and then it became the learning tool of a deeper ophthalmic quirurgic procedure (Khalifa et al, 2006).
The first training program based on simulation of neonatal resuscitation was developed in Standford University by the mid 90's (Halamek et al, 2000); then, Gaumard Scientific Company produced a mannequin of a neonate capable of simulate cyanosis.

2.2.3 Electronic simulator

A computer application was developed by the end of the 90's which enabled remote observation and control of the most relevant signals for the neonates monitoring (cardiac frequency and skin color), and also, a virtual model of the patient was implemented in which the vital signals could be controlled by an external Java application (Korosec et al, 2000).
In the year 2000, Laerdal presented SimMan®; it was the first human-scaled portable mannequin designed to train the skills and performance on resuscitation scenarios. This model also generates heart bits, mimics respiration and blood pressure and allows the trainer to develop and to edit his or her own scenarios or reuse preset scenarios (Perkins, 2007).
Then, SIMA adopted a new approach and introduced a personal computer, software, a monitor and 8 training scenarios. Currently, SimBaby® is the simulator used for training neonatal resuscitation which includes the software and a technologically advanced and interactive mannequin.
These commercial simulators have excellent quality but present some disadvantages; among them are the high cost (Halamek, 2008) and the fact that there are special training centers needed that at the same time require instruments, monitors, mannequins and technical personnel to control and supervise training (Korosec et al, 2000).

3. Proposed design for the neonatal pathologies simulator

Taking into account the characteristics of the models presented in Section 2, and in order to build a tool for both Medicine and Nursery students to acquire skills in diagnosing neonatal patients, an interactive simulator has been designed. This device consists of a screen that allows the instructor to program the health status of a patient by modifying its vital signs to create different pathologic and non-pathologic scenarios; then students are asked to define what they believe should be the appropriate treatment.
The vital signs are simulated because they are the main indicators that reflect the physiological status of vital organs (brain, heart and lungs) which immediately express the functional changes in the organism. The vital signs are the measure of different variables: cardiac frequency, pulse, respiratory frequency, blood pressure (systolic, diastolic and average) and temperature. Nevertheless, literature also recommends complementing these parameters with other useful measurements such as Pulse-Oximetry.
Acquiring the ability to interpret in an adequate and opportune way those physiological parameters (vital signs) is essential in medical training as it helps healthcare professionals and first aid personnel in selecting an appropriate treatment among the different choices. Determining and analyzing vital signs is very important during an emergency where many

patients arrive with a huge variety of clinical conditions, especially for neonatal patients whose symptomathology cannot be described thoroughly. Healthcare students must learn how to choose the correct medicine and dose according to the patient's particular symptoms. The minimum increase of a dose or the wrong medicament injection can be very prejudicial for an infant, it also can cause dead in extreme cases. Hence, a mannequin has been adapted to identify some medicines that trainees apply via umbilical vein catheterization and to show the health status after the treatment.
Figure 1 shows the graphic scheme of the neonatal pathologies simulator its main elements are: *a graphic interface* that shows the vital signs and allows selecting the medication, an RFID *medicines programmer*, a *syringe applicator*, a *mannequin that identifies medications* and a *tool to acquire data*.

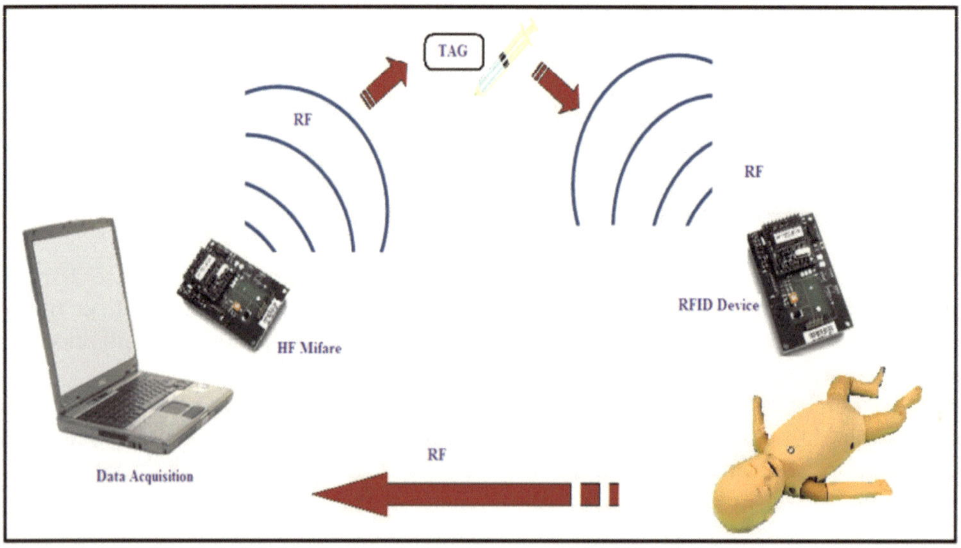

Fig. 1. Graphic of a virtual and physical simulator of a neonatal patient

In a training scenario, students and instructors must do the following: the instructor changes the vital signs of the patient (frequency and maximum and minimum values) through the graphic interface that shows the vital signs such as: ECG, pulse, pulmonary pressure and CO2 and O2 levels. In this way, the instructor can modify the health status in order to generate diverse medical scenarios. Subsequently, the student has to choose the applying medication and its dose once the diagnostic has been carried out through the same graphic interface.
The data of the medication and its dose chosen by the student for treating the patient are sent by an RFID programmer connected to the computer to the fields of an RFID Tag that is attached to the syringe (see Figure 1). In addition, the mannequin has an RFID reader embedded in its abdomen to receive the data stored in the Tag when the syringe is approached to the identifier by the student.
Once the described process is carried out, the vital signs of the patient are automatically modified by the software in the mannequin according to the chosen treatment. In this way, a new health condition is presented to the student as a feedback indicating whether the choice of medication and dose has been the correct one or not. The neonate's condition is reported continuously to the computer by using an acquisition tool implemented with wireless

communication between the emitting module in the mannequin and the receptor in the computer. In this way the patient's health is constantly monitored not only by watching the mannequin but also it can be seen in the graphic interface.
The mannequin produces cardiac sounds. It also recreates the skin flushing and, by an LCD screen, it is possible to see its rectal temperature and cardiac frequency.

4. Theoretical foundations

In this chapter is presented the previous investigation made about the vital signs which are relevant to accurately make a diagnostic over a newborn's health as it would happen in real life. Numeral 4.1 summarizes the main medical signals that were simulated: ECG, cardiac frequency, pulse, respiratory frequency, arterial pressure and levels of CO2 and O2, among others.
The selected medicines to be used in the system are shown in numeral 4.2 as well as some parameters such as the supply method and affected variables. These medicines can change the health status of the newborn which will be immediately reported to the computer where the instructor can evidence the decision made by the trainee considering the changes in vital signs and appearance.

4.1 Variable monitoring

Intensive care units were created due to the need of exhaustive and strict monitoring of patients with high risk pathologies. The current status of a patient is assessed by watching and continuously recording the physiological and pathophysiological parameters and then their evolution as result of the therapeutic applied by watching the hemodynamics. Nowadays, monitoring patients is an important part of all medical care due to it allows watching the progress of a patient and guarantees an early detection of adverse events or late recovering.
In Figure 2 the variables that were simulated in this project are presented.

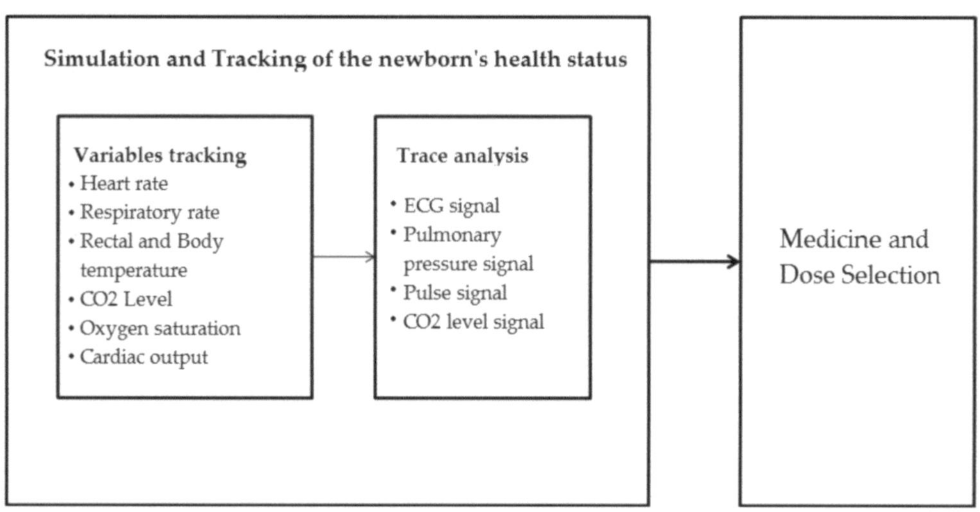

Fig. 2. Diagram of the virtual simulator of a neonate patient (Software)

4.1.1 ECG Signal y cardiac frequency

Signal morphology

The heart is the central structure of the cardiovascular system. Contraction of any muscle is associated with electrical changes called "depolarization"; those changes can be detected by electrodes located on the body surface. Although the heart has four chambers, from the electric point of view it has only two as the two auricles contract together with the two ventricles (Hampton, 2008).

The muscular mass of the auricle is smaller than the one of the ventricles and in thus, the electrical change produced by the contraction of the auricles is also small. The contraction of the auricle is associated with the "P" wave of the ECG signal. The ventricular mass is large which generates a high deflection of the ECG signal when the ventricles are depolarized; that wave is called the QRS complex. The "T" wave of the ECG signal is associated to the returning of the ventricular mass to its electrical state - repolarization (See Figure 3.a).

The diagnostic of the diverse pathologies is done based on the analysis of the following characteristics of the ECG signal (see Figure 3.b) (Resiner & Clifford, 2006):

- Cardiac frequency (Heart Rate): the number of heart bits or pulsations is commonly the ventricular frequency. The normal range for an adult is between 60-120 bpm; for a newborn it fluctuates between 100 to 160 bpm.
- Regularity: R-R and P-P intervals are analyzed in search for anomalies.
- P Waves: Size, shape and position are analyzed.
- QRS waves (complex): Size, shape and position are analyzed.
- T Waves: Size, shape and position are analyzed.
- PR, QRS and QT intervals: These are analyzed and compared to standard ranges.
- U Waves: These waves are normally invisible, that is, their presence is symptom of anomaly.

The implemented mathematical model

In order to generate the synthetic signal, the dynamic model adapted from MsSharry (MsSharry et al, 2003) was used; this model generates a trajectory in a tridimensional space (3D) with (x, y, z) coordinates. The quasi-periodicity of the ECG signal is shown by the movement of the trajectory along a limit cycle of unitary radius in the (x, y) plane. Each revolution of this cycle corresponds to a heartbeat.

The different points in the ECG (P, Q, R, S and T) are described as attractors or repulsors, positive or negative in the z direction; these are placed with fixed angles along the unitary circle given by: P , Q, R, S and T (MsSharry et al, 2003). The Dynamic equations of movement are given by a set of ordinary differential equations (Equations 1, 2, 3).

$$\dot{x} = \alpha x - \omega y \tag{1}$$

$$\dot{y} = \alpha y - \omega x \tag{2}$$

$$\dot{z} = -\sum_{i \in = \{P,Q,R,S,T\}} a_i \Delta\theta_i e^{\left(-\frac{\Delta\theta_i^2}{2b_i^2}\right)} - (z - z_0) \tag{3}$$

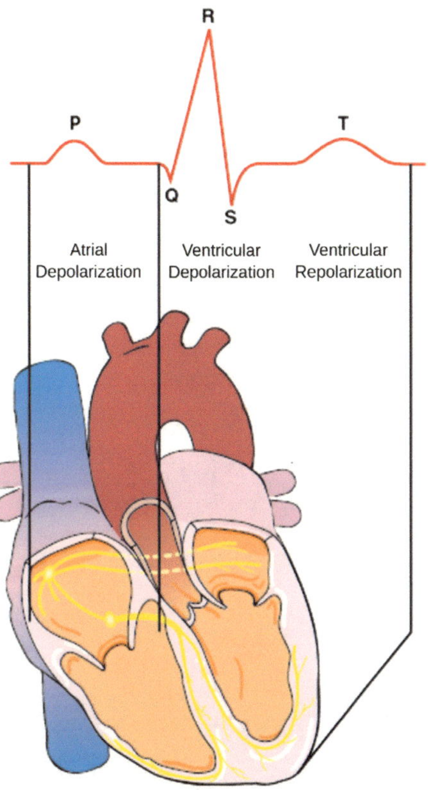

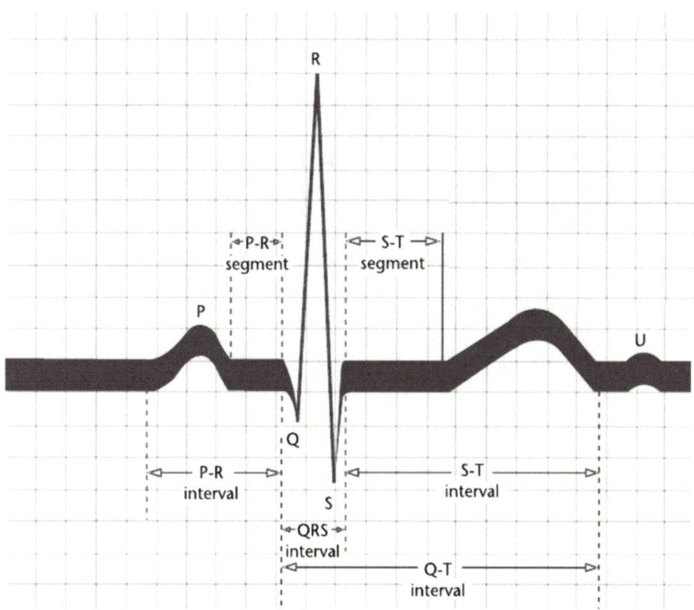

Fig. 3. a) Heart depolarization and repolarization (Jones, 2005), b) Characteristics of the ECG signal (Resiner & Clifford, 2006)

Where:

$$\alpha = 1 - \sqrt{x^2 + y^2} \tag{4}$$

$$\Delta\theta_i = (\theta - \theta_i)\mathrm{mod}(2\pi) \tag{5}$$

$$\theta = \arctan(y, x) \tag{6}$$

Yω is the angular frequency of the trajectory; time, angles, *a* and *b* values for a normal child can be found in (MsSharry et al, 2003).
Angular speed is obtained from the power spectrum of the signal given by the sum of Gaussian distributions described in the Equation 7.

$$S(f) = \frac{\sigma_1^2}{\sqrt{2\pi c_1^2}} e^{\left(\frac{(f-f_1)^2}{2c_1^2}\right)} + \frac{\sigma_2^2}{\sqrt{2\pi c_2^2}} e^{\left(\frac{(f-f_2)^2}{2c_2^2}\right)} \tag{7}$$

With $f_1 = 0{,}1$, $f_2 = 0{,}25$ and standard deviations $c_1 = 0{,}01$ y $c_2 = 0{,}01$ (MsSharry et al, 2003).
The synthetic signal was obtained in LabView ®, as can be observed in the Figure 4.

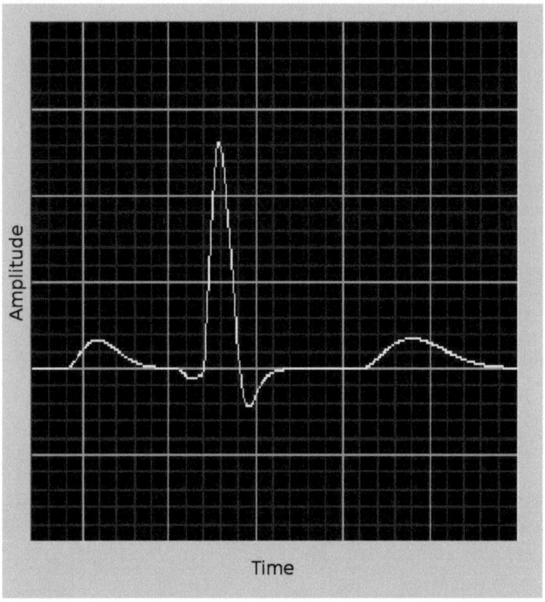

Fig. 4. Synthetic ECG signal

4.1.2 Pulse signal

Signal morphology

When the heart beats, it generates a pulse wave caused by expansion of the arteries by the circulating blood. This signal has a rounded initial peak that smoothly decreases to a sharp depression called "dicrotic notch" that occurs as a result of abrupt closure the aortic valve, finally descending to the diastole (see Figure 5.a). This particular waveform is due to the overlapping between a pressure wave, which starts from the heart to the periphery and the other, reflected at the bifurcation of the descending aorta (see Figure 5.b).

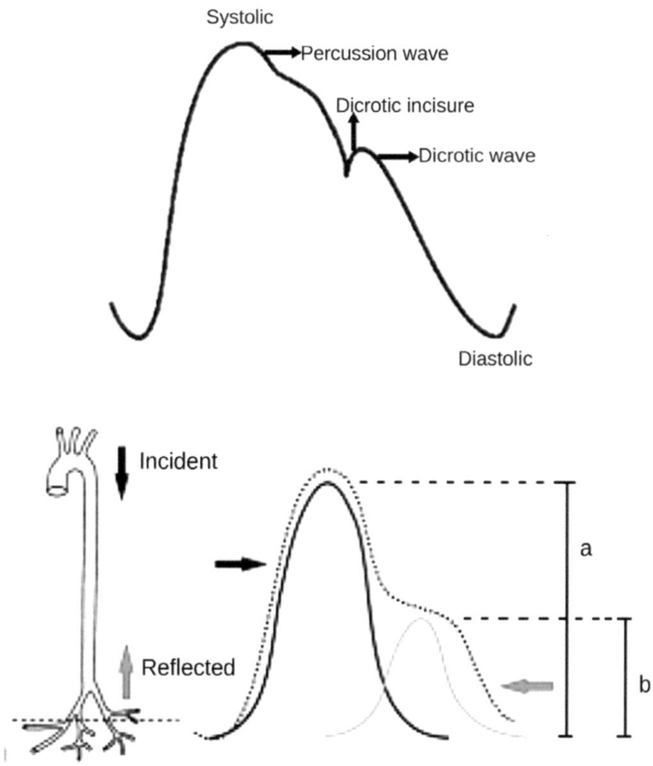

Fig. 5. a) Morphology of the pulse signal (Jones, 2005), b) Composition of the pulse signal (Vanetta & Gomez)

The Elasticity and status of arterial walls determine the size and shape of those waves. The pulse wave measures the speed at which blood travels throughout the vascular system. A slow or obstructed movement of the blood flow means slow transference of nutrients to the cell. This condition might result, among other things, in high blood pressure, lack of energy, low metabolism, loss of memory and can affect negatively the immune system.
In general, the following can be identified by analyzing the characteristics of the pulse signal:

- Premature levels of ageing and stress of the vascular system
- Efficiency of heart pumping
- Arterial elasticity and obstruction levels of large and small arteries
- Early signs of cardiac stress

The implemented mathematical model

In order to generate the synthetic signal a mathematical model was used, this model generates a trajectory in a tridimensional space (3D) with (x, y, z) coordinates. Each revolution of this cycle corresponds to a heartbeat. The waves that compose the signal are described as attractors or repulsors, positive or negative in the z direction; these are placed with fixed angles along the unitary circle The Dynamic equations of movement are given by a set of ordinary differential equations (Equations 8, 9, 10).

$$\dot{x} = \alpha x - \omega y \quad (8)$$

$$\dot{y} = \alpha y - \omega x \quad (9)$$

$$\dot{z} = -\sum_{i \in =\{R,I\}} a_i \Delta\theta_i e^{\left(-\frac{\Delta\theta_i^2}{2b_i^2}\right)} - (z - z_0) \quad (10)$$

Where:

$$\alpha = 1 - \sqrt{x^2 + y^2} \quad (11)$$

$$\Delta\theta_i = (\theta - \theta_i)\text{mod}(2\pi) \quad (12)$$

$$\theta = \arctan(y, x) \quad (13)$$

The synthetic signal obtained can be observed in Figure 6

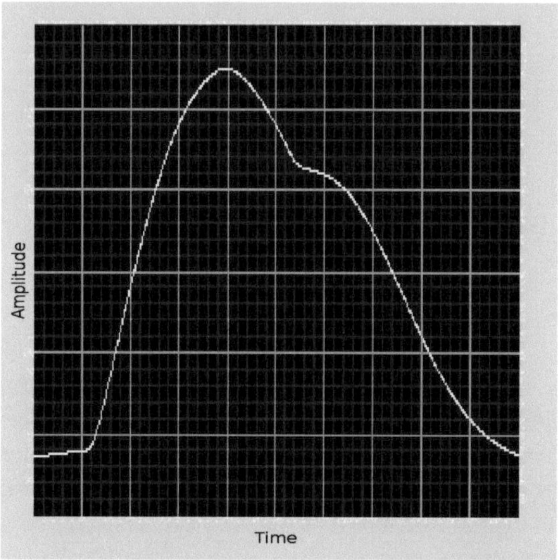

Fig. 6. Synthetic Pulse signal

4.1.3 Arterial pressure

Morphology of the signal

Blood pressure is the force that blood exerts against the arteries' walls. This variable depends on the volume of blood in the vessels and the distensibility of the walls. If the volume of blood that enters the arteries equals the exiting volume in a period of time, the arterial pressure remains constant. Nevertheless, during the ventricular systole (contractions of the ventricles) a high volume of blood enters the arteries while only a third is expelled towards the arterioles. During diastole (heart relaxing after a contraction) there is no blood entering the arteries although there is a continuous amount of blood going out caused by the elastic recoil of the blood vessel walls. The maximum pressure exert on the arteries while the blood is expelled during systole is called "systolic pressure". The minimum pressure on arteries when the blood is drained to the rest of vessels during diastole is called "diastolic pressure". The pulse pressure is the difference between the systolic pressure and the diastolic pressure; finally, the mean pressure is the average of the pressure during the whole cardiac cycle (Sherwood, 2010) (see Figure 7).

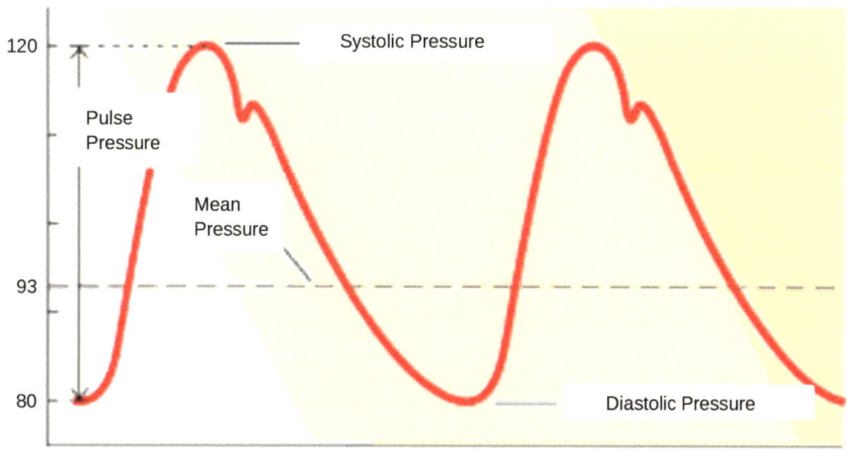

Fig. 7. Components of the arterial pressure wave (Sherwood, 2010)

In practice, arterial pressure is expressed as the systolic pressure over diastolic pressure. Values produced by those measurements and their limits (meaning hyper or hypotension) are relative and depend on each patient and their inherent factors; nevertheless, it is established that a normal reading for an adult patient could reach up to 135/90 mmHg. In contrast between 140/90 mmHg and 160/110 mmHg there would be mild hypertension. If the result is above these values, it would indicate a severe hypertension. On the contrary, values under 100/60 mmHg would represent hypotension or low arterial pressure. Values of arterial pressures in newborns vary significantly compared to those of the adults and are defined by variables such as gestational age, weight and postnatal age, among others.

The implemented mathematical model

In order to obtain the blood pressure signal, the linearized and improved cardiovascular physiology model presented by Beneken has been used (Beneken, 1965). This hydraulic model of 10 compartments describes: **systemic and pulmonary circulation** (see Figure 8). The model accepts changes in blood volume and intrathoracic pressures as inputs, and generates the pulmonary and systemic pressures as outputs. Blood pressure is calculated for the model of each compartment (Equation 14), the input flow (Equation 15) and the volume changes (Equation 16). Equations of the compartments adjust with each other as the input flow of one compartment depends on the pressure of the previous one and the changes in volume depend on the input and output flows. The expressions use elastance, resistance and volume variables.

$$p(t) = E(v(t) - UV) \tag{14}$$

$$f(t) = \frac{p_{in}(t) - p(t)}{R} \tag{15}$$

$$\frac{dv(t)}{dt} = f(t) - f_{out}(t) \tag{16}$$

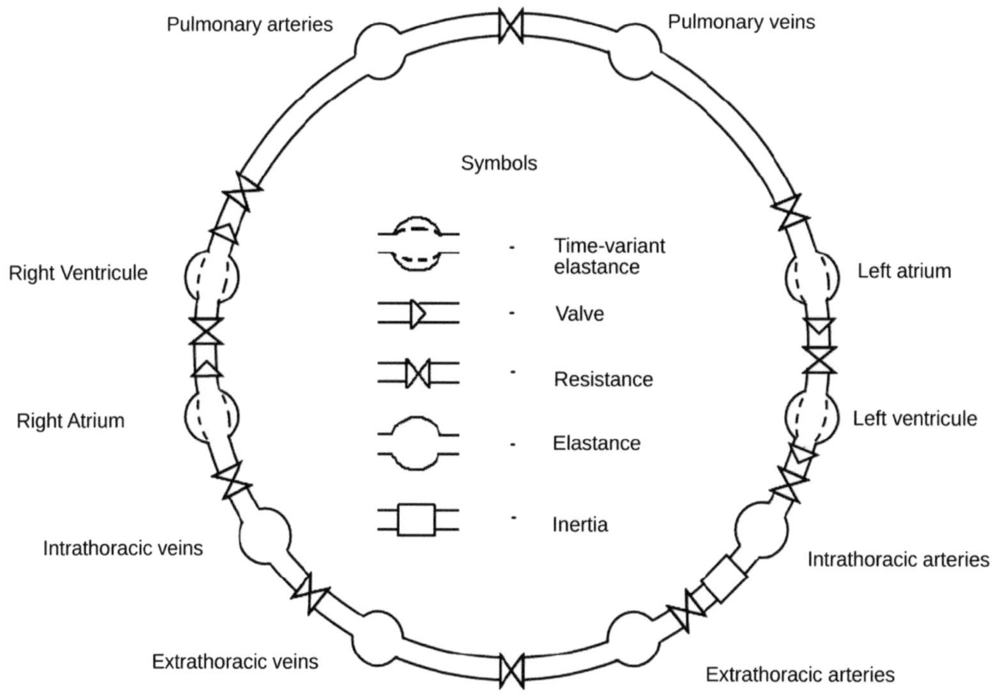

Fig. 8. Physiology Cardiovascular Model (Beneken, 1965).

The inertial behavior of the blood in the arteries is defined by the differential equation (see Equation 17).

$$\frac{df_{etha}(t)}{dt} = \frac{p_{itha}(t) + PTH - RETHAf_{etha}(t) - p_{etha}(t)}{LETHA} \tag{17}$$

Where PTH represents the average intrathoracic pressure, RETHA is the resistance of the extra-thoracic arteries and LETHA represents the inertia of the blood flow in the arteries. Data of the constants of a newborn patient were obtained from (Beneken, 1965). The synthetic signal obtained to represent the pressure is presented in Figure 9.

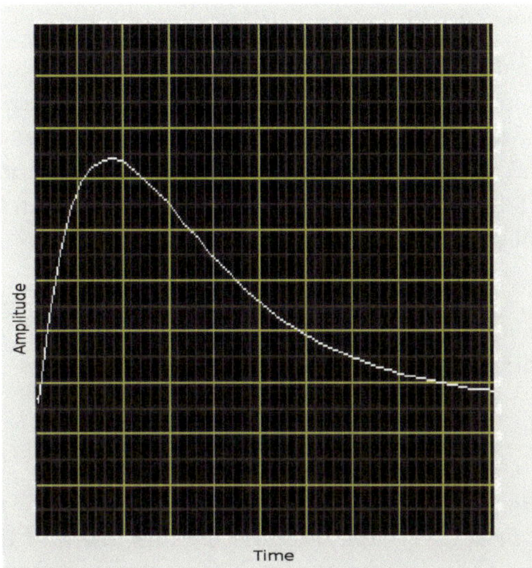

Fig. 9. Synthetic pressure signal

4.1.4 CO_2 Levels and respiratory frequency

Morphology of the signal

The concentration of CO_2 in expired gases has a close relationship with tissue metabolism, systemic circulation and ventilation. Capnography is the graphic record of instant concentration of CO_2 in gases expired during a respiratory cycle (Bhavani-Shankar et al, 1992). A capnogram is divided into four fundamental phases (see Figure 10).

The first Phase (A-B) represents the initial stage of breathing. In this phase, the gas occupies unused space, normally containing CO_2. In point B, a strong movement is shown in the capnogram which is the Phase (B-C). The slope of this movement is determined by the uniformity in the alveolar ventilation and in the respiratory emptying. In point D, the CO_2 concentration shows its highest value at the end of the respiratory cycle. When a patient initiates the inspiration, fresh gas enters and there is a significant drop of the baseline. Unless there is a re-inhalation of CO_2 the baseline approximates to zero (Barash et al, 2009).

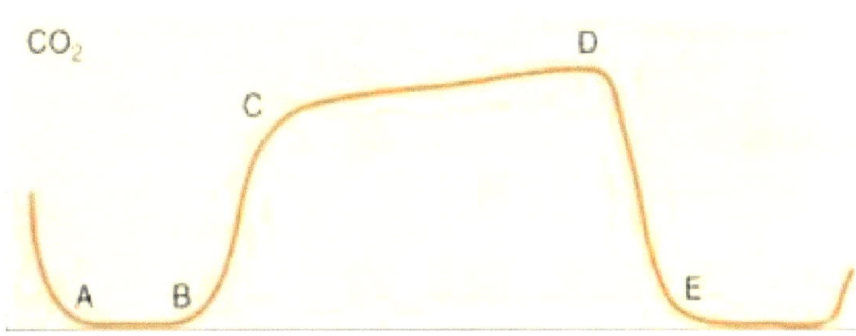

Fig. 10. Normal Capnogram (Barash et al, 2009)

The frequency of the figure above is known as the respiratory frequency or respiratory rate and corresponds to the number of respirations (inhalation and exhalation) within a period of time.

The implemented mathematical model

The capnogram is divided into four fundamental phases (see Figure 10). This wave shape can be described by decreasing exponentials that model the aspiratory and expiratory processes. The frequency of this signal is related with the respiratory frequency.
The dynamic model used describes 2 first degree differential equations; the first expression describes de aspiration (see Equation 18) and the second describes all the cycle, expiration and aspiration (see Equation 19).

$$\frac{df}{dt} = \frac{1}{\tau}(-f + \Phi) \tag{18}$$

$$\frac{dN_{CO2}}{dt} = \frac{1}{\tau_2}(-N_{CO2} + \alpha(f(t - D))) \tag{19}$$

τ y τ_2 define the time constants of the exponentials that represent the inspiration and the expiration respectively. Besides, φ and α define the baseline and the maximum CO_2 of a respiratory cycle. Finally, D is defined as the time in which the respiratory process takes place. In the Figure 11 synthetic signal obtained is shown.

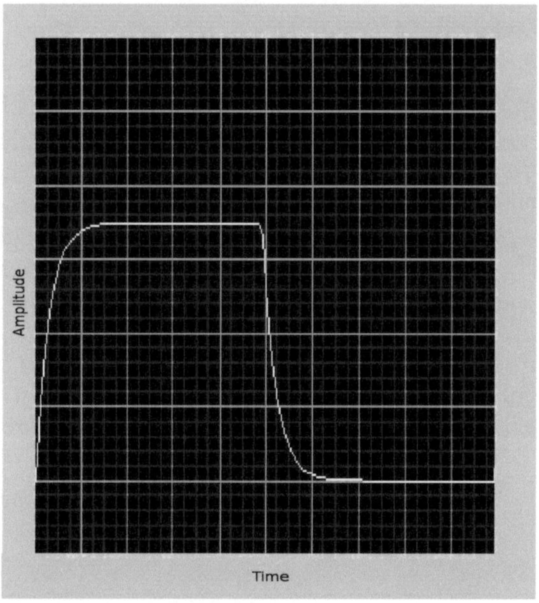

Fig. 11. Synthetic capnogram

4.1.5 Other variables

Temperature

Human beings along with birds and mammals are categorized as warm-blooded animals or homeothermic beings; that is, that despite of being exposed to a variety of temperatures, homeothermic organisms keep their temperature steady. Cells in the body perform optimally within a temperature range between 35 to 38 centigrade degrees.

The center of temperature regulation of humans is the hypothalamus; this is an area in the brain above the pituitary gland that acts as a thermostat to maintain the body's internal temperature within a range between 36,1 - 37,7 centigrade degrees.
Regarding to this measured values, if the oral temperature is above 38 °C, it can be said that the individual has a fever. On the other hand, rectal temperature is always higher than the oral one by 0,6 °C whereas axillary temperature is lower than internal temperature by 1 °C. A failure in the thermoregulatory system with temperatures equal to or greater than 41 °C would lead to a malign hyperthermia, which is characterized by a failure in the mechanisms of heat loss.
Hyperpyrexia takes place when the body temperature is 41 °C, taken as an isolated reading, or if there is an increase of 1 °C every 2 hours. This could be originated by fever or hyperthermia. On the contrary, hypothermia is the decrease in central body temperature (rectal reading) below 35° C. The most common cause is the accidental exposition to extremely low temperatures which may take place during winter, accidents in mountains and immersion in cold water.

Cardiac output

Cardiac output is composed by two main factors: the "ejected volume", which is the blood volume expelled by the heart in each heartbeat and the "cardiac frequency". The multiplication of both factors expresses the cardiac volume per minute or, what has been called "cardiac output".
Cardiac output normally decreases during normal sleep as well as under general anesthesia. Some anesthetics such as the halothane can reduce the cardiac output excessively as it reduces the sympathetic discharge in the cardio vascular system. In particular, a strong circulatory insufficiency is characterized by an abnormally low cardiac output. In chronic cardiac insufficiency, the cardiac output can be limited only during intense physical activity; nevertheless, after certain time, the reduction also takes place even during rest limiting the physical capacity .
During physical exercise, incremental cardiac output takes place; likewise, cardiac output can be greater than 50% by the end of pregnancy as well as under certain pathological conditions such as hyperthyroidism or arteriovenous fistula.

Oxygen saturation

Oxygen saturation is defined as the relationship between the amount of oxygen combined with hemoglobin present in a particular location and the maximum amount of oxygen that could be combined with the hemoglobin in the same setting. In this way, oxygen saturation indicates the amount of oxygen that is being transported by the plasma.
Under controlled conditions and constant monitoring, the saturation needed to reach and keep proper blood oxygenation can reach levels of 97% in infants; similarly, at altitudes such as that of Bogota, saturation can fluctuate between 88 to 92 % with a maximum range between 85% and 95%.

4.2 Selecting medication and dose

Once vital signs of newborns have been simulated to create different scenarios, the medicines that will be used by the simulator have to be selected in order to stabilize vital signs in case the trainee finds a pathological scenario. The following substances that are of common use in neonates were initially considered (Taketomo, Hodding, 2010):

- Cardiovascular: Adenosine, Digoxin , Dobutamine, Dopamine, Indomethacin, Terbutaline.
- Respiratory System: Aminophylline, Dexamethasone, Salbutamol.
- Central and peripheral nervous system: Phenobarbital, Phenytoin, Fentanyl, Midazolam.
- Miscellneous: Adrenaline, Atropine, Human albumin 20%, Atropine, Sodium Bicarbonate, Furosemide, Calcium Gluconate 10%, Cristalline Insulin, Physiological Serum 0,9%, Pulmonary Surfactant, Vitam K1, Vecuronium.

From the previous list some medicines that are administered via intravenous route were selected. Similarly, those that can be administered via umbilical vein were chosen since this is one of the most common ways used during the neonatal period and also because this is the place in where the identifier will be located (See Figure 1). According to these parameters, the selected medications are: Adenosine, Adrenaline, Atropine and Terbutaline. Each of them has a different purpose but physically they have a similar effect in the patient. Table 3 summarizes the selected medications with their corresponding uses and their side effects.

Medication	Use	Side effects
Adenosine	Convert tachycardia in to a sinus rhythm	Arrhythmias Redness - Flushing Bradycardia Hypotension Apnea
Adrenaline - Epinephrine	Increase heart rate	Tachycardia Cardiac arrhythmia Sudden death Hypertension
Atropine	Sinus bradycardia	Arrhythmia Fever Flushing Tachycardia
Terbutaline	Increase heart rate bradycardia	Tachycardia Arrhythmias Flushing Hypertension

Table 3. Table of medicines, their uses and side effects.

It was also important to determine the proper dose for each of the selected prescriptions. In order to find this information, the following guidelines have to be taken into account:

- Concentration in Vaccine Bottle: it is the ratio between the amount of solute (mg) and the amount of solvent (mL). It has to be specified how many milligrams of the vaccine bottle need to be administrated to the neonate according to his/her weight.
- Necessary dilutions: Dilution is the process by which the concentration of a solution is reduced by adding a solvent. Vaccine bottles containing pure medication or initial concentrations are not used in newborns due to their cardiovascular, respiratory and immune systems would not tolerate them.

- Neonate's weight: This parameter is relevant to know the dose to be administered by taking into account the weight in kilograms (Kg); along with this information, the proper dose to be given to the newborn can be determined. The proper dose has to be calculated accurately since in case of administering a wrong amount the newborn can suffer undesired side effects.

In Section 5.1.2 the appropriate dose is presented for each medicament according to the newborns' weight.

5. Implementation of the system

All the information referenced in the previous section was considered when implementing the virtual and physical interactive simulator. Vital signs simulated in a virtual way allow the instructor to recreate different medical scenarios; medications allow trainee to choose the treatment that will be applied. The mannequin reflects the health condition of the neonate which indicates the student whether the right medicament and dose have been selected. Taking these mentioned elements into account and in order to design and implement the simulator, the main system blocks and the data flow are shown in Figure 12. Each of these system blocks will be explained in this chapter along with the implementations obtained in each of them.

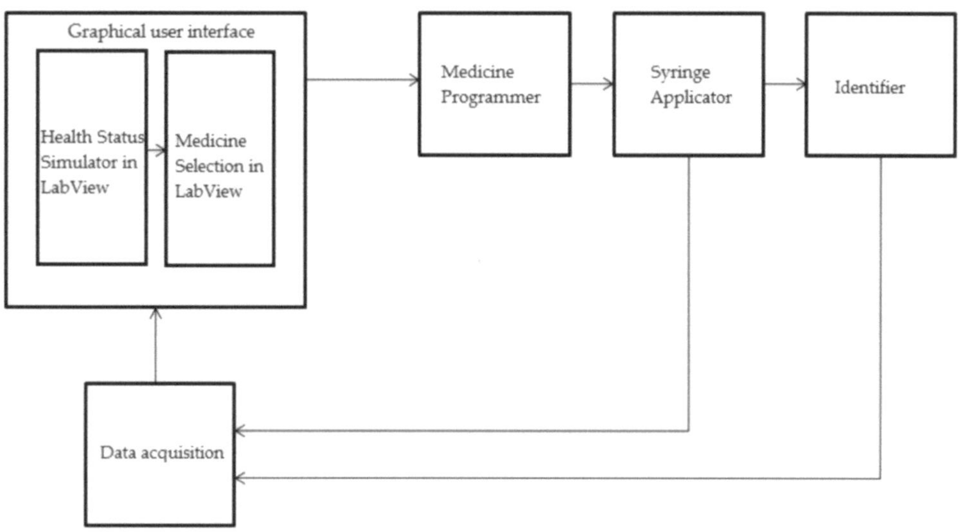

Fig. 12. System block diagram of the neonatal virtual and physical simulator

5.1 Implementing the graphic interface

Observing the vital signs on a screen is very important for both doctors and nurses during their training process because specific problems can be found through their traces, shapes, curves and their numeric values. Usually these specific problems cannot be found only by hearing the heart beats, checking the temperature or by chest auscultation. In the software application developed, different vital signs can be read and the patient can be treated according to the diagnosed pathology.

5.1.1 Simulating the health condition of a neonate in LabView ®

The models explained in the previous section are implemented and visualized in a graphic interface developed in LabView®. The result of this process can be seen in Figure 13.

The interface allows the modification of the different parameters in order to obtain a wide diversity of medical scenarios; nevertheless, the feedback variables from the mannequin are: the cardiac frequency, the respiratory frequency, the rectal temperature and skin flushing. The interface is used, mainly, to train students of the healthcare area for acquiring diagnostic skills.

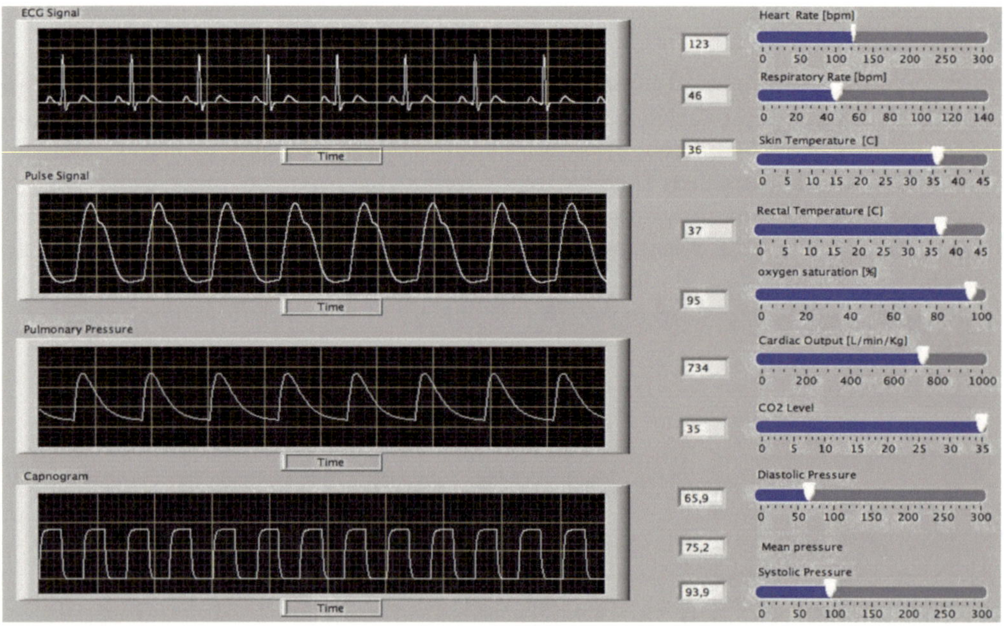

Fig. 13. Graphic interface developed in LabView ®

5.1.2 Selection of the medication in LabView ®

The correct dose is calculated for each medicament according to the drug main information. Table 4 shows the concentration of each vaccine bottle, the dilution and the dosage according to the neonate's weight. These 4 medications are available in the graphic interface of the computer according to the pathological scenario that also includes the neonate's weight that is also selected on screen. The interface of the medication programmer can be seen in Figure 14. (Young & Magnum, 2008)

	Concentration (mg/mL)	Drug Dilution	Dosage mL by weight			
			1 Kg	2 Kg	3 Kg	4Kg
Adenosine	3	1:9	0,17	0,33	0,5	0,67
Adrenaline	1	1:9	0,1	0,2	0,3	0,4
Atropine	1	1:9	0,2	0,4	0,6	0,8
Terbutaline	0,5 or 1	1:9	0,1 or 0,05	0,2 or 0,1	0,3 or 0,15	0,4 or 0,2

Table 4. Correct dosage according to the neonate's weight

As shown in Table 4, the scenarios that can be generated by the instructor are created in the virtual interface where the neonate's weight and medication are selected throughout a dropdown menu for each item; for weight selection purposes there are four options: 1Kg, 2Kg, 3Kg and 4Kg. The medications implemented are: Adenosine, Adrenaline, Atropine and Terbutaline. The selection of the dose is detailed below.

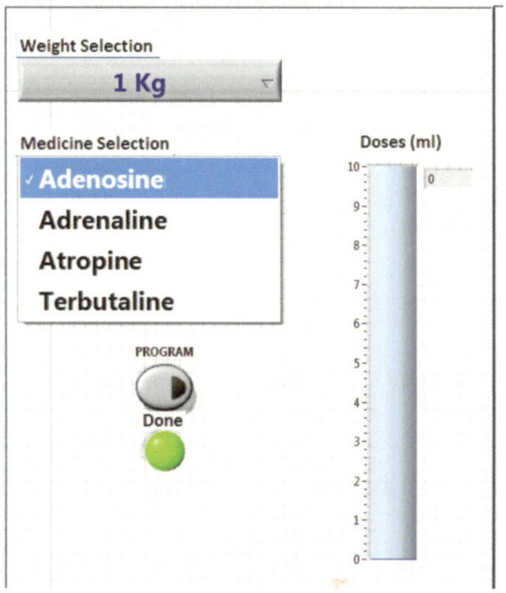

Fig. 14. Snapshot of the interface to select medication and neonate's weight

5.2 Implementation of the syringe applicator and selection of the dosage

The dosage is selected by the trainee when he or she takes the syringe or applicator and makes the load up movement. To determine the dose, a Hall Effect sensor (UGN3503) is strategically located in the rubber plunger tip while a magnet is placed in the bottom of the syringe barrel in order to measure the magnetic flux density changes while the load up movement is simulated. The sensor is a transducer that varies its output voltage when detecting a change in the magnetic field. An ATmega8 microcontroller is in charge of converting the data from analogue to digital and then codification is made. Data are sent wirelessly to the data acquisition module in the computer by using the serial communication transmitter TLP434 connected to the ATmega8 microcontroller.

The wireless communication is unidirectional. The transmitter operates in a frequency of 433,92MHz which belongs to the Ultra High Frequency (UHF) band. The receiver RLP434 is connected to an ATmega8 microcontroller that is in charge of the signal decoding. The decoded byte is sent to the Data Acquisition card of LabView ® (ADQ Labview®) (See Figure 15).

Once the data are in the computer, the information is processed and the amount of medication is shown in the screen. Dose may fluctuate between 0 mL to 1 mL with a resolution of 0,02 mL. Once the medication and the neonate's weight have been selected and the trainee has loaded up the medicament in the syringe, the information is programmed in the Tag that is attached to the syringe.

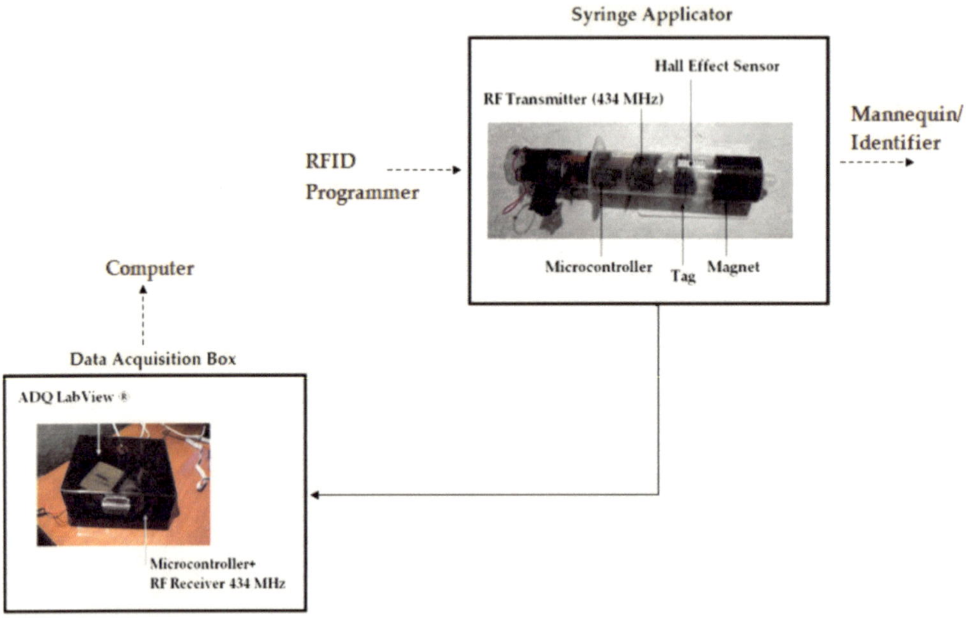

Fig. 15. Syringe applicator and wireless signals sending to the Data Acquisition Module

5.3 Implementation of the medicine programmer to write the RFID tag

Tag is programmed in order to store the information related to: the medicine chosen, the neonate's weight and dosage that will be applied to the mannequin with the intention of stabilizing its vital signs. Figure 14 shows the "Program" button that needs to be clicked to save the data in the Tag. To achieve this task, an RFID read/write device is used; this device allows writing (no contact) several types of RFID smart cards. The one used in this project belongs to the Mifare® family based upon the ISO 14443A standard.

The device has and RS232 interface which is used to establish the communication with the computer where medication, neonate's weight and dosage are selected. In order to have an adequate communication between the computer (by means of LabView®) and the writing device to program the Tag, a protocol that depends on the predetermined communication parameters of the ACG HF Mifare Easy Compact Plug and Play needs to be adjusted. Some of the device's features are: the transmission speed is 9600b/s, with 8 bits; it does not require a parity bit or handshake and it requires a stop bit. See Figure 16.

In order to access the Tag, it is necessary to send some control words by means of the reading/writing device (Three pass Authentication). The integrated circuit that is embedded in the electronic Tag has an EEPROM memory of 8192 Kbit; this circuit is organized into 16 sectors with 4 blocks of 16 Bytes each. An authentication procedure needs to be carried out for each sector to change the value of each block in the Tag. Three blocks in sector 2 are used in this particular project: block 8 stores the information about the medication, block 9 is used to store the information about dosage and block 10 stores the information about the neonate's weight. (See Table 5).

The Tag has a built-in PCB antenna that provides the typical reading range of 90 mm which implies that when the programming needs to be done, it has to be within the detection range; therefore, it is necessary to place the syringe or applicator (with the Tag attached) near the programmer so data can be sent properly from the writing device. This process is carried out when the student loads up the medication into the syringe; in that moment the applicator or syringe needs to be near the vaccine bottle where the RFID programmer is located. Simultaneously, the instructor, who should be in front of the computer at that moment, has to click on the "Program" button on the screen and the information is transmitted and stored in the Tag.

Fig. 16. Programmer of medications RFID

It is worth to mention that the memory capacity of the Tag allows the storage of a high number of medications, turning this module into a scalable project.

Tag Information				
Sector	Block	Data	Byte Value	Means
2	8	Medicine type	000000002	Adenosine
			000000003	Adrenaline
			000000004	Atropine
			000000005	Terbutaline
	9	Dosage	00000000<Byte<00110010	0 mL < dose < 1 mL
	A	Newborn's Weight	00000001	Weight=1Kg
			00000002	Weight=2Kg
			00000003	Weight=3Kg
			00000004	Weight=4Kg

Table 5. Tag information and memory uses

5.4 Implementation of the identifier (Mannequin)

As soon as the data have been stored in the Tag, the trainee has to approach to the identifier or mannequin simulating the injection of the medicine by umbilical vein catheterization, and meanwhile the Tag is read to determine whether the action carried out was the indicated one or not. In that moment, the decisions of the medicine or nursery student are evaluated. Even though the graphic interface can also be used to judge the trainee's choice by using its virtual simulation models, the physical evaluation of the patient confronts the student with reality. It is necessary to install an RFID reader/writer device in the abdomen that allows reading the information in the Tag in order to make the neonate react according to the decoded data. A microcontroller is specially programmed to receive the information from the Tag right at the moment when the trainee is about to inject the mannequin.

Once the Tag's information has been read, it is necessary to process the data to determine the reaction that should take place in the neonate taking into consideration its current health status, medication and dosage programmed in the syringe. The physical simulator will have the capability of changing its condition based on the effects that should be produced as a result of the decision made by the trainee.

Visual and audible reactions can be seen and heard in the mannequin according to the uses and side effects of each medicine shown in Table 3. Medications mainly affect cardiac frequency and temperature, therefore, the mannequin has a circuit that emulates the heart sounds (S_1 and S_2) at different frequencies such as tachycardia (high frequency), bradycardia (low frequency) and cardiac arrest (too low frequency or absent heart beating) or simply normal status. In order to make this possible, a circuit was implemented by using the LM555 timer that allows generating oscillations. The variation of those oscillations allows the simulation of the sound produced by the heart when it beats.

The systole and diastole are the movements that a heart performs and each of them produces a different sound. The first one - systole - is the one with the highest pitch at 60 Hz and the second one - diastole - produces a 45 Hz sound. Those sounds are intercalated to generate different cardiac rhythms; there is a long silence between the second sound and the first one of the following beat cycle. This silence lasts at least 50 ms, which coincides with the ventricular diastole.

Rectal temperature of the neonate can also be viewed on an LCD screen located by its side; this temperature can vary between 35,5° and 38,3°. Lights on the mannequin surface are located to emulate skin flushing.

If the trainee makes an inadequate choice to treat the mannequin, its health condition will be negatively affected showing a pathology that is reported to the instructor on the computer.

5.5 Implementing the data acquisition unit to report the mannequin's health status on the screen

Once the effects of the medication are seen on the mannequin, the information is sent to the computer so the mannequin's vital signs update automatically every time there is a change in its health condition (See Figure 17).

Signals are sent from the mannequin to indicate its reaction according to the treatment applied. Even though the communication is wireless, it is worthwhile mentioning that the identifier's transmitter works with a frequency of 418 MHz. This means that no interferences are generated in the communications between the syringe or applicator transmitter (434 MHz) and the mannequin transmitter (418 MHz).

Data strings of 8 bits (1 byte) are transmitted from the identifier; that byte contains information about the following variables: cardiac frequency, respiratory frequency, rectal temperature and skin flushing in the neonate. The cardiac frequency is equivalent to the frequency of the pulmonary pressure signal, the pulse signal and the electrocardiogram signal (ECG) which traces can be seen on the screen. In case the respiratory frequency changes, the frequency of the CO2 concentration level is affected. Rectal temperature remains 0,5°C above body temperature under normal conditions (room temperature needs to be between 20 and 24°C). Room humidity may also fluctuate between 30% and 60% and the ideal range is 50% to 55% (Bureau of Maternal and Child Health and Resources Development, 1993).

Width, shape and other features of each graph have to be analyzed thoroughly by medical personnel and nurses because trace's anomalies shown on the screen can be a sign of problems in the neonate's health.

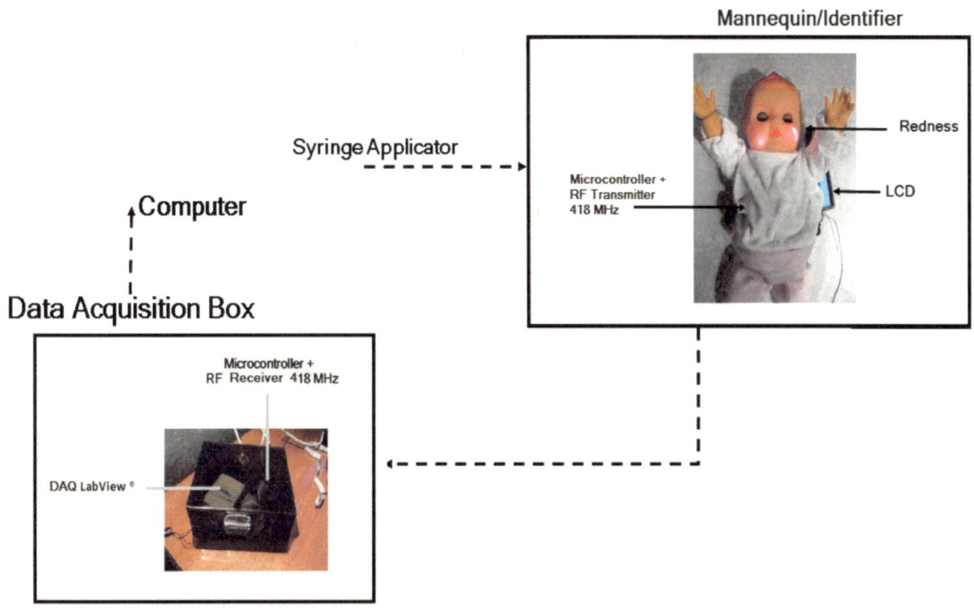

Fig. 17. Connection between the identifier and the Data acquisition Box

A noticeable message intended to catch the attention of the person who is monitoring the neonate's health condition appears on screen to report the neonate's health status.

6. Results

In order to present the most relevant results of this project of virtual and physical simulator, such results were classified into three aspects: medical scenarios, medical validation and cost analysis.

Four medical scenarios of special importance in the neonate's health condition are presented in Section 6.1; in addition, simulations of vital signs as well as some images of the mannequin are shown.

A review of the simulator was carried out by specialist in neonatology and the results are shown in Section 6.2. The results are presented as a percentage of experts that accept that the medical scenario modeled in the graphic interface is the correct representation of the real situation.

Finally, Section 6.3 shows a simplified and updated cost analysis of the virtual and physical interactive simulator.

6.1 Generating medical scenarios

The following are the four medical scenarios that can be generated in the system; different pathologies are created to try the functionality of the simulator. In addition, the way the mannequin reacts to the treatment is tested and the mannequin health status is monitored.
In order to select a medication, a student has to:

- Take the applicator
- Move the Tag near the RFID programmer
- Select the medication and neonate's weight on the screen
- Emulate the filling in of the syringe with the dosage according to the neonate's weight and pathology
- Click on the "programm" button on the screen.

In order to give the medication, the student has to:

- Move the applicator towards the mannequin's abdomen
- Wait for about 15 seconds until the mannequin reacts[1]
- Assess the mannequin's reaction
- Observe the different signal's graphs on the screen to evaluate the neonate's condition.

For each of the cases, a diagnostic is presented along with the suitable treatment to improve patient's health.

First scenario

Diagnostic: In the Figure 18 the neonate presents a normal status (Thompson & Crocetti, 1998). The cardiac frequency is 140 beats per minute, the respiratory frequency is 46 cycles per minute and the rectal temperature is 37,1 ° C which is slightly higher than the skin temperature.
Treatment: the neonate does not require any treatment.

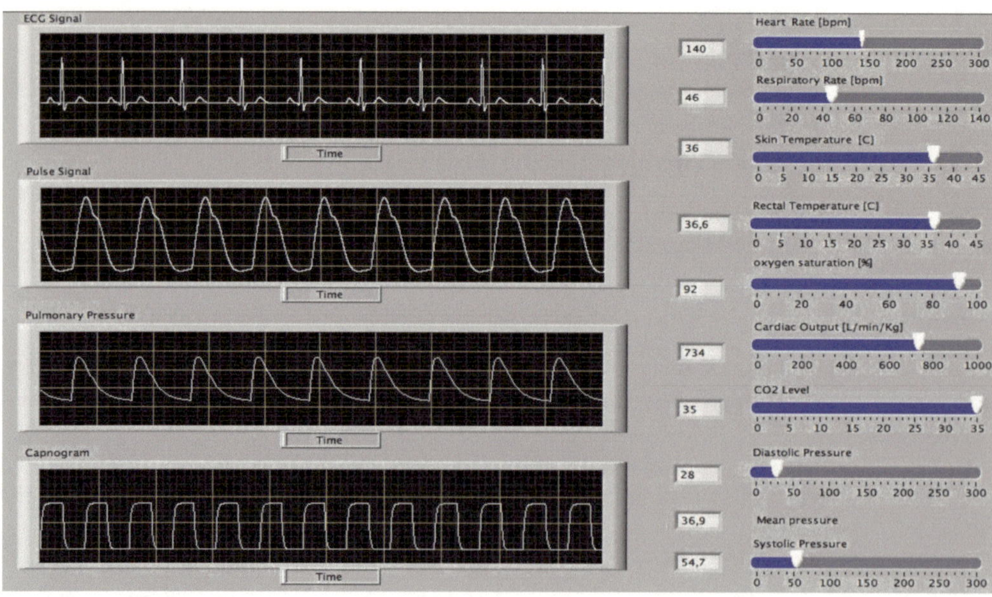

Fig. 18. Normal status

[1]In real life, times for choosing to re-administer the medicine might change due to the patient's reaction but for the model and data processing purposes, the student will only have to wait for 15 seconds.

Second scenario

Diagnostic: In the Figure 19 the neonate's weight is 3Kg and presents tachycardia, as shown in the cardiac frequency image that is at 220 pulses per minute. The respiratory frequency is 62 cycles per minute showing therefore tachypnea without fever as the rectal temperature is 36,8°C.

Treatment: The patient needs to be administered 0,5 mL of Adenosine. If after waiting for 15 seconds there is no reaction from the neonate, it is necessary to apply the medication again.

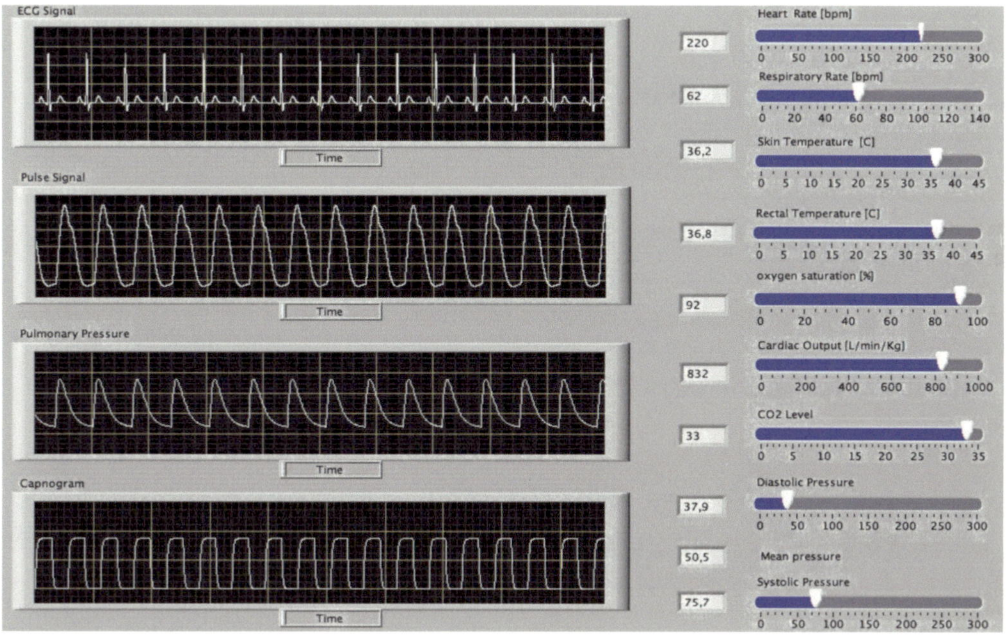

Fig. 19. Tachycardia and tachypnea

Third scenario

Diagnostic: In the Figure 20 the neonate's weight is 2 Kg. The neonate presents bradycardia as shown by the cardiac frequency of 70 bpm; the respiratory frequency is 20 cycles per minute which means there is also bradypnea and hypothermia (also shown).

Treatment: It is necessary to administer 0.4 mL of Atropine to reverse the severe bradycardia condition and wait for 15 seconds for the patient's response; in case the neonate does not show any reaction it is necessary to inject the medication again. The mannequin's skin may show some blush.

Fourth scenario

Diagnostic: In the Figure 21 the neonate's weight is 4 Kg and presents cardiovascular arrest (relative); the cardiac frequency is 24 pulses per minute and may continue decreasing (to a full cardiac arrest). The respiratory frequency is 12 cycles per minute meaning there is severe bradypnea as well as hypothermia.

Treatment: it can be administered either 0,4 mL of Terbutaline or 0,4 mL of Adrenaline, in both cases the cardiac frequency increases. If 15 seconds after there is no response from the neonate, it is necessary to inject the medication again.

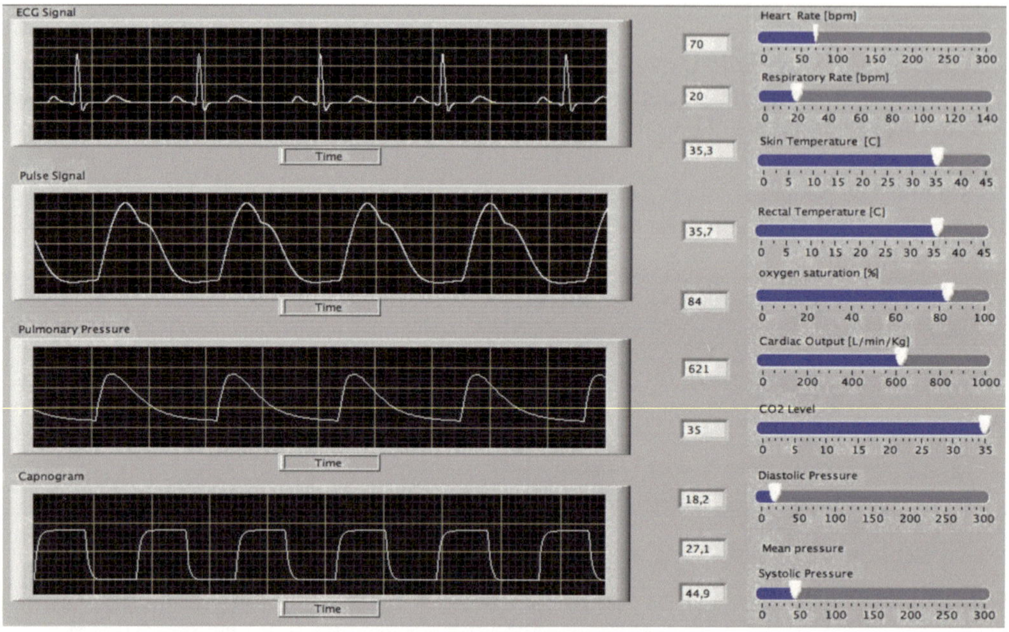

Fig. 20. Bradycardia, bradypnea and hypothermia

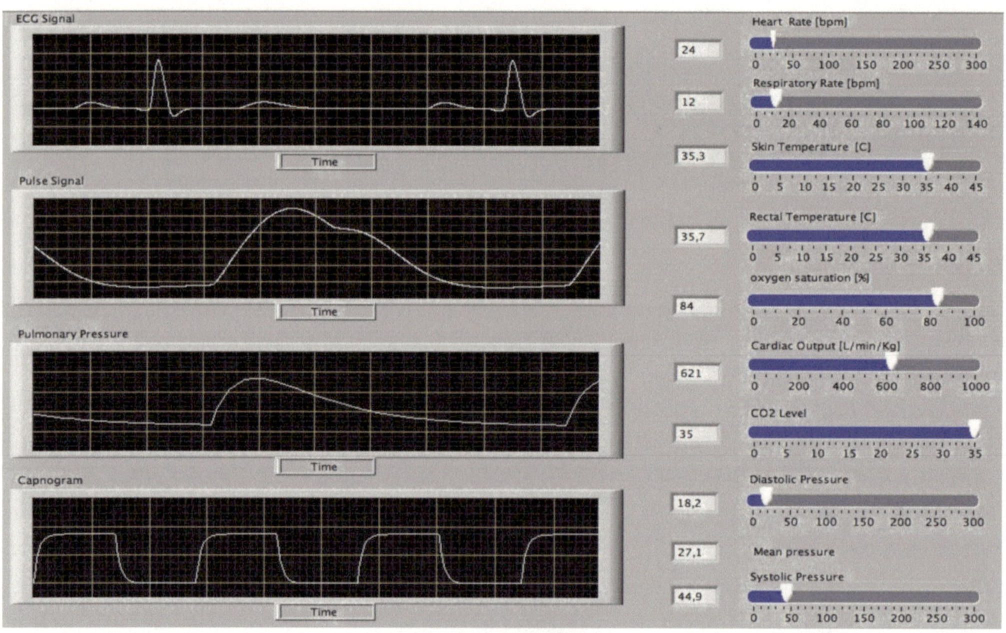

Fig. 21. Arrest (relative), bradypnea and hypothermia

6.2 Medical validation

After developing this project, a study was conducted in order to validate the usefulness of the interface in the training of personnel from fields such as Perinatology and Neonatology.

This user evaluation was a key step of this work as it allows confirmation of the veracity of the signals obtained in the interface.
A group of 16 experts in Perinatology and Neonatology was selected for this stage in order to evaluate the trustworthiness of the scenarios previously described. In this way, they evaluated the second and third scenarios described before where the neonate shows fever, tachycardia and tachypnea - second scenario - and the other where bradycardia, bradypnea and hypothermia are shown- third scenario. The constants were chosen based on expert medical advice from team members of this project.
The specialists were then presented the two scenarios in the simulator and a sheet where they wrote the set of pathologies they considered matched the represented constants.
The results (see Table 6) were highly satisfactory as the signals and, in general, the tool was considered excellent, realistic and user friendly by the consulted specialists in the healthcare area.

Fever, Tachycardia and Tachypnea		
Expected selection	100,00%	16
Unexpected selection	0,00%	0
Hypothermia, Bradycardia and Bradypnea		
Expected selection	97,00%	15
Unexpected selection	3,00%	1

Table 6. Evaluated Scenarios

7. Conclusions

The tool developed in this project consists of a neonatal monitor that shows ECG, pulse, pressure and CO2 level signals based on a physical system that simulates the use of medications with the implementation of an RFID module. This module allows wireless communication between the syringe and the dummy that cannot be found in commercial simulators.
Neonatal simulators, like the one presented in this work, are an educational tool for students of health sciences as they allow the acquisition of knowledge and skills, making faster decisions and more confidently, promoting realistic training in teams and acquiring practical clinical experience. The results of the validation of scenarios were satisfactory confirming that it is an educational tool as well as a practical and intuitive one.
The present developed tool has advantages over the commercial simulators in terms of budget needed for its implementation; the cost of the developed tool is around 7350 USD while the cost of the commercial ones, depending on their degree of complexity, range from 20000 USD to 58000 USD. This fact makes the project a viable and profitable option for training teams on neonatal care.

On the other hand, the development of a simulator that suits local training necessities provides the possibility of working in multidisciplinary research topics where knowledge from Medical Doctors, Engineers, and industrial designers, among others can be shared for successful results. In addition, it generates an environment that allows increasing the trust and experience needed in research in order to resolve multidisciplinary issues as the ones dealt with herein.

This work is the first phase of a larger project that includes a virtual simulator with the ability of generating synthetically all the signals that describe the patient's vital signs; and a physical simulator - mannequin - that exhibits the characteristics of a neonate allowing the simulation of signals that are also in the virtual simulator.

As future developments, we propose the implementation of bidirectional communication (monitor-mannequin) when transmitting all the variables that are visible in the simulator. Also, the implementation of new visible signs in the mannequin such as cyanosis, sounds, among others can be developed in the future. The simulated monitor could be enhanced with a tridimensional model of a neonate that would also allow the representation of vital signs.

8. References

Barash, P., Cullen, B., Stoelting, R., Cahalan, M. & Stock, M. (sixth edition). (2009). Clinical Anesthesia. LIPPINCOTT WILLIAMS & WILKINS, ISBN 978-0-7817-8763-5, Philadelphia, PA, USA.

Beneken, J., (1965), A mathematical approach to cardiovascular function. The uncontrolled human system. *The Netherlands:University of Utrecht.*

Bhavani-Shankar, K., Moseley, H., Kumar, A. & Delph, Y. (1992). Capnometry and anesthesia. Can J Anaesth, Vol 39, No 6, pp. (617-632)

Buck, G., (1991), Development of simulators in medical education, *Gesnerus*. Vol. 48, Pt 1, pp.(7-28)

Bureau of Maternal and Child Health and Resources Development. (1993) American Institute of Architects. Committee on Architecture for Health, United States. - pp(52)

Chopra, V., Engbers, F. & Geerts, M. (1994). The leiden anaesthesia simulator. *British Journal of Anaesthesia*, Vol. 73, No 73, pp. (287-292)

Cooper, J. & Taqueti, V. (2004). A brief history of the development of mannequin simulators for clinical education and training. *Qual Saf Health Care*, Vol. 14, No 1, pp.(72)

Currea, S., (2004). La adaptación neonatal inmediata. La reanimación neonatal. *Unibiblos*.

Gillies, G. & Williams, C. (1987). An interactive graphic simulator for the teaching of fibe-rendoscopic techniqes. Proceedings of Eurographics, pp. (127-138)

Hayes, G. J. (1994). Issues of consent: the use of the recently deceased for endotracheal intubation training. *J Clin Ethics*. Vol. 5, No 3, pp. 211–216.

Halamek, L. P., Kaegi, D. M., Gaba, D. M., Sowb, Y. A., Smith, B. C., Smith, B. E. & Howard, S. K. (2000) Time for a new paradigm in pediatric medical education: Teaching neonatal resuscitation in a simulated delivery room environment. *Pediatrics*. Vol. 106, No 4, pp. (106–110)

Halamek, L. P. (2008) The simulated delivery-room environment as the future modality for acquiring and maintaining skills in fetal and neonatal resuscitation. *Seminars in Fetal & Neonatal Medicine*, Vol. 13, No 6, (2008 December), pp. (448-453)

Hampton, J. (Seventh edition). (2008) The ECG made easy, Churchill Livingstone, *Elsevier*, ISBN:978-0-443-06817-1, Nottingham, UK.

Hoznek, A., Katz, R. & Gettman, M. (2003). Laparoscopic and robotic surgical training in urology. *Current Urology Reports*, Vol. 4, No 3, pp. (130-137)

Jones, S.A. (Second Edition). (2005) ECG Notes. F.A. Davis Company, ISBN-13:978-0-8036-2142-8, Noblesville, Indiana.

Khalifa, Y., Bogorad, D., Gibson, V., Peifer, J. & Nussbaum, J. (2006). Virtual reality in ophtalmology training. *Survey of ophtalmology*, Vol. 51, No 3, pp. (259-273)

Korosec, D., Holobar, A., Divjak, M. & Zazula, D. (2000). Building interactive virtual environments for simulation training in medicine using vrml and java/javascript. *Computer Methods and Programs in Biomedicine*, Vol 80, No 1, pp. (61-70)

Letterie, G. (2003). Medical education as a science: the quality of evidence for computer-assisted instruction. *Gynecol*, Vol. 188, No 3, pp. (849-853)

Lynöe, N., Sandlund, M., Westberg & Duchek (1998), Informed consent in clinical training - patient experiences and motives for participating. *Medical Education*, Vol. 32, No. 5, pp. 465–471

McCloy R. & Stone, R. (2001). Science, medicine and the future. virtual reality in surgery. *BMJ*, Vol. 323, No 7318, pp. (912)

MsSharry, P.E., Gari, D., Tarassenko, L., & Smith, L. A. (2003). A dynamical model for generating synthetic electrocardiogram signals, *IEEE Transactions on Biomedical Engineering*, Vol. 50, No. 3, pp (289-294)

Murphy, A., & Halamek, L., (2005). Simulation-based training in neonatal resuscitation. *American Academy of Pediatrics*, Vol. 6, No. 11, pp. (488-492)

Ostergaard, H., Ostergaard, D., & Lippert, A., (2004). Implementation of team training in medical education in denmark, *Qual Saf Health Care*, Vol. 13, Suppl 1, pp. (i91–i95)

Perkins, G., (2007). Simulation in resuscitacion training. *Elsevier*, Vol. 73, No. 2, pp. (202–211)

Rall, M. & Dieckmann, P. (2005). Simulation and patient safety: The use of simulation to enhance patient safety on a systems level. *Current Anaesthesia & Critical Care*. Vol. 16, No 5, pp. (273-281)

Resiner A. T. & Clifford G. D. (2006). The Physiological Basis of the Electrocardiogram, Advanced Methods and Tools for ECG Data Analysis. Artech House.

Sherwood, L. (Seventh Edition). (2010). Human Physiology: From Cells to Systems, Brooks/Cole CENGAGE Learning, ISBN-13:978-0-495-39184-5, Belmont, CA, USA.

Small, S., Wuerz, R., & Simon, R. (1999). Demonstration of high-fidelity simulation team training for emergency medicin. *Acad Emerg Med*, Vol. 6, No 4, pp. (312-323)

Smith, C. & Daniel, P. (2000). Simulation technology: a estrategy for implementation in surgical education and certification. *Tele-operators and Virtual Environments*, Vol. 9, No 6, pp. (632-637)

Stansfield, S., Shawver, D. & Sobel, A. (1998) Medisim: A prototype vr system for training medical first responders. ISBN: 0-8186-8362-7, Atlanta, Georgia. March 14-March 18

Spicer M. & Apuzzo, M. (2003). Virtual reality surgery: neurosurgery and the contemporary landscape. *Neurosurgery*, Vol. 52, No 3, pp. (489-497)

Takashina, T., Masuzawa, T. & Fukui, Y. (1990). A new cardiac auscultation simulator. *Clin Cardiol*, Vol. 13, No 12, pp. (869-872)

Taketomo C.K., Hodding J.H. and Kraus D.M (2009-2010), Pediatric Dosage Handbook, Lexi-Comp and APhA, 16th Edition

Thompson, W., & Crocetti, M., (1998) . The harriet lane handbook - manual de pediatria hospitalaria, *Cardiologia.*

Tsai, M., Hsieh, M. & Jou, S. (2001). Virtual reality orthopedic surgery simulator. *Comput Biol Med*, Vol. 31, No 5, pp. (333-351)

Vanetta, M., Herrera, M. & Gomez, M. (n.d.). Algoritmo de análisis de señal de pulso arterial para dispositivo portátil. *Departamento de Bioingeniería.*

Young Th, E., & Mangum, B., (2008), Neofax 2008. Thomson Reuters Health Care, 21th Edition.

Ziv, A., Wolpe, P. R., Small, S., & Glick, S., (2006) Simulation-based medical education: An ethical imperative, *Simul Healthcare*, Vol. 1, No. 4, pp. (252–256)

4

RFID Technology in Preparation and Administration of Cytostatic Infusions

Šárka Kozáková and Roman Goněc
Masaryk Memorial Cancer Institute
Czech Republic

1. Introduction

Cytostatics which are drugs used to treat oncologic diseases belong to very dangerous substances. These drugs often have very low therapeutic index, i.e. the difference between therapeutic and toxic dose is very low. Wrong dose can thus endanger the patient very easily. As these drugs are perilous also for the personnel who are handling them, Czech laws demand that the personnel concerned pass regular medical examinations. The number of shifts on sites where contact with cytostatics is possible has to be recorded. The laws order the record of basic information on preparation and administration of cytostatics; however, detailed monitoring of the drug in the course of the whole process is not required.

A state-owned medium-sized hospital with 200+ beds and more than 80 years of experience, Masaryk Memorial Cancer Institute highly specializes in the treatment of oncologic patients by surgery, chemotherapy and radiotherapy. The institute is focused on the treatment of solid tumours, which are in Czech environment represented mainly by breast cancer and colorectal cancer. For the treatment, the patients can be hospitalised, or, which is less expensive and has psychological benefit for the patient, receive their treatment at the outpatient clinic. Thus, the patient comes to the hospital, is checked by his/her physician, chemotherapy/radiotherapy is prescribed and administered/applied. On the same day, often just after a few hours, the patient is sent home.

In the pharmacy of Masaryk Memorial Cancer Institute (MMCI) we intended to implement a system that would be able to record who, where, when and how were exposed to cytostatics. Furthermore, we wanted to use the active support of preparation, i.e. introduce software that would help and navigate the personnel during the whole process, thus reducing the possibility of error. In the course of the project, we decided to include the outpatient clinic so as the administration of cytostatics could be recorded and supported, too. There are several ways to monitor the process and the possibilities of information technology can offer numerous solutions. In the end, radio-frequency identification (RFID) was chosen because it is more advantageous in some aspects than other systems.

1.1 Previous manner of prescription, preparation and administration of cytostatics

The process was standardised and consisted of several steps. The doctor prescribed the cytostatic infusions using hospital information system (HIS) and printed it in two copies, stamping and signing both of them. The prescription was a sheet of paper containing all days of the protocol and for each day individual lines with particular cytostatics and other medications. In the case of outpatients, the patient had to carry one copy to the outpatient clinic, where he had his seat reserved, and one copy to the pharmacy. In the case of inpatients, the first copy stayed at the clinic, the second was carried to the pharmacy by anyone from the personnel. Chemotherapy was prepared according to the prescription, the prescription was signed by the personnel who prepared it, and returned to the clinic. The first copy was used as administration protocol at the clinic. The doctor was limited and could not prescribe any chemotherapy - the prescription was limited by diagnosis and only treatment protocols approved by the head of the clinic could have been used.

1.2 Critical points of the previous process

In the process, there were several critical points, where an error could have occurred. Because the pharmacy runs according to quality system and is regularly inspected and audited following EN ISO 9001:2008, no significant errors occurred. There were several control mechanisms, mainly based on the principle that the personnel watched each other and on strict adherence to standard operation procedures (SOPs). Thus, the change in preparation or patients was excluded. However, the person preparing the infusion could take the necessary volume twice and so accidentally double the dose. Such an error could not have been identified.

As the patients, or their relatives, had to carry the prescription to the pharmacy in person, and sometimes did not want other people to know they were treated by chemotherapy - the prescriptions were traditionally printed on yellow sheet of paper size A4 - they folded the prescription and put it away. Sometimes, they forgot to hand it over and they themselves were the reason why they had to wait for the administration for a long time.

In some cases, it was not possible to backtrack the batch number of used drug, which is important e.g. in the case of side effects. Since the drugs have limited stability after first use, this stability was recorded by dating the particular vial. If incorrect date was written on the vial, a drug of unwarranted quality might have been used.

The entrance of the personnel in the preparation room was recorded in written form. Making regular monthly or yearly sums was difficult and any erroneous record was practically impossible to find.

1.3 RFID technology in healthcare

RFID technology is based on the communication between a unique carrier of information, i.e. a RFID tag, and a suitable reader. This technology has recently found its use in healthcare (Lahtela & Hassinen, 2009; Lahtela & Saranto, 2009; Sun, Wang, Wu, 2008). Technical report prepared by RAND (Oranje-Nassau et. al. 2009) for the European Commission describes seven cases within the European Union. In one case, the project failed completely, in two cases, the RFID technology was replaced by another technology for economic reasons. It was these two cases, where RFID technology was used in hospital pharmacies to control the preparation and administration of drugs. One of these cases was the University hospital in Geneva (Spahni et al. 2006). The RAND report praises the technology as it can lead to increase in quality of healthcare; on the other hand, the report warns against its high costs. The costs are in case of RFID technology much higher than in other technologies, e.g. barcode or its derivatives.

RFID technology was used in hospital pharmacy also in Akita University Hospital in Japan. In the Czech Republic, RFID technology is used in three hospitals, in one case for equipment, in one case for laundry and in our case for the control of preparation and administration of cytostatics. Another hospital announced its plan to introduce RFID identification in its management of blood and blood products.

2. RFID project at MMCI

2.1 Information systems

The preparation and administration of cytostatics is a matter of concern of three different information systems.

Hospital information system (HIS) contains all information on patients and their visits in hospital - reports, laboratory results, records, etc. Concerning chemotherapy, HIS contains a list of all chemotherapy protocols that are or used to be approved by the management of the clinic and are based on published information. Protocols that are not in use any more are listed only for information and the doctor is not able to load them from the system. The doctor has to use particular protocol for allowed diagnosis only. Only minor changes in protocol are possible: the dose of the cytostatic drug can be reduced (the reason has to be recorded), auxiliary therapy - antiemetics, antihistaminics, ions, liquids, growth factors - can be added or modified, and the days of the protocol can be moved slightly forward or backward.

Pharmacy information system (PIS) is standard software used in Czech and Slovak pharmacies. In this case, it is modified by adding new modules, e.g. the active support of preparation or personnel entry monitoring. Both HIS (GreyFox) and PIS (Medea) are products of Stapro, a Czech software company specialising in healthcare software.

Information system for administration of cytostatics (AIS) that is used in the outpatient clinic, and in the future possibly in the inpatient clinic, was developed solely for this purpose by IBM.

This three information systems exchange and store information allowing its backtracking or control. All three information systems are also available as "testing versions", which are used for training purposes and development of new functionalities.

2.2 General communication flow

Within the system, three different information systems communicate with each other and are connected by the means of a service bus as shown in Figure 1.

HIS (blue colour) is connected with Relational Database Management System (RDBMS) and communicates through APP Server with the service bus. The communication follows the JMS/XML format.

PIS (yellow colour) has three key modules: personnel entry evidence, storage evidence and active support of preparation. It is connected with RDBMS and communicates through APP Server with the service bus. The communication follows the JMS/XML format.

AIS (violet colour) communicates with the service bus in HTTP/SOAP/WSDL format.

2.3 RFID tags

The system is based on passive RFID tags, ISO standard 15693, working frequency 13.56 MHz. These tags are used in three different forms.

- adhesive labels for the vials, 31.5 mm x 16.5 mm
- adhesive labels for the infusion bags, 55 mm x 75 mm, on which RFID printer prints further information
- plastic ID cards.

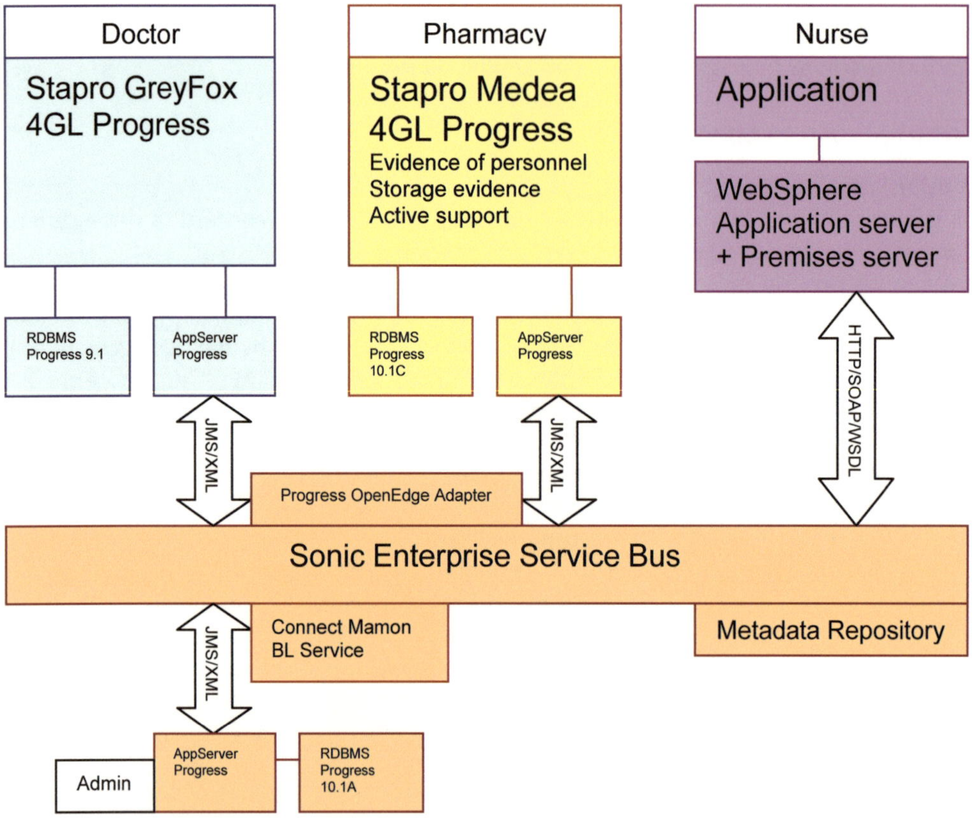

Fig. 1. General communication flow.

In the course of the project, tags of other standards, e.g. I-CODE, and other working frequencies were tested. We supposed that the evidence of personnel would be based on rings with RFID tags identified by a frame with RFID reader working with UHF frequency. This idea was abandoned. The system used now may require higher activity on the side of the personnel, on the other hand, it is clear whether the personnel is entering or leaving the room, or just checking if the reader is functional. The personnel can clearly see if their entry was recorded correctly or who is inside the preparation room without actually having to go and have a look.

For several months, the vials were labelled by ARIO-SDM70 nano-tags, which was just a tag with a small antenna covered by an adhesive. The small size of the antenna was disadvantage, as the tag and the reader had to be in close contact and in correct position, which might have been tedious. Furthermore, the small size itself made it often difficult to find the tag on the vial at all.

Even though there is evidence (Erdem et al., 2009) that interference between the tag and the infusion bag is possible because the inside of the bag is conductive we did not meet such a case. We do not have any problems with interference between RFID tags and medical equipment (infusion pumps) either. Such interferences are known with other frequencies than 13.56 MHz (van der Togt et al., 2008). In Japan, there is the shortest allowed distance between a 13.56 MHz tag and medical equipment - 22 cm; however, testing showed that the risk is significant in high-output antennas only. Actually, these are not used very often (Hanada & Horigome, 2008).

2.4 Technical solution – pharmacy

Figure 2 shows a simplified ground plan of the pharmacy. Turquoise colour corresponds to clean area class A, blue colour clean rooms class C. Light blue colour indicates material and personnel entrance hatches with material/personnel flow direction indicated by arrows. Red squares stand for RFID readers on the premises. Preparation room 1 is purposed for cytostatics, preparation room 2 for auxiliary medications.
The core of the system is a standard WLAN network, to which a service bus is connected. The network connects standard PCs in the Storage, in the small storage beside Preparation room 2, in Goods Intake room and in Completion room. These PCs are equipped with dial readers (RFID + barcode). In personnel entry, there is a small tablet PC with fixed RFID reader. In Preparation room 1, there are three isolators. An isolator is a special box, in which there is clean area Class A inside, their front side has two arms with gloves through which the personnel works. The backside of isolator is equipped with a touch screen and fixed RFID reader. The PCs themselves are positioned outside the preparation room, where they are easily accessible and not exposed to aggressive disinfectants. In Preparation room 1, there is also an industrial PC and a printer with RFID module. The printer is put in a pressure box where it is protected from disinfectants. In the pharmacy, there is also other necessary equipment, e.g. barcode printers. The system of work requires that RFID reader, industrial PC and RFID printer have to be available in at least two specimens. In case of maintenance or fault any equipment can be replaced immediately. There are three isolators, only two of which are in use, the third one is a reserve.

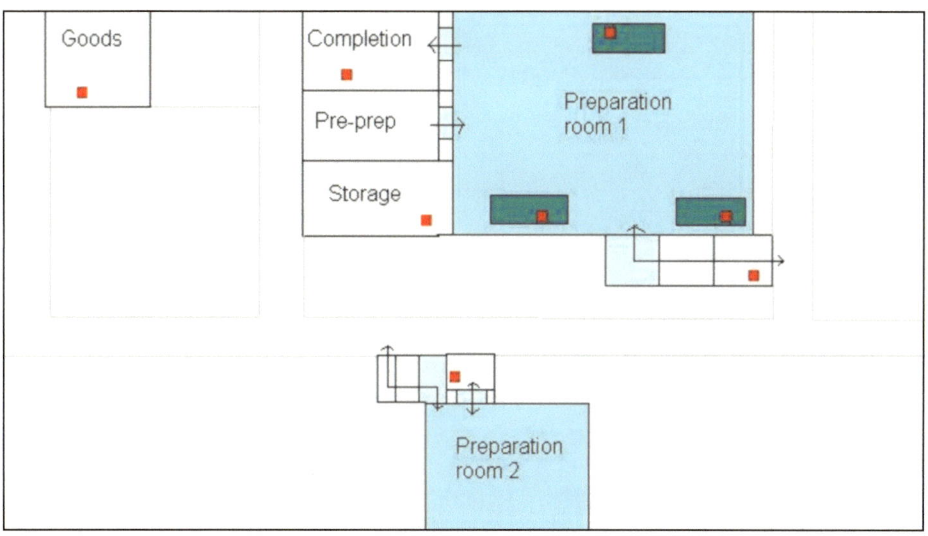

Fig. 2. Simplified ground plan of the pharmacy

2.4.1 Communication flow

RFID reader is connected with the PC through USB port that behaves as a serial port. RFID agent takes over the data by the means of RFID adapter and via a Message Queue (MQ)

client sends the message to the Premises Server, where the data go through App Server (via Message Driven Bean, MDB) and are sent by the means of Sonic MQ to the service bus.
Initially, RFID readers were connected by the means of Premises server. This solution was not fully reliable and in final solution is used in one case only. The readers are now connected through an USB port directly to the PC with the running application (Stapro MEDEA).

2.4.2 Readers

RFID reader reads the identification number (electronic product code, EPC) from the RFID tag. Obtained information is transferred along an USB connection to a PC, where it is processed further. In the course of the project, four different types of RFID readers were tested and used.
RightTag reader was initially used in the completion room where it was connected to a tablet PC. This reader reads RFID tags only. Because the completion process requires the use of both RFID tags and barcodes, this reader is now not used.
TagSys reader was placed initially in the room between Good Intake Room and the Storage; in these two rooms, there were reading frames connected to the reader via a coaxial cable. The reader itself was connected with the PC by a WiFi connection. Even though it was possible to work with one reader in two rooms, it was not possible at the same time. As there is neither visual nor audio possibility to communicate between both rooms and the WiFi connection was not reliable, we do not use this reader any more. Later, the whole hospital was covered by WiFi and so there was not any interference.

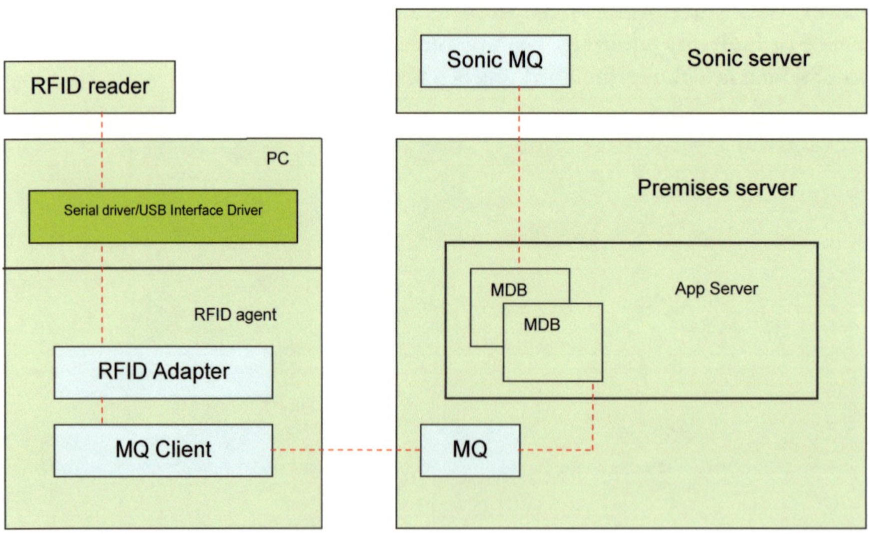

Fig. 3. Pharmacy. Communication flow

ACG readers are placed on goods intake, in personnel entry and in isolators. In the first two cases, the reader is connected with adjacent PC through an USB port. Should the reader or tablet PC in personnel entry break down (such a case happened) the particular part of the application can be run on the PC in the room of goods intake. In the isolators, the readers are placed in special covers that protect them from disinfectants and that are glued on the backside of the isolator. These readers are connected to particular PCs via an USB port (for details see chapter 2.4.7).

CPT8000 reader is a dual reader, working with both RFID tags and barcodes. These readers are used in the storage (RFID), auxiliary medication storage (barcode), and completion room (RFID+barcode). These readers are used also on other sites within the pharmacy.

2.4.3 RFID printer

In the preparation room, there is a printer SATO CL408E with RFID module. We use self-adhesive labels with in red pre-printed warning *cytotoxic substance*. The RFID tag itself is glued to the underside of the label. The printer couples the information on the preparation with the RFID tag and receives a confirmation that the coupling was successful. Only then is the label marked as usable. If the coupling fails, a new label has to be printed. The printer has to be protected from disinfectants and therefore it is placed in a special custom made pressure box. Next to the printer, there is an industrial compact PC with touch screen, which is used to run the application controlling the printing of RFID labels, a sub-module of the active support. The printer used to be connected with the computer though WiFi but after several months, this connection was replaced with standard Ethernet, which proved to be more stable and reliable.

2.4.4 Serial driver/USB driver

This driver serves the RFID reader on a low level and forms a virtual COM port, through which the communication with the RFID reader is channelled.

2.4.5 RFID agent and premises server

Console application serves the RFID reader and communicates with RFID Premises Server. Initialising and configuring when the RFID reader is switched on, RFID agent converts the protocols from RFID readers into the form of standardised messages. It shows up as an icon in the status area of the task-bar (green rectangle if running correctly, otherwise yellow or red).

Premises Server is a server application and connects communication buses IBM MQ and Sonic MQ. Each reading corresponds to one record in this format: *[2/26/09 15:31:34:734 CET] 00000027 SystemOut O EventTagMDBean CONSUMED MESSAGE: null*

2.4.6 Communication protocol

The protocol manages the transfer of information on the EPC of a RFID tag from the reader to the service bus SonicMQ, where it is picked up by the HIS for further processing. The message contains information listed in Table 1.

The protocol contains other types of messages that enable the beginning and the end of the reading process at particular location and find out the status of reading location. These types are not in use at present.

2.4.7 Connection of isolators

Technology in isolators is connected with particular PCs that serve it by the means of USB/Ethernet converter. This method was chosen as the distance between an isolator and its PC is as much as 8 m. Such a distance can't be bridged reliably by a mere USB cable.

In isolators 1 and 2, the LCD touch screen is connected via a VGA cable with the PC. LCD touch screen and RFID reader are connected with the PC through USB cable, USB/Ethernet converter and switch. The switch is connected by Ethernet with an IP Watchdog.

Tag Message - TagEvent	
Published to	MOU.RFID.Event.Tag
Message Property	**Value**
RFIDEvent	TagEvent
RFIDEpc	EPC, e.g.: E07576576576AD
RFIDLocation	Name of the location, e.g.: Sklad.Cyto.Reader.1 *(Reader in Storage room)*
RFIDTagType	Type of the Tag, now only ISO15693 tags are used
RFIDTimestamp	In the format used byMQ

Table 1. Tag event

In isolator 3, the LCD touch screen is connected with the PC via a VGA cable as well as via a serial port, Ethernet and switch. The RFID reader is connected with the PC via an USB cable, converter, and switch. The switch is connected by Ethernet with an IP Watchdog.

The conversion from USB to Ethernet is managed by UBox2. This device enables the prolongation of USB through a LAN local computer network (Ethernet, Internet). USB-connected devices at the UBOX are accessible for more users at the same time. In contrast with traditional USB cable, the UBOX connection is not limited by the distance from the PC. Two USB devices can be connected at the same time (full-speed 12 Mbps). The energy supply of the USB device is standard (up to 500 mA for each device). UBOX ports appear in the operating system among other USB ports; however, they are only *virtual USB* ports, redirected to the UBOX port.

IP Watchdog GIOM 1200 is an automatically controlled socket which behaves according to user-defined rules. The socket can be used to switch on and off various devices, watch IP devices, reset servers, etc. In our case, the Watchdog is used to watch automatically if the communication is running correctly.

2.4.8 Software solution

Technologically, the solution is based on the integration of applications from two information systems that were already in use at MMCI. This software was modified or newly developed to suit the new demands. Following applications have to be changed:

StaproMEDEA Logistcics - pharmacy information system that manages complex logistics of drugs, or other stored commodities. Based on the demands of the project, the system was updated with a module using RFID identification of drugs, a module using RFID identification of personnel and a module supporting the preparation of cytostatics. Electronic communication with other systems was modified and improved, too.

StaproGreyFox - hospital information system supporting many inpatient and outpatient processes. Based on the demands of the project, the whole chemotherapy module was transformed. The medication is now supported more effectively and the communication with adjacent systems is improved. The system represents a sole information system that enables the doctor to have access to all necessary information without a need to search in other systems.

All communication XMLs have the main element named "chemo". This can have different attributes, in which the particular XMLs differ.

The receiving of XML have to be confirmed by the other side. If the "return" value equals "OK", the communication happened correctly. If the "return" value differs from "OK", an error occurred and the user is informed.

2.5 Technical solution – outpatient clinic

The outpatient clinic is used by those patients, who only visit the doctor, prescribed intravenous chemotherapy is administered to them and they go home, spending only several hours in the hospital. This is repeated for all the days of their chemotherapeutic cycle. Outpatient clinics are more convenient for several reasons but two of them are prominent. Firstly, such a system of care is much cheaper than standard care when the patient spends in the hospital several days. The second reason is psychological as a short visit in hospital does not stress the patient so much as a long stay.
RFID support of the administration at the outpatient clinic is processed through the co-operation of several components. These are HIS, PIS, PDAs used by the outpatient clinic nurses and the application in these PDAs, and the application on the Premises Server that communicates both with the PDAs and the HIS.

2.5.1 Hardware

Premises Server components and the HIS communicate with each other by the means of message exchange through MQ bus. This is the only possible way of communication between these components. PDA Sockets SoMo650 are used, the CF (Compact Flash) RFID reader for PDA is inserted in a slot in the PDA. In the outpatient clinic, there is also a WiFi access point LinkSys WRV200. The number of used PDAs is equal to the number of used seats.
Outside the outpatient clinic, there is an application running on IBM Premises Server that is placed in the server room of MMCI. This application holds status information on running application programmes and enables the users to use any PDA for particular record of administered dose. Thus, the PDA do not contain any long-term status information or any data on the patient or his/her application programme. The architecture has three levels and PDAs represent pure presentation level.
PDAs are used by nurse to record the steps taken when administering each drug to the patient. PDA serves as a tool to read the identificators - RFID tags and barcodes - and guides the nurse through the process of drug administration, using graphic user interface. PDAs communicate through a coded wireless network with IBM Premises Server where the server part of the application is run.
The WiFi access point is located in the nurses' room at the outpatient clinic. PDAs are placed at the door to the nurses' room in their chargers. When the nurse is going to administer a dose, she picks up the PDA from the charger and when working with the patient she has the PDA at her person. When returning to the nurses' room, she puts the PDA back to the charger. Because the PDAs are expensive and there are a lot of people coming in and going out of the outpatient clinic, a safety frame was installed at the door to the clinic. The communication is protected by the means of WPA-PSK (Wi-Fi protected access pre-shared key) and set rules at the MMCI's firewall.

2.5.2 Communication flow

The communication flow at the outpatient clinic follows the scheme pictured in Figure 4. The piece of information is read by the RFID reader, via its driver the information is send to the PDA and on the presentation level it is sent to Premises Server. Premises Server (WebSphere Sensor Events) is a middleware mediating the communication between RFID readers and Sonic ESB. Premises Server is connected to Sonic data bus, through which the information is sent to the HIS. WebSphere application server is an application server, on which the server part of application is run.

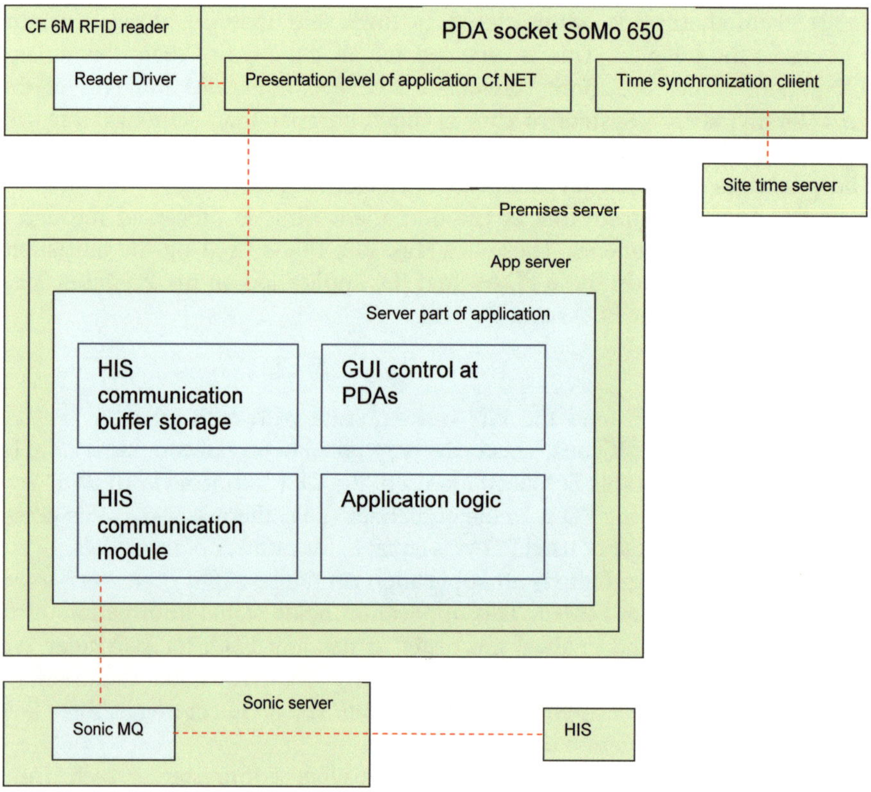

Fig. 4. Outpatient clinic: Communication flow. (MQ = message queue, GUI = graphical user interface)

Examples of communication between HIS and AIS are shown in Table 2.

Name	Direction	Queue	Structure
Query	AIS → HIS	MOU.RFID. Stacionar.doNIS	dotaz
Answer	HIS → AIS	MOU.RFID. Stacionar.doNIS	odpoved
Administration programme	HIS → AIS	MOU.RFID. Stacionar.zNIS	program
Administration protocol	AIS → HIS	MOU.RFID. Stacionar.doNIS	aplikace

Table 2.

2.5.3 Software

The application in the PDAs support the process of administration of cytostatic and auxiliary drugs. The application uses information on the patient and prescribed drugs that it receives from the HIS. This information is used to check the process of administration. The application was developed solely for this purpose.

2.6 Evidence of personnel

On the entrance to the preparation room, the personnel have to identify themselves with their personal RFID ID card. In this way, the evidence of personnel who prepare cytostatic drugs is recorded. Obtained data are exported monthly and stored in defined folder where they are accessible for internal and external audits. The number of entries as well as their total length is recorded. The former meets the requirements ordered by the law; the latter is more quantitative and can be better related to any incidence of industrial disease. There is also another important quantitative value - the number of preparations per each person that shows direct participation in the preparation process and not only the presence in the room. Figure 5 shows the number of preparations for individual employees in 2010.

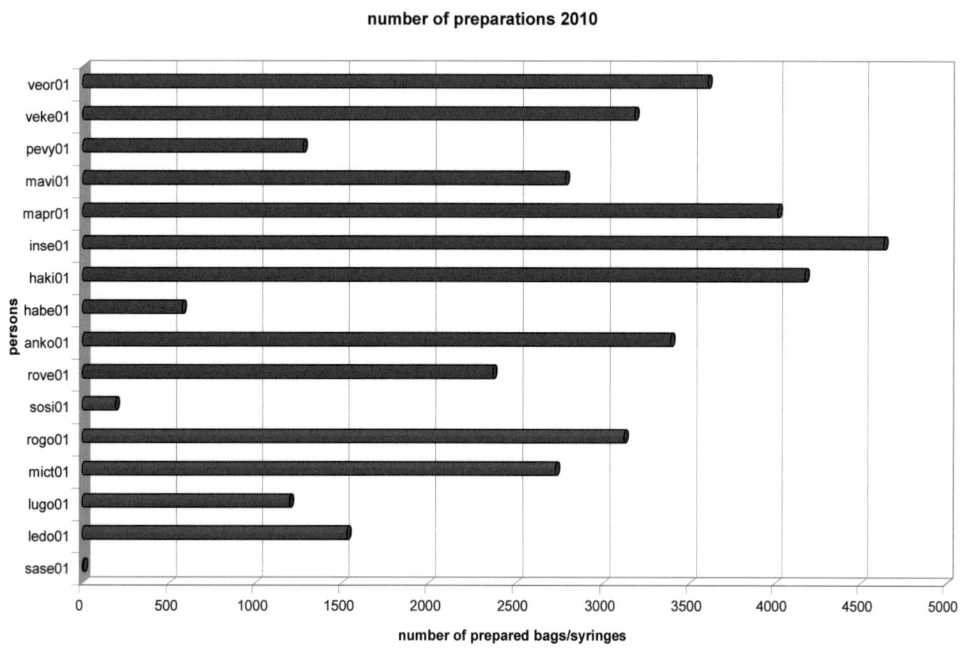

Fig. 5. Number of preparations. Employees with university degree in red colour, technicians in violet.

2.7 Preparation of chemotherapy

2.7.1 Entry of goods

When goods arrive at the pharmacy, they are put in the entry room where they have to stay until the entry of goods is finished. Cytostatic drugs can't be moved to the storage sooner than all vials are labelled with RFID tags. The tag with its antenna has a form of self-adhesive label. The PIS couples particular RFID tag with information on the vial: name, strength, ATC code, batch number, expiry date, price, VAT, supplier, etc. In this way, unambiguous identification of each cytostatic vial is certain. Cytostatics can be moved between storages or moved to a bill only following the rules of RFID identification and only with the use of the tags. There are several kinds of bills - preparation, move to other storage, return to supplier, stock-taking - and all of them demand the use of RFID identification.

2.7.2 Electronic prescription of cytostatics

The doctor creates the electronic prescription of cytostatics in the HIS. He/she selects the patient and for his/her diagnosis can choose only from the list of approved chemotherapeutical protocols. The protocols are approved by the head of the clinic and are based on published information only. The doctor has the possibility of some modifications in the protocol. The individual days of the protocol can be moved slightly forward or backward, e.g. to avoid the weekends. The dose of a cytostatic can be reduced or particular cytostatic completely removed from the protocol. Such a reduction or removal has to be justified (e.g. serious adverse event or side effect) and the reason recorded. Auxiliary medications (antiemetics, antihistaminics, ions, liquids, growth factors) can be added, removed, or changed freely. Auxiliary medications can also be used as separate prescriptions. The prescription has usually one to ten lines. If the protocol consists of more days, each day has its individual prescription. Then the prescription is signed electronically - either all days of the protocol at once or only the first day, e.g. when there is a high risk of an adverse event.

If day 1 of the prescription equals to the actual date, the prescription is sent immediately to the pharmacy. The doctor has 15 minutes to recall the prescription, e.g. if he/she receives some new information on the status of the patient and needs to cancel or modify the prescription. Only after this interval the prescription can be processed in the pharmacy (the prescription is visible during this interval). If the doctor wants to recall the prescription after this interval, it is possible but the pharmacy does not guarantee that the prescription was partially or completely processed (prepared drugs won't be administered to the patient but they will be charged to the clinic).

If day 1 of the prescription does not equal to the actual date, the signed prescription awaits the right date. In case of inpatients, the prescription is sent to the pharmacy automatically at 5:45 in the morning (configurable time). In case of outpatients, the patient has to come in person to the outpatient clinic and the nurse sends only then the prescription to the pharmacy. The prescription for the actual date can't be sent to the pharmacy when the pharmacy is closed unless permitted by the pharmacy.

2.7.3 Identification

The electronic prescription arrives in the pharmacy. The system records the receipt (reads the heading), divides the prescription into individual lines and after 15 minutes informs the user by increasing the number of received unprocessed lines on the taskbar where there are three numbers - cytostatic, auxiliary, unknown. The prescription is not a fully functional bill, it is just a prescription stored in the data structure as a heading carrying all information that arrived from the HIS. The heading also bridges all bills that are created later and belong to this prescription (at least one bill for each line of the prescription).

When the user sees that there are unprocessed lines, he/she double clicks the taskbar and thus the CytoEvidence window is opened. This window has three sub-windows. In window 1 the user processes the unprocessed lines, in window 2, all received lines are visible, in window 3, all bills related to received prescriptions are visible. Windows 2 and 3 can be used only for viewing. In window 1, each line is attributed a bill of material, i.e. list of material that is necessary for the preparation and will be charged automatically (infusion bags, infusion lines, syringes, stopper, needles, etc.). This attribution is done according to the ATC code (defines the active substance of the drug) of the line and the way of administration. The same drugs can be often administered in different ways - e.g. intravenous infusion or intramuscular bolus syringe; different ways of administration demand different bills of material. If the system is unable to find a bill of material, i.e. it is not able to recognize both the ATC and the way of administration, the line is listed as unknown and the whole preparation can't be processed until a new/correct bill of material is defined. The bills of material are defined in the code-list of the system and are divided to two categories: RC (cytostatics, follows RFID process) and RS (auxiliary, follows barcode process).

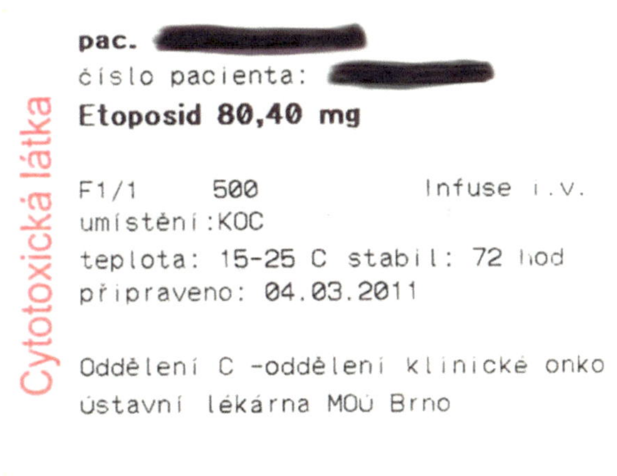

Fig. 6. RFID label used on the infusion bags (actual size).

Next step is the identification of the drug. Here, the system checks if the ATC code read from the vial is equal to the ATC code in the prescription. In cytostatics, RFID tag is read (identifying single vial), in auxiliary medications, barcode is read (identifying stock item). If the drug has not defined volume and amount of active substance that are necessary for the active support, the user is warned. After the identification, the user is allowed to print a label with all necessary information (name and ID of the patient, active substance and its dose, way of administration, medium and its volume, date of administration, + barcode in case of auxiliary medications). Vials, labels and all material necessary for the preparation are thoroughly disinfected and sent to the preparation room.

The system allows for the individual lines to be marked as priority, these appear in later stages of preparation in red colour and remind the users to process them as quickly as possible.

2.7.4 Preparation

The application "waiting room" is run on the computer with the RFID printer in the preparation room. When identified, individual lines are listed in the waiting room in order corresponding to the time when the prescription arrived in the pharmacy. The lines can be listed either according to surname of the patient or according to the ATC code. Both ways can be convenient under certain circumstances. Lines, where RFID label was already printed but the preparation was not processed yet, remain in the waiting room and are in green colour. In such lines, the repeated printing of RFID label is possible, e.g. if the label was damaged before the preparation. Lines, which have been prepared but not completed yet, remain in waiting room, too, they are in grey colour and they can't be processed any more. In the application, the protocol can be viewed by the user if necessary.

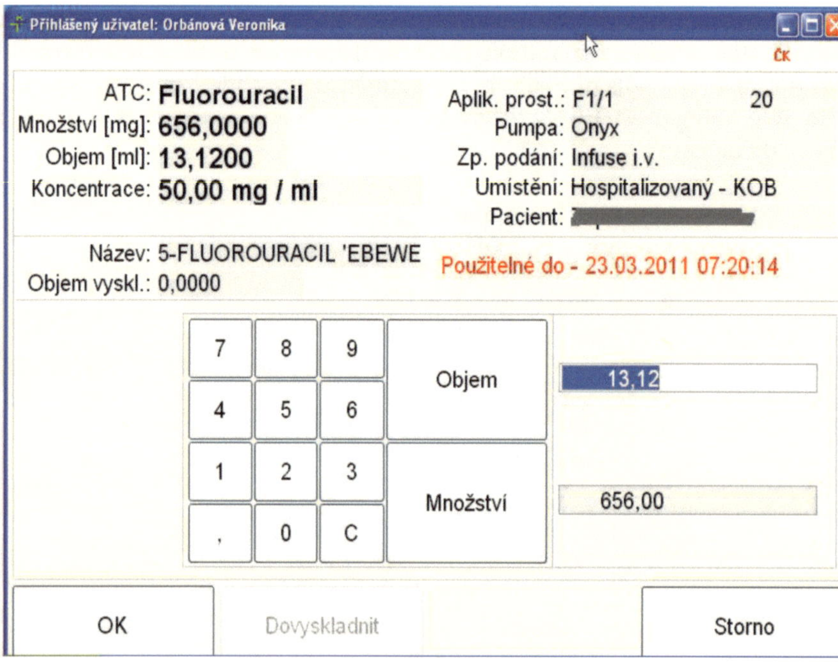

Fig. 7. Isolator. Vial window.

The user chooses a line and prints the RFID label (Figure 6.). Thus, the direct relation between the EPC of the tag on the label and the particular medication for particular patient is established. If the cytostatic is going to be administered in intravenous infusion, the RFID label is attached directly to the infusion bag. If the cytostatic is going to be administered as intravenous bolus, i.e. in a syringe, the RFID label is only put beside the necessary material. Because the syringes are too small, the label can't be used directly and is attached to the secondary packaging only. The label contains (from the top): identifiers of the patient - first name and surname, ID number, INN name of the drug and dose, used medium (here: saline) and its volume in ml, way of administration, location of the patient, temperature at which the bag has to be kept and its stability, date of the preparation, location where the medication was prescribed and the name of the pharmacy.

The user in the isolator identifies himself/herself (by his/her personal RFID ID card, time lock is configurable and set to 5 min, repeated identification is usually not necessary) and then reads the RFID label on the bag or the RFID label for the bolus. Then the user reads the first vial. The system automatically suggests necessary volume (either the full volume of the vial or a part of it). The user can change this volume according to actual volume found in the vial. This is important, as two significant cases occur naturally. Firstly, if the vial contains the precise volume of the drug and if the vial is used for more preparation, losses can't be avoided and the last user is not able to find suggested volume in the vial. Secondly, some vials contain higher volume - as much as 105 % - and this volume can be used, of course. In both cases it is essential that the user puts in the system the actual volume that was taken from the vial. Then, the user takes further vials until required volume of the drug is taken.

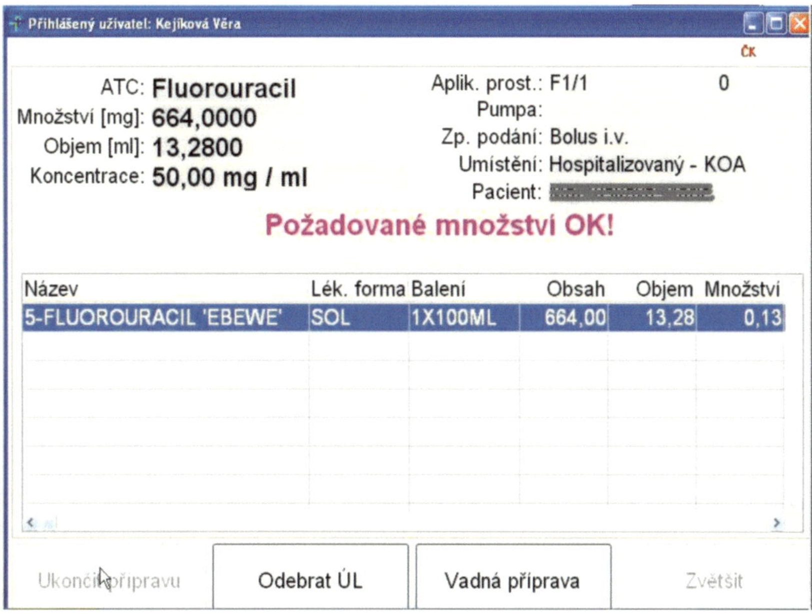

Fig. 8. Isolator. Preparation window

In figure 7 we can see the window as seen by the user in the isolator (the name of the user can be seen on top). In the top part of the screen, the user can see basic data on the preparation - INN name of the drug, dose, volume, concentration, medium (here: saline) and its volume (here: 20 ml), used pump, way of administration, location of the patient and his/her name. In the frame below, there is information on the used vial: name of the product, the volume that was already taken for the preparation (here: zero), and the date and time limiting the use of the vial. The dials serve for putting in of the volume that was taken (the input can be either volume or amount/dose). Buttons at the bottom are used to confirm that the step was finished or for return to previous step. In figure 8 we can see the moment when the preparation was just confirmed as finished. This windows lists in the tablet used vials and when the correct volume was taken, it enables the user to finish the preparation.

The system checks if the user takes the right drug (the ATC code on the vial has to the same as on the bag) and that the user does not take a drug whose stability after first opening has already expired. This stability has to be defined for each stock item and has to be watched because it not rare that the drugs containing the same active substance have significantly different stability after the opening. If the system recognizes the vial as past its usability it is charged to individual bill.

During the preparation, the system automatically creates particular bills (either a standard bill or an invoice) that contain used vials or their parts and material that is defined by the bill of material. If the drug was stored in the Study-storage, two bills are formed, one for each storage. In clinical studies, the pharmacy sometimes uses own cytostatic drugs but these are not charged to the hospital, respectively to the health insurance company, but invoiced to the sponsor of the clinical study. This can be easily configured in the system.

When the infusions bags/syringes are prepared, they are sealed in a foil that acts as secondary packaging and first barrier in the case of an accident. In case of boluses, on this foil the user attaches the RFID label.

2.7.5 Completion

In the completion room the prepared infusion bags and syringes are controlled by another employee. All lines of the prescriptions are controlled here. The user reads the RFID label or barcode and visually checks that the preparation is not damaged, that there are not any leaks, etc. If everything is correct, the user confirms this fact in the system. If the last line was confirmed as checked the system suggests that the prescription should be completed. The message on completion is sent to the HIS. The completion has to processed also with prescription that was recalled after the limit and was at least partially prepared because the cycle has to be closed; in such a case no message is sent to the HIS.

2.7.6 Repeated use

In some cases, the preparation is not administered to the patient, usually because of sudden change in health status, e.g. allergic reaction. If the personnel at the clinic think it is possible to return the unused infusion bag to the pharmacy, it can be done. Firstly, the doctor or the nurse has to mark the bag as unused in the application programme in the HIS. The, the bag is carried back to the pharmacy. Here, the user in Cyto-storage room chooses suitable recipient - a line in the prescription that has identical ATC code and way of administration. Moreover, the dose has to be the same or higher than the dose of the returned bag, as the drug can be added to the bag but not removed. The user then couples the returned bag with the new preparation. This requires the user to identify himself/herself with RFID ID card. The stability of the returned has to be respected, too. However, it is not necessary to use the returned bag on the same day. The user now destroys the original labels on the bag and sends the bag with material, including more vials of the drug if necessary, to the preparation room.

In the preparation room, a standard RFID label is printed and attached to the old bag. Nevertheless, when the RFID label is read in the isolator, it already contains the drugs that are in the bag. If the dose is the same, the user just confirms that the preparation is finished, otherwise he/she adds required amount of the drug.

2.7.7 Repeated preparation

At most steps of the process, there is a risk of an accident and damage of the infusion bag. The bag can be dropped to the ground and burst, cut with sharp instruments, or, which is most frequent, the bag just leaks. In all cases, repeated preparation of the bag is necessary. Should the damage occur in the pharmacy, the user can mark the bag as damaged either in the isolator or in the completion room. Should the damage be found in the clinic, the nurse marks the bag as damaged in the HIS and the HIS sends a message to the PIS that the bag has to be prepared again. In all cases, the particular line of the prescription appears back in the CytoEvidence window. The whole cycle of preparation has to be processed again. Damaged bag can be charged either to the pharmacy or to the recipient.

2.8 Outpatient clinic

The administration of drugs is guided by application software that was developed solely for this purpose and is described in previous chapters. The steps that the nurse has to perform are described in Table 3. This process is repeated for each line (=medication) of the protocol.

Step	Process
1	The nurse reads her ID card with the PDA.
2	The system verifies the ID. If the ID is listed as nurse, the system shows information on this nurse (first name and surname).
3	The nurse reads the ID of the patient.
4	The system verifies the ID. If the ID is listed as patient, the system shows information on the patient (first name and surname) and asks the nurse to confirm the identity of the patient.
5	The system finds the actual application programme of the patient.
6	If there is not any actual application programme for the patient available, the system reports an error and returns to the Step 3, enabling the nurse to start working with another patient.
7	The system finds out if the patient had his/her blood pressure and pulse measured on that day. If yes, step 8 follows, if not, the nurse has to measure these values and put them in the application.
8	The system shows the actual application programme of the patient.
9	If the patient is in his/her first cycle and was not educated by the nurse, the nurse is asked to educate the patient and confirm the education in the system.
10	If the patient has in his application programme unconfirmed oral medications, these medications are listed one after another and have to be confirmed as the patient takes these medications.
11	The system finds out if an infusion pump, or two infusion pumps in case of parallel administration, is necessary for the administrations. If yes, the nurse has to identify an infusion pump. Identified infusion pump is shown on the screen.
12	The system finds out the actual status of the patient.
13	There are 4 possible statuses: Free, Drip, Pause, and Wait
13a	Free - the administration of the actual item of the application programme can be started. The system checks if the right dose is going to be administered.
13b	Drip - an item of the application programme is just being administered. The administration can be interrupted but the reason of the interruption has to be recorded (e.g. in case of a 3-hour infusion a visit to the toilet or a short walk).
13c	Pause - the administration was interrupted. The pause can be finished and the administration restarted.
13d	Wait - the administration of the item of the programme can be started only when the defined waiting time passes (e.g. some treatment protocols have necessary pause between two subsequent items).

Table 3. Administration of at the outpatient clinic

2.9 Further data

The system produces various data, as almost any operation is to some extent recorded. This enables retrospective control and traceability of who, when and how performed a particular step. These data can be further processed and analysed and the results can be used to improve the process.

As example, the comparison of at which time the preparation is performed most frequently can be shown. The working time is divided in 4 shifts as shown in Figure 9. The amount of work is not divided evenly. Most preparations are processed in the morning. This is caused by three significant facts. Firstly, the prescriptions of inpatients, whose protocols cover several days, arrive in the pharmacy first thing in the morning. Secondly, the outpatients who come only for the administration and not for a check-up by their physician tend to come around 9 o'clock. Thirdly, the first wave of patients who come also for a visit at the physician have their blood results ready also around 9 o'clock. Therefore, most inpatients want their infusions to be prepared in a very short period, flooding both the pharmacy and the outpatient clinic with their requirements. An unwritten agreement between the pharmacy and the outpatient clinic states that the patient should wait for his medications one hour on average (including the 15-minute recall period reserved for the physician).

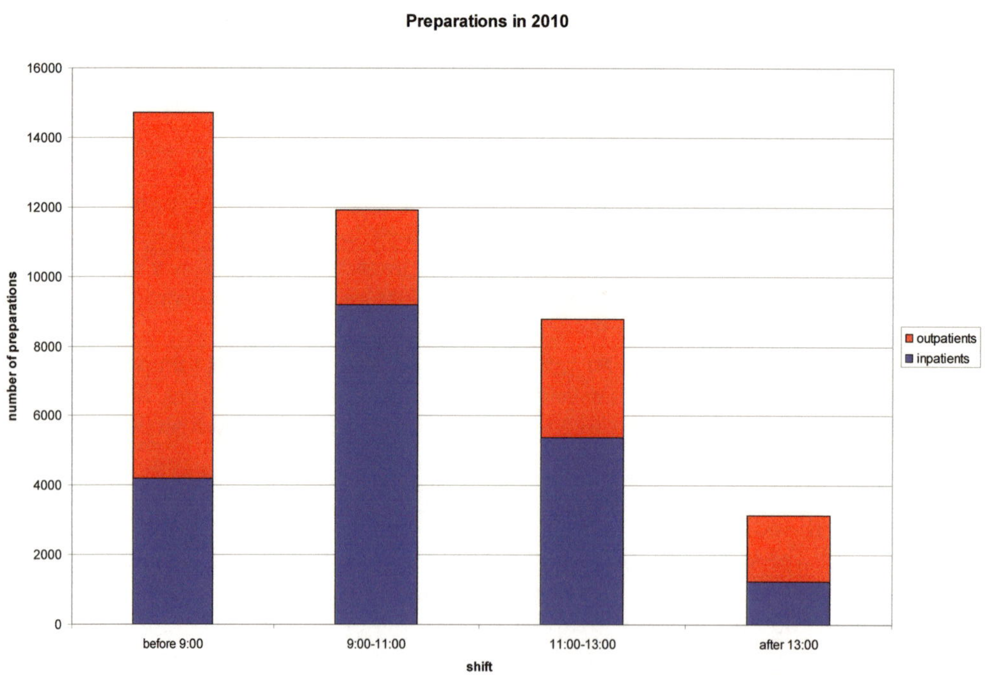

Fig. 9. Preparations during the day

In table 4, preparation times are compared. The value represents net preparation time in the isolator, i.e. between reading the RFID label and confirming the preparation as finished. The most frequent preparations in 2010 are listed, the lowest number of preparations in this set being 367 in bleomycin. The preparations were divided in three groups according to the amount of necessary steps. Group A lists preparations where the required amount is loaded in the syringe from the vial and then either the syringe is stoppered, or the content is first diluted with saline to required amount and then stoppered. Groups B and C include preparations that are administered in infusion bags. During the preparation, the infusion

line has to be connected to the infusion bag and prefilled with the medium, what increases significantly the time needed for the preparation. Group B includes preparations that are taken from the vial directly. Group C includes preparations that have to be first dissolved/diluted and only then they can be used.
Average and median times are very similar, SD is quite high. Once the content in the vial is dissolved, diluted or does not require such a treatment, the preparation time depends only on two physical factors - the volume that has to be taken and the viscosity of the solution. The data show how convenient it is when a preparation that used to be supplied as lyophilized powder or concentrate that had to be dissolved/diluted is suddenly available in dosage form for direct use. This is the case of topotecan in three last months of 2010. This fact is one of the causes why the data for topotecan are so low. The data show how inconvenient it is when only vials of low strength are available in the market. This is the cause of cetuximab exhibiting so high numbers in Group B. Cetuximab is available in 100mg strength only, requiring 4-10 vials for each preparation. On the other hand, the numbers for cisplatin and etoposide, which require 1-2 vials for the use, are low. However, in 2011, significant increase in cisplatin times is expected, as the product is now available in twice lower concentration - the volume that has to be taken is now twice so high.

Preparation	Average (min:s)	Median (min:s)	SD (min:s)	Number of preparations
Group A				
5-Fluorouracil bolus	1:03	0:57	0:36	4728
5-Fluorouracil onyx	1:04	0:58	0:37	3953
Vinblastin	1:46	1:29	1:10	631
Bleomycin	1:54	1:47	1:19	367
Group B				
Cisplatin	1:45	1:29	1:13	5009
Paclitaxel	2:40	2:32	1:08	2962
Cyclophosphamide	2:30	2:09	1:32	1900
Bevacizumab	2:34	2:34	1:06	1378
Etoposide	1:24	1:21	0:42	1305
Carboplatin	2:26	2:19	1:01	1196
Oxaliplatin	3:16	3:24	1:28	1090
Irinotecan	2:35	2:25	1:16	990
Doxorubicin	2:29	2:04	2:34	931
Epirubicin	2:50	2:48	0:54	925
Cetuximab	5:08	4:30	2:28	684
Panitumumab	3:21	3:07	1:58	397
Group C				
Trastuzumab	3:43	3:32	1:44	2357
Docetaxel	3:49	3:32	1:54	1146
Gemcitabin	4:17	4:11	1:54	1025
Topotecan	2:06	1:58	0:59	1004
Iphosphamide	3:50	3:09	2:54	687

Table 4. Preparation time

The table illustrates that the preparation time is 1-5 minutes. In the preparation of cytostatics, automated preparation performed by robots has been in the vogue recently. In spite of the fact that the producers of the robots promote them as extraordinarily convenient, the fastest robots need at least 5 minutes for each preparation. The robots decrease the human factor in the preparation but the benefit on the side of efficiency is still awaited as well as the patient waits until his medication is prepared.

3. Conclusion

The project was implemented in full run in October 2009. Neither erroneous preparation nor administration has been recorded. Within the Czech Republic, the RFID technology in preparation and administration of cytostatics is used solely at MMCI. It is an example of a multidisciplinary solution which was tailored to the needs of the MMCI. The goal was to reduce the human factor in the whole process.

The solution leaves two important challenges for the future. Firstly, the administration of cytostatic and auxiliary medication to the inpatients will be included, as well. Inpatients receive 40-50 % of preparations processed in the pharmacy. This will require minor investments in hardware and thorough education of nurses in inpatient clinics. Secondly, unless the prize of RFID tags decreases significantly in three years, we will be forced to adjust the system to another type of unique identificator. The system was developed to be suitable for unique identification in general and only some changes in hardware will be necessary.

4. References

Erdem, E., Zeng, H., Zhou, J., Shi, J., & Wells, DL.: Investigation of RFID tag readability for pharmaceutical products at item level. Drug Dev Ind Pharm. 2009 Nov;35(11):1312-24.

Hanada, E., Horigome, S.: The safety of medical equipment near data communication devices that use RFID with magnetic coupling in the 13.56 MHz band. The Journal on Information Technology in Healthcare 2008; 6(6): 421–428.

Lahtela, A., Hassinen, M.: Requirements for radio frequency identification in healthcare. Stud Health Technol Inform. 2009;150:720-4.

Lahtela, A., Saranto, K.: RFID and medication care. Stud Health Technol Inform. 2009; 146:747-8.

Oranje-Nassau, C. van, Schindler, H.R., Valeri, L., Vilamovska, A.-M., Hatziandreu, E., & Conklin, A.: Study on the requirements and options for Radio Frequency Identification (RFID) application in healthcare. RAND 2009.

Spahni, S., Lovis, C., Ackermann, M., Mach, N., Bonnabry, P., & Geissbuhler, A.: Guarantying the quality of chemotherapy: from the order to the administration. Stud Health Technol Inform. 2006;124:241-6.

Sun, P.R., Wang, B.H., & Wu, F.: A new method to guard inpatient medication safety by the implementation of RFID. J Med Syst. 2008 Aug;32(4):327-32.

Van der Togt, R., van Lieshout, E.J., Hensbroek, R., Beinat, E., Binnekade, J.M., & Bakker, P.J.M.: Electromagnetic interference from radio frequency identification inducing potentially hazardous incidents in critical care medical equipment. JAMA, June 25, 2008 – Vol 299, No. 24.

5

Application of RFID Technology in eHealth

Cristina Turcu, Tudor Cerlinca, Marius Cerlinca and Remus Prodan
Stefan cel Mare University of Suceava
Romania

1. Introduction

Today's hospitals are particularly interested in increasing the quality and efficiency of *patient identification* and *monitoring procedures*. While most patient health records are stored in separate systems, there is still a huge stack of paperwork left for health-care providers to fill out in order to comply with different regulations. Since many health care errors occur when important patient information is missing or simply not available, the electronic medical records (EMRs) may easily alleviate the distress of most doctors and nurses working in today's care system.

The next step beyond the EMR is to connect and provide medical information to primary care physicians, medical and surgery specialists, anesthesiologists, nurses, assisted-living staff, patients themselves, patient's family and so on.

However, each hospital may use a different system, store data in many ways and even decide upon its own data format. Furthermore, file system access and data retrieval are often governed by inconsistent parameters. Thus, the availability of medical information is seriously affected and effective communication among physicians is not achieved.

Within this framework, the present chapter focuses on how RFID technology can be used in order to solve the problems eHealth is dealing with. After defining the EHR and EMR terms, we shall focus on presenting an RFID-based system (named SIMOPAC) that integrates RFID and multi-agent technologies in the field of health care. The purpose of this system is to make patient emergency care as efficient and risk-free as possible, by providing doctors with as much information about a patient as quickly as possible. SIMOPAC could be used in every hospital with the existing systems in order to promote patient safety and optimize hospital workflow. The SIMOPAC system will assure information exchange with electronic health record (EHR/EMR) (Smaltz & Berner, 2007; Hallvard & Karlsen, 2006) systems set up in healthcare units. This information exchange will be in accordance with the HL7 standards specifications. In the present chapter, we will focus on the RFID technology and how it could be used in emergency care in order to identify patients and to achieve real time information concerning the patients' biometric data, which might be used at different levels of the health care system (laboratory, family physician, etc.). Within the SIMOPAC system, the access to medical information is granted by an electronic memory-based chip (RFID tag or transponder). This tag, named Personal Health Information Card (CIP, in Romanian) (Turcu & Turcu, 2008), allows patient information storage (Jonathan, 2004). We describe a general purpose architecture and data model that is designed for storing and presenting clinically significant information to the emergency care physician. Also, we present the strengths and weaknesses of this RFID-based systems used in eHealth.

2. EHR vs. EMR

The most common method currently used by physicians in hospitals to record patient data is paper-based, a method considered low cost and easy to use. But there are various disadvantages concerning this practice, especially when health records must be stored for a long period of time. One disadvantage concerns the storage space: paper-based records require a significant amount of storage space in comparison with electronic/digital records. Another one refers to the costs involved: electronic storage media is cheaper then traditional storage media. More problems can occur when a patient's paper records are stored at different levels of several health units: the process of collecting patient's information by a health care provider proves very difficult and time consuming. In 1990, a study commissioned by the Institute of Medicine (IOM) highlighted the strength and weakness of the traditional paper-based health records. Some identified weaknesses were the disorganization, the illegibility, and the short availability (Van der Meijden, 2001). In order to eliminate the mentioned disadvantages and weaknesses the use of electronic medical records becomes imperative (I. de la Torre et al., 2010).

Electronic Health Record (EHR) and Electronic Medical Record (EMR) are two terms often used in the last years. Even though related to the same field and considered interchangeable, each of these terms describes a completely different concept and is defined in various ways. Recently, National Alliance for Health Information Technology (NAHIT), a leadership organisation that aims at enforcing the use of IT health systems in order to improve the US healthcare system has defined EMR and HER as follows:

> **EMR:** The electronic record of health-related information on an individual that is created, gathered, managed, and consulted by licensed clinicians and staff from a single organization who are involved in the individual's health and care.
>
> **EHR:** The aggregate electronic record of health-related information on an individual that is created and gathered cumulatively across more than one health care organization and is managed and consulted by licensed clinicians and staff involved in the individual's health and care.

A more elaborated definition of EHR was delineated by the Healthcare Information and Management Systems Society (HIMSS) - a professional member organization exclusively focused on providing leadership for the optimal use of healthcare information technology:

> **EHR:** A longitudinal electronic record of patient health information generated by one or more encounters in any care delivery setting. Included in this information are patient demographics, progress notes, problems, medications, vital signs, past medical history, immunizations, laboratory data and radiology reports. The EHR automates and streamlines the clinician's workflow. The EHR has the ability to generate a complete record of a clinical patient encounter - as well as supporting other care-related activities directly or indirectly via interface - including evidence-based decision support, quality management, and outcomes reporting.

By considering these definitions one can conclude that an EHR is an EMR with interoperability. This attribute was also highlighted by Justin Barnes (Chairman Emeritus of the HIMSS) who believes that "the future of healthcare IT is interoperability".

But selecting an EMR or an EHR software proves a difficult task for a health care services provider. Mr. Barnes considers that there are three criteria to be taken into consideration while choosing between EMR and EHR systems:

- Current-year interoperability certification standards (CCHIT- Certification Commission for Healthcare Information Technology, HL7 – Health Level Seven);

- A unique workflow that matches your practice and specialty;
- Excellent usability at the point of care.

Electronic medical records (EMRs) may easily alleviate the distress of most doctors and nurses working in today's care system. Besides improving the degree of data availability among physicians and patients, they certainly increase the traceability of numerous medical details so deeply buried in traditional records. Still, despite these potential benefits that cannot be disregarded, doctors have been reluctant in adopting electronic health records.

Most hospitals have improved patient care by reducing wait times in the emergency ward when they decided to replace their paper-based process for emergency ward admission with a solution based on informatics systems. With these solutions in place, hospitals save minutes each time they admit a patient because doctors and nurses no longer fill out forms manually and improve healthcare outcomes. Thus, it has been estimated that 15 to 18 per cent of US physicians already use electronic health records (Tucci, 2008).

"Instant access to patient information is key to lifesaving care, especially in the emergency room and intensive-care unit, where delays may mean the difference between life and death", Dr. Mark Smith said (Microsoft, 2006). Currently, Emergency Medical Service (EMS) providers rely completely on personal and medical history information provided by patients or family members. It is common knowledge that stress, physical and mental discomfort prevent most patients and family members to impart vital medical information. Problems may also arise if there are no family members around or if the patient is unconscious, incoherent or unable to talk or communicate (e.g. language difficulties).

The next step beyond the EMR is to connect and provide medical information to primary care physicians, medicine and surgery specialists, anesthesiologists, nurse practitioners, assisted-living staff, patients themselves, patient's family and so on.

However, each health unit may use a different system, store data in many ways and even decide upon its own data format. Furthermore, file system access and data retrieval are often governed by inconsistent parameters seriously affecting the availability of medical information. Hence the inefficient communication among physicians.

Microsoft's Feied, a pioneer in medical training computer programs and medical intelligence software, said physician collaboration is the critical element for improving health care. He offered an impassioned testimonial. An emergency room physician who estimates he treated 80,000 patients "with my own hands", Feied said the thing that stuck out as he looked back on his career was how many times he was put in a position of "guessing over and over", "flying solo" in an information vacuum. In situations where people "die right in front of you", he said he often felt he was "one data element away" from stopping a patient from dying (Tucci, 2008).

The market for bringing healthcare data from disparate sources into one view is growing by leaps, according to a new study from KLAS, a healthcare research firm based in Orem, Utah (Klas, M., 2009).

For example, through Microsoft HealthVault and Google Health, Microsoft and Google have a common goal of managing vast quantities of personal health information to benefit end users. Thus, these systems encourage and support healthcare patients/consumers to control and account for their own and family health records.

According to (Impact, 2008), data integration - the automated aggregation and consolidation information from a variety of disparate systems and sources - across sites of care (inpatient, ambulatory, home), across domains (clinical, business, operational), and across technologies (text, video, images) - is the Holy Grail of healthcare information technology.

Still, it is necessary to find a way to get the vital medical data into the hands of those who can use it to save lives in emergency medical services, even when there is no connection to the Internet or the server is down.
Health systems and health policies across the European Union are becoming more and more interconnected, and also more complex. The European Commission aims at improving the safety of care for patients in all EU Member States through sharing information and expertise (EC, 2006). But healthcare is provided through different systems that run at the national level.
For the moment, introducing informatics systems within the Romanian healthcare proves to be relatively difficult, as patients data has not been shared at the level of medical entities, the medical records are not unitary and complete, and cannot be accessed online by the medical staff, when needed.

3. Healthcare in Romania

Medical care in Romania is not up to European standards and medical supplies are limited, especially in some rural areas. In larger cities, there are hospitals and private clinics, but in some small cities or village areas, quality health care level is low.
Hospitals are organized on geographical criteria at the regional, district and local level. Tertiary care is provided in specialized units (specialized hospitals, institutes and clinical centres) and in a number of cardiovascular and other surgery departments of teaching hospitals. Inpatient care is also provided by long-term care hospitals (for patients with chronic diseases who require long-term hospitalization), medicosocial care units (institutions under local authorities that provide both medical and social care), sanatoriums (units that besides usual treatments provide natural therapies) and health centres (inpatient units that assure medical services for at least two specialties) (C. Vlădescu et al., 2008).
The fact that primary care is provided especially by family doctors is an indicative of the low efficiency and underuse of primary and ambulatory care services. It is also a proof of the fragmentation of services and insufficient development of different levels of care.
Emergency care is provided through a network of emergency centres with a territorial dispatch system connected to hospital wards specialized in dealing with emergencies. Each district has an emergency dispatch system with a number of ambulances located in hospitals or dispatch centres and emergency wards at designated hospitals. All hospitals have to be prepared to receive emergencies, but not all of them are properly equipped for this purpose (C. Vlădescu et al., 2008).
The emergency system is based on the traditional ambulance system and SMURD (Romanian acronym for Mobile Emergency Service for Resuscitation and Extrication) as a complementary service, with a lot of bases in the whole Romania, still expanding.
Today's Romanian medical sector has not fully embraced the gains and benefits of information systems. Thus, the medical staff is faced with endless amounts of paperwork of former and present patients. But, healthcare costs are expected to grow due to the aging of the population and the increasing demand on health systems. Considering this, one of the Romanian government's priorities is controlling public spending on healthcare, in part, by IT investments.
In this context our research team proposed an integrated system for identification and monitoring of patients - SIMOPAC (C. Turcu & Cr. Turcu, 2008). This system was designed to integrate within the distributed medical information system, and privately, to solve the

issues related to patient identification and monitoring. The SIMOPAC system will assure the information exchange with electronic health record (EHR/EMR) (Smaltz & Berner, 2007; Hallvard & Karlsen 2006) systems set up in healthcare units. This information exchange will be in accordance with the HL7 (HL7, n.d.) standards specifications. Within the SIMOPAC system, information needed in medical services is stored and can be accessed by means of a Personal Health Information Card (CIP, in Romanian) (C. Turcu & Cr. Turcu, 2008). This card will be implemented by using the RFID technologies (Jonathan, 2004), where information carrier is represented by a transponder (tag).

4. SIMOPAC system

In order to provide high-quality medical services to all its citizens, EU has recently proposed the interconnection of all health and medical care systems and services. Thus, this proposal aims at creating a large continental medical service space available to all European citizens and authorized medical personnel. Unfortunately, the major challenge of implementing e-Health applications in Europe is the lack of interoperability of European medical systems and services. In our attempt to address this complex issue, we have proposed an integrated system for the identification and monitoring of patients, a system that suits the Romanian medical environment and allows further adaptations to any medical environment.

Today's Romanian medical sector has not fully benefited from all gains and advantages of information systems. Patient-related information is scattered among various medical units, the patients' charts have no standardized form or content and are seldom complete or up-to-date; moreover, if needed, they cannot be accessed online by the medical staff. Considering these major inconveniencies, we have devised an RFID-based system, called SIMOPAC, for the distributed medical field. Employing the latest Radio Frequency Identification solutions, the system permits the real time patient identification and monitoring, ensures the collaborative problem solving in distributed environment (multi-agent technologies) and provides the communication infrastructure with multi-point connections to the medical information within the system.

4.1 SIMOPAC objectives

The research's main objective was the implementation of an integrated system using RFID technologies, agents and web services in order to identify and monitor patients. Delivering multi-source real time medical information, the SIMOPAC system aims at optimizing medical decision by increasing the quality of patient-oriented medical acts.

The major objectives of the research were:

a. increase the efficiency of medical information management;
b. increase the quality of medical services by adopting advanced information technologies;
c. build and expand the Romanian health information system in accordance with EU requirements in the field of health and medical care;
d. eliminate all physical constraints of hardcopy documents and to grant immediate access to medical charts or patient records;
e. establish partnerships among research units in different fields and motivate them to cumulate their experience and expertise in joint health projects;
f. give assistance in providing citizens with comprehensive and reliable information.

The specific objectives of the research were:

a. implement several RFID software applications aimed at patient identification by using Personal Health Identity Cards (CIPs) that allow the extraction of vital data in medical care and emergency situations and strengthen patients' trust in medical treatment as by considerably reducing medication errors;
b. implement a high-speed communication system that secures the access of the medical staff to the electronic medical records (bi-directional access) and thus allows all medical and patient-related information to be shared by all parties involved in health and medical care;
c. improve the communication among all health-service providers: family physicians/specialist physicians, hospitals, medical laboratories, etc.

4.2 SIMOPAC facilities

The SIMOPAC system allows:

a. access to medical services via RFID Medical ID Cards;
b. sharing of patient-related information and development of databases containing patients' electronic medical records;
c. secure access to medical information databases (for both medical staff and patients), as well as the complete and speedy bi-directional transfer of information;
d. quick and accurate information gain on the medical status of patients transported in emergency units (ambulances) and requiring appropriate medication;
e. enhanced communication among all health and medical care services: family doctors, specialists, hospitals, medical laboratories, pharmacists;
f. automated information-flow in the medical system.

4.3 Standards and technologies employed

SIMOPAC employs the latest technologies and software solutions. Widely used in a variety of other applications, RFID technologies have proved considerable advantages for the medical environment. Efficient patient identification solutions have already been reported by many European and American hospitals. However, according to recent surveys, the implementation of RFID solutions in healthcare is still in its infancy. The application of this technology in hospitals is part of the view that in the hospital of the future the patient's life will not be saved by the latest medicine, but by computer systems.

Within the next ten years, multi-agent systems will trigger major transformations in health and medical care. The decision to integrate this technology in our SIMOPAC system was taken after a close consideration of its major advantages such as intelligent, adaptive and decentralized coordination-solutions and data availability in fragmented and heterogeneous environments. Our major aim was to design and develop software agents which could dynamically extract patient-related information from heterogeneous environments within a distributed communication structure.

4.3.1 RFID technology

RFID technology has been considered one of today's "hottest" technologies due to its specialized capacity to track and trace objects in real time (Castro & Wamba 2007). RFID technology is classified as a wireless Automatic Identification and Data Capture (AIDC) technology that uses electronic tags to store identification data and other specific information, and a reader to read and write tags (Mehrjerdi 2007). Tags are small chips with an antenna. There are three different types of RFID tags: passive (uses the reader's signal to be activated), active (battery powered) or semi-passive (battery-assisted, activated by a signal from the reader).

RFID technology is also providing a high level of security and has various important advantages over similar technologies, such as barcodes. It has been successfully implemented in a variety of areas, such as: logistics operations, inventory and materials management, industrial automation etc.

Healthcare industry can also benefit from the RFID technology. Although most of the current RFID healthcare applications and systems are just in some experimental phases, the future looks promising. Thus, some studies estimate that the market for RFID tags and systems in healthcare will rise from $90 million in 2006 to $2.1 billion in 2016 (RFIDUpdate 2008). The RFID-based systems can provide a number of benefits to the healthcare industry. By attaching RFID tags to different entities in healthcare industry (people and objects), RFID technology can ensure the following: identification, tracking, location and security. These capabilities directly affect the major issues currently experienced by healthcare organizations while helping to drive down costs (RFIDHealthcare, n.d.).

The main idea of any RFID healthcare system is to tag patients. Thus, an RFID tag attached to a patient needs to store some of the patient's relevant information, such as: identification data, a list of chronic diseases the patient is suffering from and the most significant data of patient's medical history. But, the common problem of any memory based system has always been that no amount of memory is ever sufficient (Peacocks, n.d.). On the other hand, it is well known that RFID tags with large memory capacity are too expensive to be used in a system with thousands of patients and the only way to keep costs low is to use passive tags with reduced memory capacity. But it is obvious that a tag with a reduced memory capacity cannot store all the relevant information related to a patient. This problem can be solved by storing the vital information on the RFID tag and the additional information into a central database, based on a tag template. The IP address of the database server could also be stored on the RFID tag, so that the additional information could be accessed by the medical staff over the Internet. This way, all relevant patient-related information will always be available for the medical staff.

Another important feature that an RFID healthcare system should provide is the ability to integrate and exchange information with similar systems. This could be achieved by using HL7 standards. HL7, an abbreviation of Health Level Seven, regards the information exchange between medical applications and defines a specific format for transmitting health-related information. Using the HL7 standard, information is sent as a collection of one or more messages, each of which transmits one record or item of health-related information. The HL7 international community promotes the use of such standards within and among healthcare organizations, in order to increase the effectiveness and efficiency of healthcare delivery for the benefit of all (HL7_1, n.d.; Iguana & Chameleon n.d.; Shaver, 2007).

4.3.2 The HL7 standard

What is HL7? HL7 (Health Level Seven) is a non-profit organization that is a global authority in the field of interoperability of health information technology (*, HL7). HL7's more than 2,300 members represent approximately 500 corporate members, which includes more than 90 percent of the healthcare information systems vendors (Ehto, n.d.).

Furthermore, HL7 "is a standard series of predefined logical formats for packaging healthcare data in the form of messages to be transmitted among computer systems" (OTech, 2007).

Why HL7? Because "HL7 is the most widely used standard that facilitates the communication between two or more clinical applications. The prime benefit of HL7 is that it simplifies the implementation of interfaces and reduces the need for custom interfaces. Since its inception in the late 1980's, HL7 has evolved as a very flexible standard with a documented framework for negotiation between applications. The inherent flexibility makes deploying HL7 interfaces a little more challenging at times." (Mertz 2010).

The HL7 messages are in fact clinical information and not only collections of data used to send information about some events in some healthcare enterprise.

Originally developed in 1987, HL7 Version 2.x is now in use in more than twenty countries around the world. It contains messages for almost every conceivable healthcare application area, including the following: registration, orders (clinical and other), results and observations, queries, finance, master files and indexes, document control, scheduling and logistics, personnel administration, patient care planning, network synchronization, laboratory automation (OTech, 2007).

In order to acquire all these, the HL7 standard includes: conceptual standards: RIM (Reference Information Model), document format standards: CDA (Clinical Document Architecture), clinical application standards: CCOW: (Clinical Context Object Workgroup) and messaging standard.

But the use of intelligent agents reduces the need for knowledge about HL7 and interfaces, and thus reduces the barriers to entry for the introduction of HL7 (Long et al., 2003).

Thus, ontology-based multi-agent systems provide a framework for interactions in a distributed medical systems environment without the limitations of a more traditional client server approach (Orgun & Vu, 2006).

We consider agents (Turcu et al., 2009) that cooperate with each other in order to manage the information flow between local EMR database applications and HL7 message templates.

4.3.3 Multi-agent technology

Agent technology is an emerging and promising research area, which increasingly contributes to the development of value-added information systems for different applications. An agent is a small, autonomous or semi-autonomous software program that performs a set of specialized functions to meet a set of specific goals, and then provides its results to a customer (e.g., human end-user, another program) in a format readily acceptable by that customer (Wagner, n.d.). For example, agent technology has been applied in the area of gathering information from World Wide Web heterogenous data sources. The performance evaluation of the agent-based system versus traditional systems (client-server and relational database based systems) was undertaken by some researchers (Yamamoto & Tai 2001; El-Gamal et al. 2007). The tests reveal that the agent-based systems provide better times of response as well quicker notification processing.

Healthcare systems are characterized by a wide variety of applications working in autonomous and isolated environments. The use of agent technology in healthcare system has been increasing during the last decade. Multi-agent systems become more and more important in the field of health care as they significantly enhance our ability to model, design and build complex, distributed health care software systems (Nealon & Moreno 2003)).

4.4 SIMOPAC architecture

In the last few years, most world-wide medical bodies and healthcare units have shown an increased interest in the employment of Healthcare Information and Management Systems and Electronic Medical Records (EMRs). Nevertheless, there are still many problems to be tackled upon, such as the case when patient information is not available because the unit which is supposed to offer medical assistance does not own the patient's medical record. Furthermore, it is imperative to eliminate the duplication of medical services (e.g. laboratory tests) so that physicians may easily obtain any patient-related information that is stored in different databases within different EMR systems. Our research team developed a distributed RFID based system for patients' identification and monitoring, named SIMOPAC. This system enables real time identification and monitoring of a patient in a medical facility, on the base of CIP. A CIP is a passive RFID tag that is storing relevant medical information regarding its carrier. The CIP provides a quick access to the actual health state of a patient and helps the medical staff in taking the best decisions, especially in case of an emergency. Thus, the risk of administrating wrong medication is highly reduced. The system is also able to integrate and exchange information with other HL7 and even non HL7 based clinical applications already developed by other companies or organizations. The presented system provides an interface to different areas of healthcare, such as: emergency services, medical analysis services, hospital services, family medicine services, etc.

The different components of this scalable and robust distributed system are depicted in figure 1.

The *Personal Electronic Health Identity Card* (PIC in English, CIP in Romanian) is a prerequisite of patient identification. SIMOPAC CIPs are designed to store patient personal data, minimum general health data, as well as other vital information indispensable in emergency situations. Employing the Domain Name System (DNS), the RFID tag permits patient identification in a SN@URI format, where SN represents the tag series corresponding to the patient's CIP. The CIPs store the following data:

a. emergency medical information (blood type, RH, allergenic substances, HIV/ AIDS and any other chronic or transmissible diseases, etc.);
b. patient ID + URI server keeping the medical chart;
c. values of 1 and 0 corresponding to a template defined within the system by the medical staff.

SIMOPAC offers reliable solutions for the distribution of patient-related information among several medical units. The system requires that all medical units own EHR/EMR information systems to store patient electronic medical records. Moreover the information systems must be compatible with 2nd version of HL7 standard. Whenever a member of the medical staff needs to consult a patient's medical record, the multi-agent system allows the gathering of patient-related information, regardless of the patient's location.

Related to SIMOPAC architecture we can assert that this RFID-based system includes the following main modules:

- User management;
- EMR viewer;
- Tags management;
- HL7 server.

These modules are shortly described in following subsections.

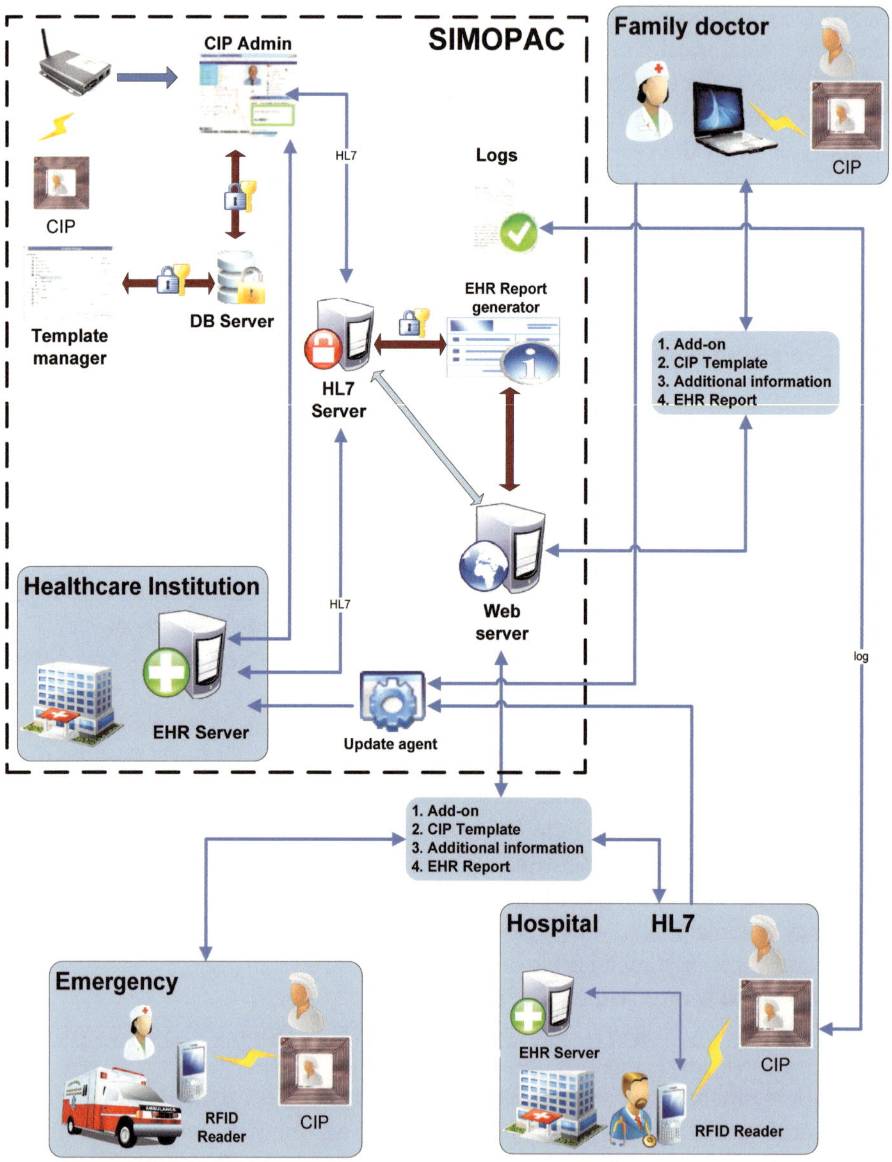

Fig. 1. SIMOPAC System Architecture

4.4.1 User management

Needless to say, security is one of the main aspects that should be taken into consideration when implementing such a distributed system. User management is a critical part of maintaining a secure system. Ineffective user and privilege management often lead many systems into being compromised (Teambusiness, n.d.).

The User Management module was designed as a generalized system that enables the management of all users and users groups within a distributed system. It consists of different modules, each of them with its own list of entities and rights.

Within the framework of SIMOPAC system, the User Management module provides the following main facilities:

- password based access to the User Management application;
- data encryption with the TripleDES algorithm for all important information transferred over the Internet and stored into the central database (e.g.: user names, passwords, access rights);
- support for different levels of access rights. This implies that users are granted different rights to the system's features;
- management of system registered users (users visualization, adding or removing of certain users, profile modification, granting/revoking user privileges, etc.);
- modules and entities management.

Figure 2 exemplifies the process of granting/revoking user privileges for different modules and entities of the SIMOPAC system.

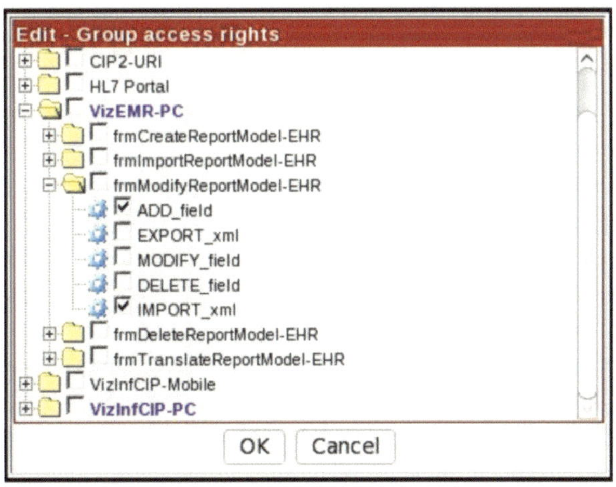

Fig. 2. Granting/revoking user rights

4.4.2 EMR Viewer

This module, generically named VizEMR-PC, allows patient identification based on his own CIP and displays some pre-configured information from the electronic health record of that patient. The patient identification is based on the patient's identifier that is stored on his RFID tag and printed on the CIP. VizEMR-PC module also displays patient information in the language requested by the user.

This module can be used when the CIP is read at a medical unit and the medical staff wants to obtain more information about the patient. VizEMR-PC provides the following main facilities:

- a specialized editor that allows the design (configuration) of a report template. This template will be used for the interest information from the electronic health record of the EHR/EMR system that is integrated with SIMOPAC. The report template is created only once by skilled health personnel and contains all or only some fields of the electronic medical/health record. This template can then be translated into several foreign languages in order to facilitate cooperation between medical units from different countries and assure a good care for a patient from another country.

- a report generator that will be responsible with the completion of the following tasks:
  - filling the report template fields with information taken from the electronic medical/health record of a patient by using HL7 dedicated commands;
  - generating a custom report in different formats (XML, CVS, MDB, etc.), using the language specified by the user.

In order to have access to VizEMR facilities, authorized users must first login to the application by entering their username/password. The client-server communication is secure; all the passwords that are sent over the Internet are first encrypted on the client-side. Also, the access to various facilities offered by VizEMR-PC is granted in accordance to the rights previously set for the registered user. Access rights are established by the User Management module.

4.4.3 Template management

This component of SIMOPAC system is mainly focused on the designing of the templates used for information structuring on patients' CIP sheet and stored on a Web server. The patient's CIP sheet contains two different areas, each of them storing specific information about the patient. The first one contains clear-text information that is needed especially in emergency situations. This information uniquely identifies a patient and specifies if he/she is suffering from any serious illnesses. The second section of the CIP contains data that can be interpreted only with the same template that was used for writing the information into the RFID tag. This template will be available for download at an URL written on the CIP. The medical staff can have quick access to the information written on the CIP by downloading (from the same URL address) a specialized add-on application that is mainly used to communicate with the RFID reader. Moreover, the medical staff can obtain a translation of this information, if it has been previously translated by the person created the template and the CIP sheet. This translation, available in an XML format, could be easily transferred and read. On the base of these templates, the medical staff can create the CIP sheet that corresponds to one or more patients.

One of the main advantages of template based information structuring is the fact that in order to be included on the CIP, information is translated only once. Other advantages are listed below:

- the use of a single template for a specific target group (because everyone will have the same type of data included in the CIP);
- allows a better organization of data to be included on the CIP.

A template consists of a list of user defined fields. Each field is defined by name and data type. The basic data types are shown in figure 3, to which more types can easily be added.

ID	Type	Display format
1	String	(A/_)
2	Integer	[+-](9)
3	Real	[+-](9)[.(9)]
4	Data	yyyy-mm-dd
5	Time	hh/nn/ss
6	DateTime	yyyy-mm-dd hh/nn/ss

Fig. 3. Common data types

As seen in figure 3, each data type has been associated with a display format that will be used by a plug-in module for the correct displaying of the information stored on the CIP. The display format can be interpreted as follows:

- (A/_) - letters (A. … Z) and other displayable ASCII characters;
- [+-](0...9) – the symbol + or - (optional), followed by digits;
- [+-](0...9)[.(0...9)] – the symbol + or - (optional), followed by digits. The decimal point is optional and it is used for floating point numbers representation;
- yyyy-mm-dd – standard representation of dates (y - year, m - month, d - day);
- hh/mm/ss – standard representation of time (hh - hour, mm - minutes, ss - seconds);
- yyyy-mm-dd hh/mm/ss – standard representation of date-time values.

When the system contains at least one CIP sheet associated with a particular template, the template cannot be edited anymore, but another one could be built on the base of the first one. After building the template, the next phase is the translation of the fields; this translation will be saved in an XML format and then stored into the central database. There is no restriction related to the number of translations that can be done. When a doctor consults a patient's CIP sheet, he is granted access to the structured information as well. Regarding to the translation of the template's fields, the medical staff can choose between an automated translation (performed by the plug-in application, based on localization) and a translation that was downloaded once with the template associated to the patient's CIP sheet (see figure 4).


```
- <grammar xmlns="...">
    <name
      type="simopac.template">Sablon
      Cardiologie</name>
  - <languages>
      <RO>romanian</RO>
      <EN>english</EN>
    </languages>
  - <language name="RO">
      <field idx="1">Nume
        persoana</field>
      <field idx="2">Data nastere</field>
      <field idx="3">Grupa
        sanguina</field>
    </language>
  - <language name="EN">
      <field idx="1">Person name</field>
      <field idx="2">Birth date</field>
      <field idx="3">Blood type</field>
    </language>
  </grammar>
```


Fig. 4. SIMOPAC - CIP sheet

The template is automatically accessed through the add-on module downloadable from the official site of the SIMOPAC system. The URL is printed on the label of the RFID tag (see figure 5). After being downloaded and launched, the add-on module will perform the following actions:

- tries to find an RFID reader recognized by the system;
- if such a reader has been found, the add-on module accesses the SIMOPAC's database and downloads the template and its translation;

- based on this information and using the localizing function, the add-on displays the translated template filled with all data extracted from the patient's CIP (local RFID tag);
- after patient investigation, the add-on module sends all the results/findings to the logs' area of the SIMOPAC server.

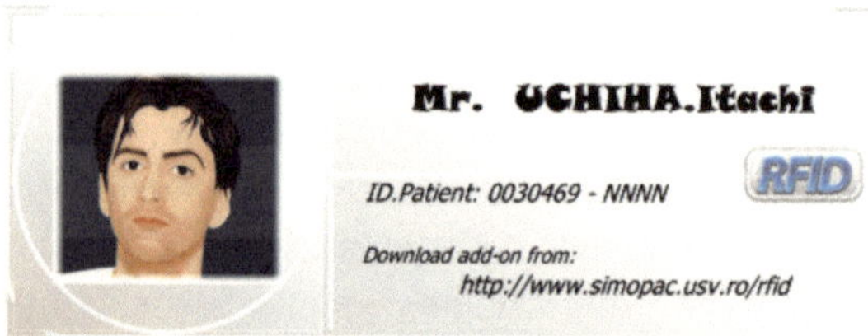

Fig. 5. An example of printed label of a patient's CIP

The filling-in of the patient's CIP sheet, along with the creation/administration of the template(s) is to be performed by the treating doctor. If the medical unit does not use such an EMR system, it is still possible to use the SIMOPAC system, but without the facilities of an EMR system (e.g.: direct import of patients' related data).
Generally, the memory space on RFID tags is limited to about 1-2 Kbytes. Thus, an efficient data compression method is needed when working with large amount of data. In order to reduce the amount of memory needed to store the structured information on RFID tags, we have designed and developed several techniques of data representation, as follows:

- representation of Floating point/Integer numbers on subintervals *[a, b]*, with step specified. This achieves a reduction in the number of bits needed for representation;
- representation of Date, Time and DateTime values by setting the startup date/time value;
- specifying the list of possible values for the fields using small sets of values;
- Huffman encoding of fields that frequently use the same numerical values.

When representing numerical values on subintervals, the template will store some additional information, as a 3-tuple *(left borderline, number of values, [step])*. If the distance between two consecutive values is different from 1, then it must be specified in the template, in the optional section *[step]* (see figure 6).

Field name	Field type	Left	Dim	Step
'weight'	INT	48	128	1

Fig. 6. Internal representation for floating point/integer numbers

When working with a Date field, the user can specify (in the template) the date from which the actual encoding within that field begins. Thus, the value 0000 corresponds to the start date. This start date will be specified as a 3-tuple *(year, [month] [day])*, *year* being the only mandatory. If *month* is missing, it is assumed to be January. When *day* is missing, it is assumed to be the first day of the month. The value stored in such a field represents the number of days elapsed from the start date (see figure 7). Time fields will be handled in a similar manner. The value stored in such a field represents the number of seconds elapsed from the start date (see figure 8).

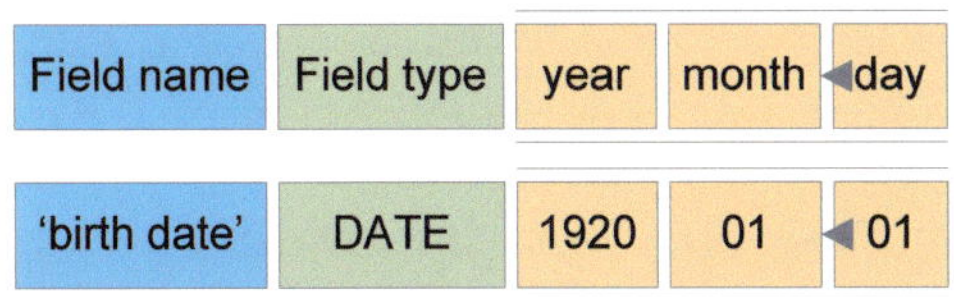

Fig. 7. Internal representation of date values

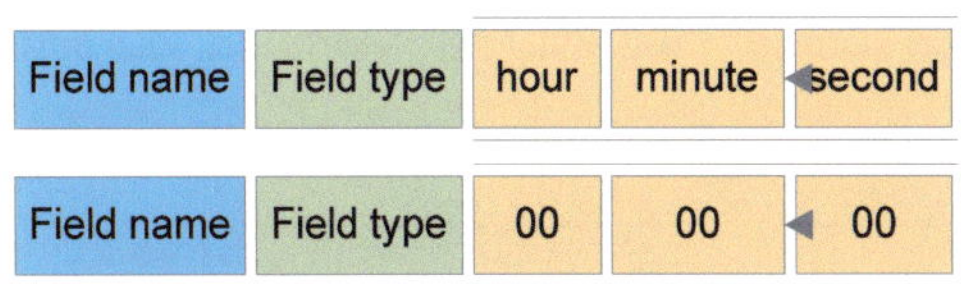

Fig. 8. Internal representation of time values

Huffman coding, a variable-length coding method, was used to allow a substantial compression ratio of the data encrypted on the RFID tag. Thus, certain fields encode information such as "diseases" and some of them may occur more often on patients' tags than others. Figure 9 presents an example of a Huffman coding tree.

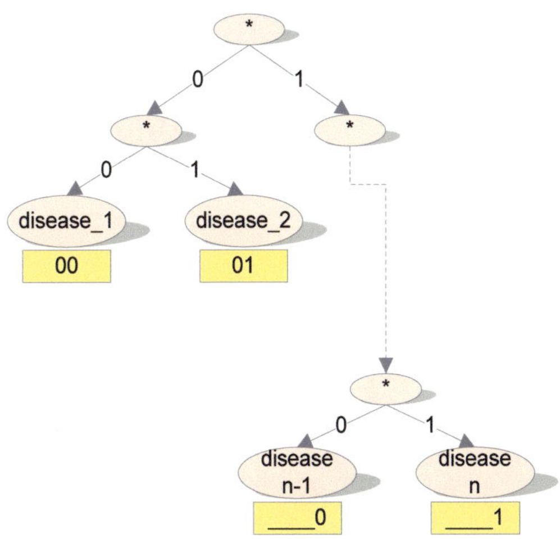

Fig. 9. Coding tree

4.4.4 HL7 portal

Our research team designed and developed a HL7 portal to integrate the SIMOPAC system with other clinical applications/systems already developed by other companies or organizations. Thus, the main purpose of this server is to acquire clinical data about patients (from different servers and applications) by using the HL7 messaging protocol. Within the framework of SIMOPAC system, the HL7 server will be primarily used to obtain the EMR of a patient that was identified by his RFID tag. There are two different ways of getting clinical data (Cerlinca et al., 2010):

- using the standard HL7 messaging protocol our HL7 Messaging Server connects to a list of medical applications and requests patient's related data;
- using simple and intuitive ASCII commands any non-HL7 application can connect to the Messaging Server and request data about a patient.

The main objective of the HL7 portal is to ensure safe and standardized communication between aware and non-aware HL7 applications and SIMOPAC system modules (Figure 10). Other objectives that we had to accomplish are:

- easy integration with other modules of the system such as: plug-ins, PDA software, software agents;
- compatibility with Linux, Windows 7/XP/2000 and Windows Mobile operating systems;
- secure data exchange using HL7 CCOW standard authentication and encryption algorithms.

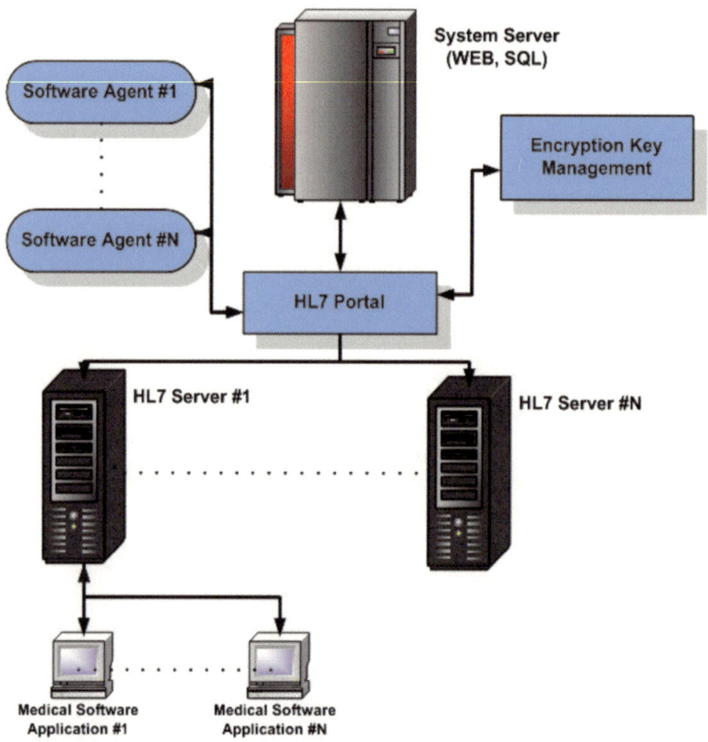

Fig. 10. HL7 Portal integration

The HL7 Portal should provide a secure and sustained flow of medical data between various system modules, regardless of whether they support or not the messaging standard. The modules we developed for this portal are (Figure 11):

- Medical Data serving module: HL7 V2/3 messaging which provides standardized communication between system's modules and also with external medical software applications. This module can be integrated anywhere HL7 data messaging is used. The design and the implementation of this sub-module are compatible with Windows 7/XP/2000, Linux and Windows Mobile operating systems. Also it will provide, if and when needed, additional clinical information, other than the one stored on the RFID tag;
- Authentication and data encryption according to HL7 CCOW, and providing the necessary confidentiality elements regarding the flow of medical data. This sub-module works only with HL7-aware applications;

- Login and transfer module that uses HL7 V2 messaging in order to transfer clinical data between external HL7 servers and SIMOPAC applications. This method involves creating a TCP/IP socket connection that will connect to another socket (IP address: port) on a server, and providing thus the medical data flow. Typical connection used is HL7 LLP (Low Layer Protocol);
- Data interpretation and translation module, the core of the entire portal;
- Encryption key management which keeps and distributes all keys inside our software system; it also provides safe exchange of keys between the system's modules. Furthermore, this sub-module keeps and distributes authentication and encryption algorithms types used by every module and by each partner module/external application.

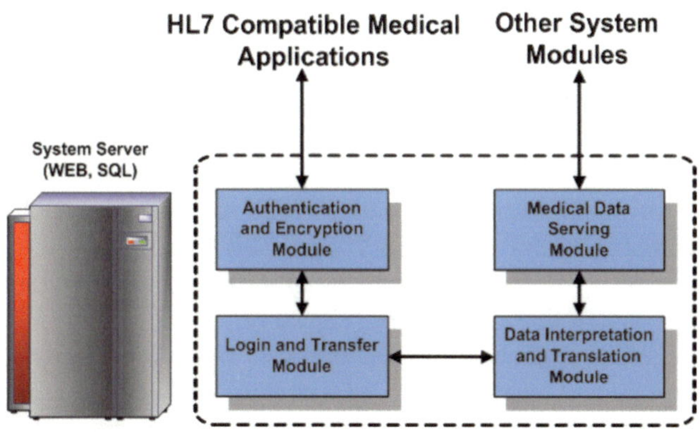

Fig. 11. HL7 Portal Architecture

HL7 Portal Facilities

The main requests covered by the HL7 Portal are:

- compatibility with Windows 7/XP/2000 and Linux operating systems;
- use of HL7 connection and authentication standards;
- acquiring clinical data regarding patients by using safe HL7 connections;
- encrypted exchange of data in all cases;
- translation of HL7 formatted data as close as possible to the natural language;
- the system architecture enables translation from/in an unlimited set of languages, as long as standard ASCII characters are used;
- ensuring connection to and authentication of an unlimited number of concurrent client applications requiring patient information from HL7 medical data servers;
- supported command set designed to provide complete support for gathering relevant medical data;
- storage of all connections, received commands and answers in an encrypted log file.

A language barrier between patients and healthcare providers is a major obstacle in providing quality care, according to (Bischoff et al., 2003).

The elements of originality of the HL7 portal are:

1. Translation of HL7 messages parts in various foreign languages;
2. Enabling partial interpretation and translation of data from HL7 segments from and in any language;

3. Providing a simple mechanism to add new languages for data interpretation;
4. Providing means to obtain and process HL7 format data into non-HL7 applications;
5. There is no other portal that has the same functionalities as the SIMOPAC portal, designed and developed by our research team.

Even if our main goal was to provide a solution for healthcare language issues, there are some aspects that our system, in its current state, cannot solve: translation of descriptive fields, translation of doctor observations, etc. To this extent, more research is needed on EMR translation systems.

4.4.5 Data interpretation and translation module

This module provides the following features:

- allows the interpretation of clinical data in different languages;
- allows users to customize interpreted messages;
- new languages can be dynamically added and then used for data interpretation;
- executes commands received from client applications, and returns the corresponding clinical data, if any;
- allows connections from an unlimited number of clients.

In order for this module to be fully functional, the following steps must be completed:

- read the *languages.txt* file that contains all supported languages;
- read all files used in data interpretation in different spoken languages and data initialization for each of these languages (Figure 12);
- create a TCP/IP server socket for the connection of potential external application;
- wait for connections and create one server socket for each client;
- reply to each client for received messages and return requested data using the appropriate foreign language;
- disconnect the client on request and close the corresponding thread/socket pair.

The *languages.txt* file is used in order to find out the available interpretation languages. Thus, this file contains all available interpretation languages identified by name and also indicates the associated language abbreviation used to find specific files. For example, the *languages.txt* file can contain: English (en), Francais (fr), Romana (ro).

The files needed to interpret clinical data in English are:

1. HL7 specific messages files: EVN.en, MRG.en, MSA.en, MSH.en, MSH-EventType.en, MSH-MessageType.en, OBR.en, OBX.en, ORC.en, PID.en, PV1.en, QAK.en, QPD.en, RCP.en, ZDS.en.
   We choose these file names because we needed to interpret the most used HL7 messages:
   - MSH (Message header);
   - MSA (Message Acknowledgement);
   - OBX (Observation);
   - OBR (Observation Request);
   - EVN (Event Type);
   - PID (Patient Identification);
   - PV1 (Patient Visit),
   - etc.
2. Files with translated error messages: -none-.en,
- files not found-.en, -not present-.en.

Table 1 presents the command set for English and the corresponding returned values.

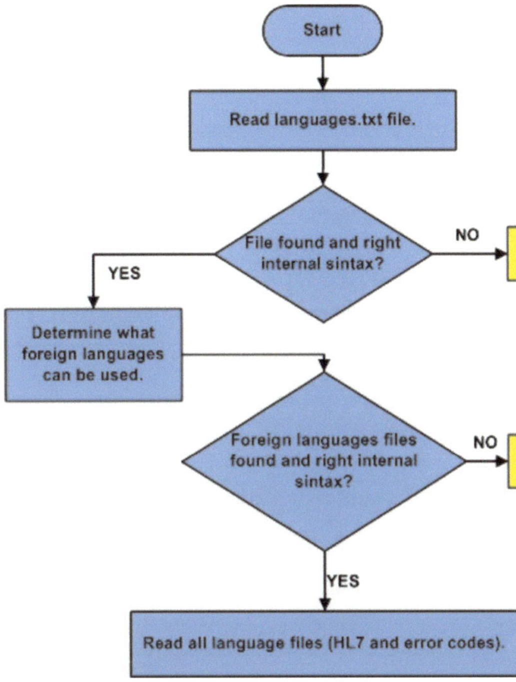

Fig. 12. Module initialization by reading languages files

English Command	**Returns**
• login(IP, port, user, password)	• command sent by the client in order to connect through the portal to a HL7 server; • ***OK*** if successful or ***NOK*** if the connection failed;
• usePatient(SSN, language)	• command that will set the current patient; all subsequent commands from the current client will receive data on this patient; • ***OK*** if successful or ***NOK*** if the connection failed;
• getExternalID()	• external identifier associated with the current patient (external to HL7 application questioned); • ***none*** if there is no external ID;
• getInternalID()	• internal identifier associated with the current patient (internal on HL7 application questioned); • ***none*** if there is no internal ID;
• getAlternateID()	• alternative identifier associated with the current patient (alternate to HL7 application questioned), etc. • ***none*** if there is no alternate ID.
• getName()	• returns the name of current patient;
• getMotherMaidenName()	• returns the patient's mother maiden name, this may be important in order to get all information about family health history. Also may be used to distinguish between patients with the same last name.

Table 1. Command set for the English languages

This solution facilitates the addition of a new language support. The interpreter has to follow these steps:

1. adding a new line in the *languages.txt* file (e.g. Espanol (es));
2. creating the following files:
   a. *–files not found-.es,* file with the following content: *"!Archivos no encontrados! !Elija por favor otra lengua!",*
   b. *-none-.es* file with the following content: *"Ningunos",*
   c. *-not present-.es* file with the following content: *"No presente";*
3. translation into Spanish of all HL7 specific files.

The data will be interpreted once commands are received from external applications.

Testing

In order to test the module, we developed a prototype for a generic client that executes all the commands described in the previous section. HL7 portal runs on a Linux machine, while the client is using a Windows platform. For testing the HL7 portal, we used three different applications, compatible with the HL7 standard: PatientOS, AccuMed EMR and the Mirth HL7 messaging server.

All tests proved that our system complies with the specified requirements and can be successfully used to provide accurate health information in different spoken languages.

From the performance point of view, our design and implementation meets all requirements of typical client/server software systems. Performance testing proves that there are no significant delays and the server response time is more than acceptable.

4.5 SIMOPAC novelties and benefits

SIMOPAC proposes a novel approach in patient identification and ensures the interoperability of HL7 medical information systems. Its implementation does not require any change in or re-design of existent information systems. SIMOPAC can easily integrate any other existing solutions in today's medical establishments and provides a reliable way of identifying patients by using the latest RFID technology. Allowing the integration of other current technical solutions available in today's medical units, SIMOPAC contributes to a considerable reduction of implementation costs. It also eliminates the import of patients' electronic medical records into other EMR systems. Furthermore, the members of the medical staff do not need to be trained how to use the information system in order to store their patients' medical records.

SIMOPAC permits the interoperability of medical information systems at an international level and especially among EU countries, irrespective of their centralized or decentralized health system organization:

a. when a traveling citizen gets sick in some other EU country and requires an emergency service;
b. when a citizen travels to an EU member state in order to benefit from some requested medical service available in the visited country;
c. when diagnostic requests are electronically posted by individual citizens or by members of the medical staff in real-time store-and-forward telemedicine.

Since SIMOPAC does not substitute the existent information systems, it represents a viable solution for the reduction of costs involved in acquiring infrastructure components of medical information system and services. Economically, the level of interoperable health information-exchange among medical institutions is expected to reach considerable values.

The actual increase of international medical contacts and the real need to exchange patient-related information in cross-border contexts pave the way towards the implementation of such systems.

SIMOPAC offers the following major *benefits*:

- *clinical benefits:*

a. the members of the medical staff can securely access medical data stored in patients' electronic records;

b. the patients' health histories are made available to authorized staff;

c. the members of the medical staff can better coordinate the provision of health services by providing accurate information about their patients' health, the history of their medical visits at any time and in any location where the system is operational;

d. the system stores and distributes upon request a whole-range of patient-related information;

e. patient-related data can be obtained on-line;

f. the paper consumption for keeping hardcopy documents may be considerably reduced or eliminated;

g. the system reduces medical errors and increases patient safety.

- *administrative benefits:*

a. on-line access to information;

b. efficient management of medical information;

c. health care providers may be connected internally and externally;

d. the system eliminates the need to re-register patients and keep multiple healthcare records in several medical information systems.

5. Conclusion

Many errors in health care relate to lack of availability of important patient information. The use of information technology (IT) and electronic medical records (EMR) holds promise in improving the quality of information transfer and is essential to patient safety (Bates & Gawande, 2003). While the adoption of Information Technology in individual medical institutions is growing rapidly, interoperability is still a major challenge, and reaching agreement over the appropriate approach to a national EHR system has proved difficult. Thus, despite the fact that most hospitals store patient electronic medical information, these data cannot be easily shared among all healthcare systems because of its discordant formats.

The continuous decrease of costs in RFID technology will soon enforce its use in everyday life. In this chapter, we have focused on the RFID technology and how it could be used in emergency care in order to identify patients and to achieve real time information concerning the patients' biometric data, which might be used at different points of the health system (laboratory, family physician, etc.).

Also, this chapter describes an RFID-based system (named SIMOPAC) that integrates RFID and multi-agent technologies in health care in order to make patient emergency care as more efficient and risk-free, by providing doctors with as much information about a patient as quickly as possible. The proposed RFID-based system could be used to ensure the positive patient identification (PPI) in a hospital. The SIMOPAC goal is to extend the procedure of patient identification beyond the hospital and country boundaries. Thus, our RFID-based system could be considered an open-loop RFID application, functioning across global hospital boundaries. The CIP will allow the identification of patients, and this RFID

card will provide access to an ambulatory EMR, namely a data repository devised as a subset of a longitudinal health record. Furthermore, the CIP could be used to allow physicians to connect to the SIMOPAC server. In order to link patient identifiers to patient information, the SN@URI approach has been proposed, SN being the CIP serial number.
The interaction of the afore mentioned system with another full EMR system will assure optimal integration. The HL7 server we have designed and developed can be used to obtain the EMR of a patient that was identified on the base of his RFID tag. Clinical data can be acquired from different servers and applications. In addition, any non-HL7 application can connect to our HL7 server and request data about a patient. Our system is able to integrate and exchange information with other HL7 and even non-HL7 based clinical applications already developed by other companies or organizations. Using multi-agent technology reduces the need for knowledge about HL7 and interfaces.
Through the use of tag templates that can be translated into several foreign languages our system facilitates the cooperation between medical units from different countries in order to assure a good care for a patient from another country.
Every hospital could use SIMOPAC with their existing system in order to promote patient safety and optimize hospital workflow. We have described a general purpose architecture and data model that is designed for both collecting ambulatory data from various existing devices and systems, and storing clinically significant information in order to be accessed by the emergency care physician. The SIMOPAC complexity is further amplified by the fact that most individual electronic health record systems are packaged products supplied by a variety of independent software providers and run on different platforms.
Through the use of the RFID technology, the system we have developed is able to reduce medical errors, improve the patients' overall safety and enhance the quality of medical services in hospitals and other medical institutions. For example, the risk of administrating wrong medication in case of emergency is highly reduced.
Our future research will focus on the development of various software modules that will use the medical information collected via RFID in order to optimize the patients' treatment process.

6. Acknowledgment

This work was supported in part by the Romanian Ministry of Education and Research under Grant named "SIMOPAC - Integrated System for the Identification and Monitoring of Patients" no. 11-011/2007.

7. References

Bates, D.W., Gawande, A.A. (2003). Improving safety with information technology, *The New England Journal of Medicine*, 348:2526-2534

Bischoff, A., Perneger, T., Bovier, P., Loutan, L., Stalder, H. (2003). Improving communication between physicians and patients who speak a foreign language, *British Journal of General Practice*, 53 (492), p. 541-546, Available from http://www.ncbi.nlm.nih.gov/pmc/articles/PMC1314645/pdf/14694667.pdf

Castro, L., Wamba, S., (2007). An inside look at RFID technology, *Journal of Technology Management & Innovation*, Volume 2, Issue 1.

Cerlinca, T., Turcu, Cr., Turcu, C., Cerlinca, M. (2010). RFID-based information system for patients and medical staff identification and tracking, *Radio Frequency Identification Fundamentals and Applications*, IN-TECH

Ehto (n.d.), *About HL7*, Available from http://www.ehto.org/ikb/standards/news.html

El-Gamal, Y., El-Gazzar, K., Saeb, M., (2007). A Comparative Performance Evaluation Model of Mobile Agent Versus Remote Method Invocation for Information Retrieval, *Proceedings of World Academy of Science*, Engineering and Technology, vol. 21, ISSN 1307-6884

European Comission (2006). *Medical Errors*, Special Eurobarometer, 241/Wave 64.1 /64.3 - TNS Opinion & Social

Hallvard, L., Karlsen, T., (2006). Effects of Scanning and Eliminating Paper-based Medical Records on Hospital Physicians' Clinical Work Practice, *Journal of the American Medical Informatics*, pp. 588–595.

HL7, (n.d.). *Health Level Seven*, Available from https://www.hl7.org.

HL71, (n.d.). Health Level 7, Available from http://en.wikipedia.org/wiki/Health_Level_7

I. de la Torre; Díaz, F.J.; et al. (January 2010). Choosing the Most Efficient Database for a Web-Based System to Store and Exchange Ophthalmologic Health Records, *Journal of Medical Systems*, Springer Science+Business Media

Iguana & Chameleon, (n.d.) iNTERFACEWARE, HL7 Overview, Available from http://www.interfaceware.com/manual/hl7.html

Impact, (2008). Healthcare Data Integration Market Overview, *Impact Advisors*, LLC

Jonathan, C., (2004). RFID Remedy for Medical Errors, RFID Journal, 2004, Available from http://www.rfidjournal.com/ article/view/961/1/1.

Klas, M., (2009). Beyond the CIS: Why are hospitals buying aggregation solutions, *Report*, http://www.klasresearch.com/

Long, A.J., Liu, C.T., Chang P. (2003). An Open Source Intelligent HL7 Agent for Integrating Laboratory Information, *The Journal on Information Technology in Healthcare* 1(5), pag. 357–368

Mehrjerdi, Y.Z. (2007). *RFID-enabled systems: a brief review*, Available from http://www.emeraldinsight.com/Insight/ViewContentServlet?Filename=Published/EmeraldFullTextArticle/Articles/0330280307.html

Mertz, J. (2010), HL7 Interface - An Overview, *Corepoint Health*

Microsoft, (2006). Available from http://www.microsoft.com/presspass/press/2006/jul06/07-26azyxxiacquisitionpr.mspx

Nealon, J., & Moreno, A., (2003). Agent-Based Applications in Health Care, *Applications of Software Agent Technology in the Health Care Domain*, Whitestein Series in Software Agent Technologies, Birkhäuser Verlag (2003), pp. 3-18.

Orgun, B., Vu, J. (2006). HL7 ontology and mobile agents for interoperability in heterogeneous medical information systems, *Computers in Biology and Medicine* 36, pag. 817 - 836, Available from www.intl.elsevierhealth.com/journals/cobm

OTech, (2007). *What Is HL7 Version 2.x*, OTech Industries, June 7th, 2007

Peacocks, (n.d.). *About RFID*, Available from http://www.peacocks.com.au/about-rfid.htm

RFIDHealthcare, (n.d.). The Benefits of RFID in the Healthcare Organization, *RFID Solutions for the Healthcare Industry*, Available from http://www.rfidhealthcare.com/

RFIDUpdate (2008). *Healthcare RFID Worth $2.1B in 2016*, Available: from http://www.rfidupdate.com/articles/index.php?id=1105

Shaver, D., (2007). HL7 101: A Begunner's Guide, For the Record, *Great Valley Publishing Co.*, Inc., vol. 19, no. 1, p22, January 2007, Available from http://www.fortherecordmag.com/archives/ftr_01082007p22.shtml

Smaltz, D. & Berner, E., (2007). The Executive's Guide to Electronic Health Records, Health Administration Press, pp. 03.

Teambusiness, (n.d.). Networks, Applications and User Management, Available:from http://teambusinesssolutions.com/solutions.asp?sid=82&clicked=79

Tucci, L., (2008). Electronic health record adoption an issue for health care CIOs, Compliance Management, 2008, Available from http://searchcompliance.techtarget.com/news/article/0,289142,sid195_gci1339665,00.html

Turcu, C. & Turcu, Cr. (2008). Sistem informatic integrat pentru identificarea si monitorizarea pacientilor - SIMOPAC (in Romanian), An integrated system for identification and monitoring of patients - SIMOPAC, vol. *Sisteme Distribuite - Distributed Systems*

Turcu, Cr., Cerlinca, T., Turcu, C., Cerlinca, M., Prodan, R., (2009). An RFID and Multi-agent Based System for Improving Efficiency in Patient Identification and Monitoring, *WSEAS Transaction on Information Science and Applications*, Issue 11, vol. 6, pag. 1792-1801, ISSN: 1790-0832

Van der Meijden; Tange M. J.; et al. (2001). Development and implementation of an EPR: How to encourage the user, *International Journal of Medical Informatics*, 64:173–185

Vlădescu, C., Scîntee, G., Olsavszky, V. (2008). Romania-Health system review, *Health Systems in Transition*, Vol. 10, No.3, European Observatory on Health Systems and Policies

Wagner, D., (n.d.). Associates, Inc., *Software Agents*, www.wagner.com/ technologies/ softwareagents/softwareagents.html

Yamamoto, G., Tai, H., (2001). Performance evaluation of an agent server capable of hosting large numbers of agents, *Fifth International Conference on Autonomous Agents*, Montreal, Canada, pp 363-369, May 28-June 1, 2001

6

RFID Technology and Multi-Agent Approaches in Healthcare

Felicia Gîză, Cristina Turcu and Cornel Turcu
Stefan cel Mare University
Romania

1. Introduction

Every year, thousands of people die because of medical errors. For example, in 2004 it was estimated that each year, more than 98,000 people die because of medical mistakes in the U.S., according to the Institute of Medicine, while in the United Kingdom the number is 40,000, according to the British Medical Journal. In 2005, according to a European Commission report, the number of deaths due to medical errors in the U.S. was higher than the total number of persons who died of breast cancer, AIDS or car accidents. A study conducted by the Institute for Safe Medication Practices in the United States indicated that approximately 25% of hospital patients had adverse reactions to medications; in many cases they could have been prevented or alleviated. Also, such side effects are registered in patients undergoing primary care, but there are not too many studies in this direction. In its 2008 annual report to Congress, the Agency for Healthcare Research and Quality reported that preventable medical injuries are growing each year by 1 percent (Crowley & Nalder, 2009). An investigation conducted by Hearst Media Corporation showed that nearly 200,000 people die each year from medical errors and hospital infections throughout the U.S (Hearst, 2009). Many of these errors can be avoided by using information technology. But in 2004 only 3% of the 64,000 U.S. hospitals had integrated a hospital information system (Hospital Information System - HIS) to allow the management of patient records.
The medical history of a patient is very important for his diagnosis and for setting an appropriate therapy. Unfortunately, for the moment, in many countries, keeping a patient's medical records is carried out at the general practitioner's level and healthcare units in which the patient has performed medical examinations. So, there is no complete data set comprising all the medical information about a patient and allowing quick access to the patient's complete medical history. In certain situations, for example, whether the patient has suffered an accident and he/she is unconscious, the emergency medical personnel do not have access to medical information concerning that patient. RFID technology provides a solution for enabling the access of medical personnel to the patient's medical history, by using a device (RFID tag) that allows storing relevant medical information related to its carrier, which provides a quick access to the actual health state of a patient and helps the medical staff to take the best decisions, especially in case of emergency. Thus, the risk of administrating wrong medication is highly reduced. Also, multi-agent systems offer the framework for the collection and integration of heterogeneous information distributed in different healthcare specific systems to get access to the patient's complete medical history.

This chapter provides a structured enumeration of the most notable recent attempts to use RFID technology and multi-agent systems for healthcare. Next, the authors propose an RFID-based system (named SIMOPAC) that integrates RFID and multi-agent technologies in health care in order to make patient emergency care as efficient and risk-free as possible, by providing doctors with as much information about a patient and as quickly as possible. Thus, this system enables real time identification and monitoring of a patient in a medical facility, on the basis of passive RFID tag, entitled CIP (Personal Electronic Identity Card). The system is also able to integrate and exchange information with other HL7 (Health Level Seven) and even non-HL7-based clinical applications already developed by other companies or organizations. All hospitals can use SIMOPAC with their existing system in order to promote patient safety and optimize hospital workflow. We describe a general purpose architecture and data model that is designed for collecting ambulatory data from various systems, as well as for storing and presenting clinically significant information to the emergency care physician.

2. Applying RFID technology in healthcare

Currently, RFID technology is successfully applied in many fields. In this section, we will consider the integration of RFID technology in healthcare systems. The major challenge comes from the possibilities to incorporate RFID into medical practice, especially when relevant experience in the field is relatively low. By attaching RFID tags to persons (patients or healthcare staff) and objects (medical equipment, medical dressing, blood transfusion bags, etc.) this technology enables the identification, tracking and tracing of entities, security, and other healthcare specific capabilities (Figure 1) (Iosep, 2007).

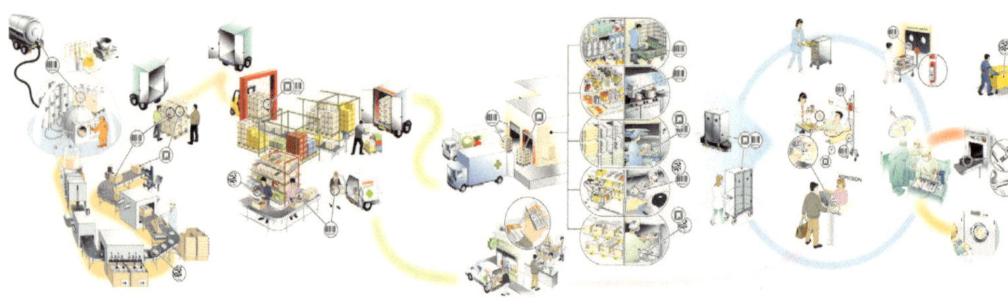

Fig. 1. RFID technology use for patient care (Iosep, 2007)

RFID tags can be used in the medical field in the following ways (BioHealth, 2007): identification of a patient in emergency situations; patient vital signs measurements (for example, for patients with chronic diseases); recording significant medical information and their transfer to an electronic monitoring device; monitoring the elderly, even at their home; monitoring of goods and equipment; controlling drugs administration and blood transfusions, thereby reducing medical errors in hospitals.

Internationally, at present, the following main areas benefit from the application of RFID technology in healthcare (Table 1):

1. Management of medical articles - The fast tracking of mobile medical articles ensures a better use of them, which reduces losses and, consequently, new acquisitions, while considerably reducing the amount of time wasted by medical staff searching for equipment;

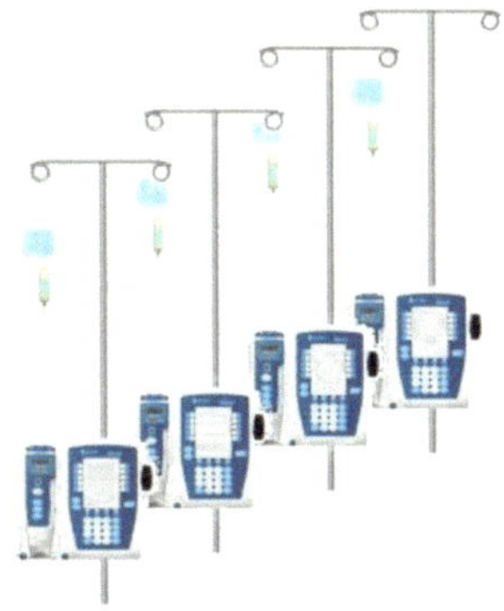

a)

2. Patient care - Correct identification of patients and their location at all times may lead to increased security (for example, in case of patients suffering from Alzheimer's disease), but also better management of hospital beds within a medical unit;

b)

3. Management of drugs and dangerous medical substances - Drug traceability is fundamental to eliminate counterfeit drugs. A significant decrease in the number of errors in patient medication administration can be achieved through quick and accurate drug identification, thus also ensuring the checking of prescribed dosage for a particular patient.

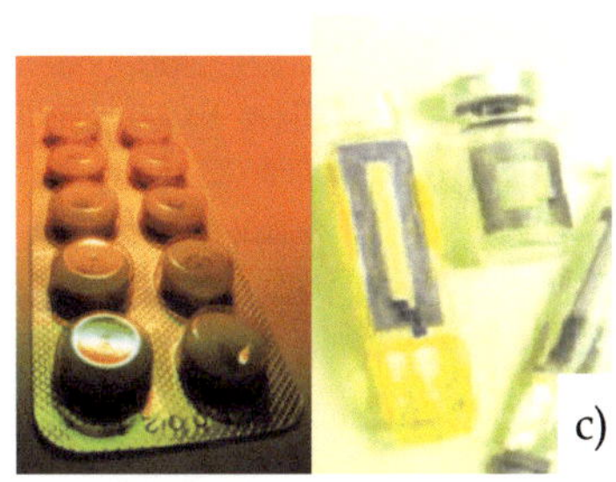

c)

4. Inventory Management - Early identification of inventory items and rapid inventory achievement may result in the elimination of '0 stock' situations and optimization of current inventory etc.

d)

Table 1. Examples of applying RFID technology in various areas of medical fields

But RFID is, also, an option for patients who are not hospitalized in a medical institution and who, for example, undergo medical treatment.
Various studies (e.g., BRIDGE project (BRIDGE, 2007)) estimate a significant increase in the coming years in the use of RFID technology in medical field (Table 2, Table 3).

Millions RFID tags items associated with	2007	2012	2017	2022
Medical equipment	2	98	190	320
Laboratory samples	1	8	30	40
Drugs	5	246	1500	6380
Total	8	352	1720	6740

Table 2. Estimating the use of RFID tags in the medical field

Use of RFID readers	2007	2012	2017	2022
Locations with RFID readers	110	2770	11900	40600
Total number of RFID readers	180	12600	70200	208000

Table 3. Estimating the use of RFID readers in the medical field

For example, in May 2008, an RFID-based system to be used in surgery rooms was implemented in San Jose, California. ClearCount Medical Solutions has chosen RFID technology to automate the process of tracking surgical dressing. The system uses passive tags, 13.56MHz, with a 2 Kb programmable memory (figure 2). Surgical dressings with RFID tags used in a surgery cost about $ 35-50. This system has been approved by U.S. organisation: FDA (Food and Drug Administration) and FCC (Federal Communications Commission).

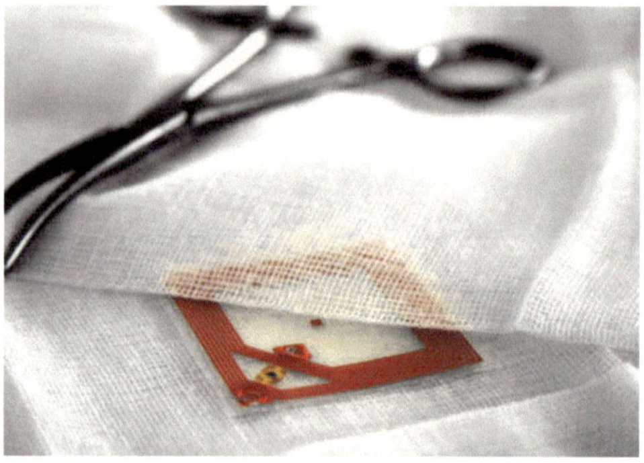

Fig. 2. RFID for tracking surgical dressing

Even if the labeling of hospital objects (such as surgical dressings, medical equipment etc.) submits a development potential for the RFID technology, patient labeling involves far more issues. Janz and others studied the impact of introducing RFID-based application in the emergency department of a hospital and found that the information collected from patient's tags has been particularly useful, especially in decision making process and resource management (Janz et al., 2005).

In 2003, at the Taipei Medical University Hospital (TMU) in Taiwan, a platform that exploited RFID technologies was implemented, due to the need to handle cases of bird flu that ravaged Taiwan that year. Thus, the implemented system used RFID technology to monitor the body temperature of medical staff and hospital patients, to allow the identification and monitoring of bird flu virus carriers. According to a report, in 2003, 94% of Taiwan's SARS victims were infected in hospitals. The implementation of this RFID-based system targeted a more rigorous control in hospitals, so that the danger of SARS disease or other transmissible diseases could be considerably reduced.

(Chung et al, 2009) proposed the Medicare-grid system (grid-based e-Health System) to facilitate the process of retrieving and exchanging patient's EHRs (Electronic Health Records) among hospitals and medical centers. Grid and peer-to-peer technologies were used to develop an EHR center as a decentralized database to store and share EHRs among participating hospitals and medical centers. In addition, they also integrate computing resources provided by hospitals, to form a computational grid for medical-related applications. Based on computing resources and a data grid platform, they developed medical related applications to improve the in-hospital medical services:

1. a data warehouse for medical decision support system; they use data mining techniques for analyzing patients' EHR information;
2. an RFID-based mobile monitoring system to identify people or items accurately;
3. an wearable physiological signal measurement system that monitors the health condition of a patient.

But it should be noted that some researchers warn that RFID technology in hospitals can influence the optimal operation of medical equipment. According to a study published in June 2008 in The Journal of the American Medical Association, RFID systems can cause random incidents over medical devices in hospitals. This study, however, is not confirmed by researcher around the globe and rather asserts that RFID technology can be used in hospitals and other patient care institutions. 25 common medical devices were tested in this study, 1,600 tests being considered. In all cases, the devices worked at standard parameters and no interference from passive RFID devices was observed. The report concluded that the RFID solutions can be applied to inventory monitoring, entities traceability etc. without adverse effects on the equipment. Therefore, passive RFID tags can be used safely in hospitals.

The price of integrating RFID technology in medical systems is the most important impediment to the adoption of this technology in the medical field. Currently, implementation and use cost of RFID systems is higher than the cost of any bar code system on the market. This is, mainly, due to the higher cost of tags production. But a decrease is foreseen over the next years in the price of tags because of the growing scope of RFID applications and, implicitly, because the number of these products is increasing.

3. Multi-agent system developed for the medical field

The medical field is characterized by information, data, knowledge and even distributed competence. Moreover, the three components (data, information, knowledge) may be of different types: natural language descriptions, images, measured signals, the results of various tests and measurements (usually lists of numbers). They are stored under different shapes: sheets of paper, photos, slides, electronic files, books (if we consider the "classical" knowledge) and sometimes private discussions. Usually they are not available in one place at a time. Therefore, this distribution is a major problem when decisions must be made very quickly.

Modern medical systems include many specialists whose practice is limited to a particular branch of medicine or surgery. Complex examination of a single patient involves several consultations by medical specialists and, also, laboratory tests. Medical knowledge, examinations and treatments are distributed geographically and temporally. Therefore, there is a need for a consistent flow of information and trust among all involved subjects to meet the global target - a patient's enhanced health. But the necessary information flow is not predictable in content and structure, and it is evolving and changing over time due to new knowledge and reactions. To meet these demands and to provide appropriate decision support, intelligent software applications have to be used.

Generally, as shown in several studies on health system design for distributed heterogeneous environments (Laleci et al., 2008) the best-suited method of implementation is the use of multi-agent systems. Those systems include independent components that communicate in a reactive way, and some of these components should be instantiated and removed dynamically on demand.

An agent is a software component that has a well-defined role in the operation of a system. Also, an agent must have the ability to communicate with other agents or human users. A multi-agent system is a collection of such entities that cooperate with one another. By using the multi-agent technology in the system implementation, the following advantages could be obtained (Bouzeghoub & Elbyed, 2006):

High performance: agents can run in parallel. They can be cloned when their tasks and goals are very important;

High flexibility: an agent can be developed for any context, providing the interface for different ontologies;

High modularity: the number of connected sources can increase practically without limit.

In the medical field, multi-agent systems can provide services that facilitate decision-making process for medical staff, providing a larger volume of information about certain situations and reducing the number of operations performed by the human operator. So far, worldwide, several multi-agent systems have been developed in the medical field. These systems provide:

- Patient monitoring and, in some cases, generation of automatic prescription;
- Automatic information extraction from medical databases and fast information analysis;
- Efficient patient scheduling in medical offices;
- Critical drugs scheduling;
- Doctors' access to patient's medical information, the information being distributed in heterogeneous databases;
- Medical images processing;
- Patient access to own medical information.

This chapter will describe a few examples of multi-agent systems developed to allow quick access to the complete medical information of a patient. The solution of developing some large centralized databases to store information about all patients is difficult to achieve. Currently, in most cases, medical information on a patient is stored in databases of the healthcare unit, where the patient resides or where the patient underwent medical investigations. The heterogeneity of the information stored in different medical information systems used in healthcare units hinders easy access to comprehensive medical information of a patient.

For the integration of heterogeneous data from several health care units, (Schweiger et al, 2007) proposed the concept of Active Medical Document (AMD). These documents are compiled at runtime and can be prepared according to user's needs. The Active Medical Documents contain agents offering internal services (access control, appointments monitoring) as well as coordinative, administrative, and medical data (patient medical records). Their built-in agents offer additional services such as information retrieving and processing. The advantages of using such documents cover, among others, the possibilities for decentralized, adaptive and intelligent coordinating, ensuring, above all, availability of heterogeneous data sources.

In order to achieve full medical information about a patient several systems have been designed to provide advanced capabilities search for this kind of information in the medical information systems of different healthcare units. For example, agent-based systems, such as MAMIS (Multi-Agent Medical Information System), developed by Fonseca et al. (2005) and eMAGS (electronic Medical Agent System), implemented by Orgun et al. (Orgun et al, 2006), enable the competent information search in a community of autonomous healthcare units and provide physicians and surgeons with easily accessible information. In MAMIS system, each medical unit must share, on request, a limited set of information about a patient. In this direction, the authors propose a common database architecture to be implemented by each healthcare unit from the considered community. This database is a supplemental database, developed in addition to their existing private databases. This database stores a limited set of information about patients and will be available within the community. The eMAGS system described by Orgun et al (2006) proposes a multi-agent architecture that uses an ontology based on the HL7 standard (Health Level Seven) to facilitate the flow of information about a patient within a healthcare organization. In the proposed model, several healthcare applications are tied together through servers of agents, one for each medical application registered in the network, a broker for agents and an ontology server.

Another solution allowing the access to complete medical information about a patient lies in placing the responsibility of sending the results of the medical test to the medical unit to which the patient belongs, in the hands of the staff from the medical unit where these tests are performed. To achieve this automatically, one should consider the multi-agent approach. Nguyen et al. (2008) made the first step in this direction, developing the MEDIMAS system, which is a multi-agent system for the transmission of test results carried out in laboratories within a medical unit to the healthcare professional who requested them. The considered medical unit already has a database and runs an application for recording and managing the performed analysis. The multi-agent system is designed and implemented in order to extract the necessary information from the existing database and to notify the appropriate medical staff within the unit.

Within a national project, Laleci et al. (2008) have developed the SAPHIRE multi-agent system, used for monitoring patients with chronic diseases both in hospital as well as at the place of residence. Based on the information provided by monitoring systems and that existing in the patient's electronic medical records, the system is able to deploy and execute clinical guidelines in a care environment that includes disparate medical units with heterogeneous information systems. As a result of conducted research, the team members have chosen to establish a semantic interoperability environment to enable communication with different heterogeneous health care systems; the considered solution is the adoption of a multi-agent system as the basic structure of SAPHIRE system.

4. Integrating RFID and multi-agent technologies

The research performed over the years has shown that RFID and multi-agent technologies can provide solutions for problems in various fields. Thus, for example, in the supply chain of companies, Dias et al. (2008) propose an intelligent transportation system that integrates both technologies. Lebrun et al. (2010) present a model of a multi-agent system dedicated to the management of objects identified by RFID tags that users shall move on a flat surface, with a first application for the study of road traffic.

An RFID Identification System (IRS) commonly uses passive information about a particular entity, such as the identification and description of the information stored in the RFID tag, and chooses a set of actions based on already established rules stored in a database (Chen et al, 2010). Since this database is static, it cannot be updated in a timely manner for new types of objects or according to the dynamics of the environment, thus creating synchronization problems. Chen et al. (2010) propose an RFID system based on a code (CRS - Codecentric RFID System) as a solution to these problems. The RFID tag encodes a mobile agent that contains up-to-date service directives realizable by an intelligent handling of the dynamics of various networks.

In the medical field, Bajo et al. (2008) presents a multi-agent architecture, the Geriatric Residence Multi-agent System (GR-MAS), developed for facilitating health care services in geriatric residences. GRMAS contains different types of agents and takes into account the integration of RFID technology, Wi-Fi technologies and portable devices. The core of GR-MAS architecture is a deliberative planning agent, aimed to optimize the visiting schedules for medical staff. This agent, which can learn and adapt to new circumstances, was designed to schedule the nurses' working time dynamically, so that patients receive proper care. Also, this agent will keep track of the standard working reports on medical staff activities. This multi-agent system used RFID technology to facilitate location and identification of patients and medical staff.

5. SIMOPAC solution for accessing full medical information about a patient

Diagnosing and setting proper medical treatment for a patient inevitably involves consulting the patient's medical history by medical specialists. Unfortunately, the access to the patient's medical history is not always possible, which may lead to errors in diagnosing or in setting a treatment, sometimes with adverse consequences for patients.

Currently, in Romania, the degree of computerization of the health system is relatively low; the information about patients is to be located in different healthcare units; patient medical records are neither consistent nor complete, and cannot be accessed online by the medical staff if necessary. In this context, our research team has designed and implemented an integrated information system for identifying and monitoring patients - SIMOPAC. The system aims to operate in the medical distributed environment and particularly, to solve the problems of identifying and monitoring patients based on the most recent technologies in the field: radio-frequency identification, collaborative problem solving in a distributed environment (intelligent multi-agent technology). Altogether, it aims to provide communications infrastructure in order to enable multi-point access to medical information conveyed in the system.

Our team designed the SIMOPAC system to use an RFID-based card (named CIP) for each patient. This card must contain patient personal information such as name, birth date, identification number and medical information considered critical, such as, blood type or certain chronic diseases. In addition, if the patient has carried out medical examinations in other medical units, another RFID card (named CIP2URI) is considered to store the EHR server addresses used in the medical units where the patient was consulted. Then, through SIMOPAC, the patient's physician can use this card to get the results of the medical investigations suffered by the patient. This information will also be stored in the SIMOPAC database.

Some of the medical informatics systems of medical units implement HL7 standard, while others do not. In the first case, medical information may be retrieved based on the HL7 standard. For healthcare information systems that do not follow the HL7 standard (hereinafter referred to as non-HL7 servers), a partnership agreement shall be previously performed, with the details of communication protocol to be used in the SIMOPAC system to allow the retrieval from their database of the information relating to a patient.

To enable access to a patient's medical history, within the SIMOPAC project, a multi-agent system was designed and implemented to provide retrieval of information of interest from different servers of healthcare units where the patient was consulted. Adopting agent technology does not require major changes in terms of software resources available in the healthcare units. Thus, existing software systems compliant HL7 can be integrated directly with the multi-agent system and healthcare information systems that do not meet this standard are interfaced through specific agents.

Figure 3 shows the flow of information within the multi-agent system which allows the update of the patients' medical records with information retrieved from HL7-compliant servers and non-HL7 partner servers, as well as the notification of the general practitioner where the patient is primarily registered.

Figure 4 shows the agents considered in the multi-agent system of the SIMOPAC platform (SMA-SIMOPAC):

Supervisor Agent - the core agent within the platform. It acts as coordinator and mediator of other agents' actions. Some of the most important responsibilities carried out by this agent are:

- the encapsulation of database connection details;
- creating HL7 and non-HL7 agents to communicate with the EHR servers of medical units where a patient carried out medical investigation, in order to retrieve medical records;
- notification of family physicians on investigation results received from other medical units.

HL7 Agent - an agent specifically designed for communication through HL7 messages. The agent relates to a specific HL7-compliant server and provides appropriate HL7 commands to retrieve the patient's observations file.

Integration Agent - an agent who mediates information gathered from non-HL7 servers. In order to get data from medical units that do not have HL7-compliant medical informatics systems, a partnership should be previously agreed upon. Thus, a protocol that indicates the server address and the exact name of the DB-Server-type agent running on the server is set. To avoid overloading the platform, the Integration Agent will be responsible for getting information from all non-HL7 servers of medical units where the patient carried out medical investigation. Essentially, its task is confined to sending REQUEST messages to specific DB Agents of partner medical units and then to processing the responses.

DB-ServerX Agent - an agent implemented at the partner medical unit system, which knows the login details and the structure of this database medical unit. This agent extracts relevant information about patient's medical investigations and sends it to the Integration agent that

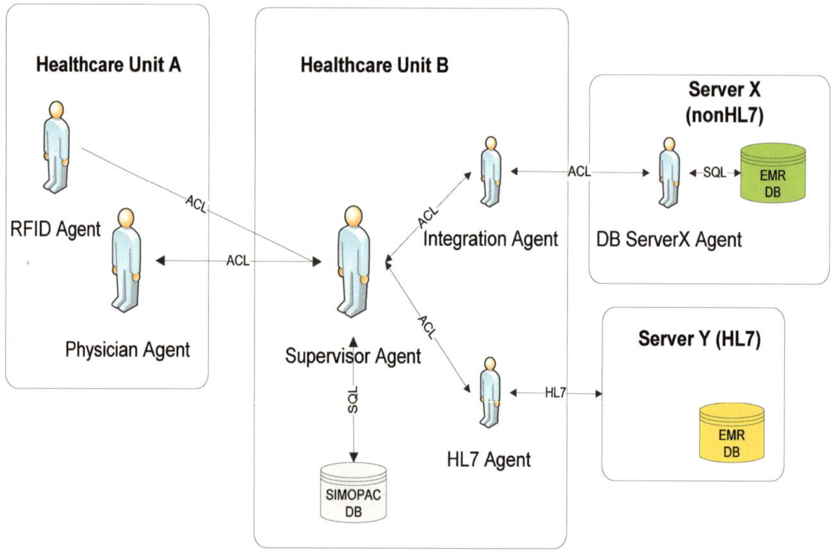

Fig. 3. Updating patient's electronic records with information from HL7 and non-HL7 servers

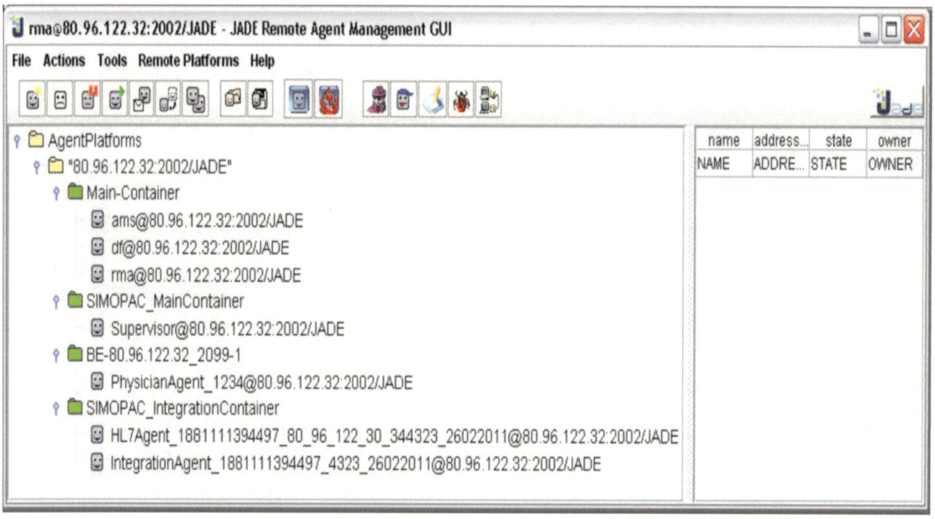

Fig. 4. Agents considered in SMA-SIMOPAC

initiated the request. The DB-ServerX agent development is based on clear specifications regarding the response to requests from Integration Agents of various medical units. Thus, this agent receives the patient's identification number, extracts data from the database, transforms the data into a message expressed in the particular ontology developed within the project and then sends the message to the Integration agent.

Physician Agent - the agent that uses the services provided by the SIMOPAC multi-agent system, namely: requesting complete electronic patient medical records, initiating a process to update the information if it has not received the results of some medical investigations, viewing the notifications received from the Supervisor agent regarding the new results received.

RFID Agent – is the agent specifically created for reading/writing RFID tags (CIPs). When reading a tag, according to the data retrieved from it, this agent performs the appropriate operations, i.e.: if the tag belongs to a family doctor/general practitioner, it creates the proper physician agent or, if the tag identifies a patient, it displays its own medical records. This agent is used for the authentication of multi-agent system users.

The update of the patient's electronic health records with information from HL7-compliant or non-HL7 servers is performed automatically at a particular time set to the Supervisor Agent. To achieve this task, the Supervisor Agent extracts from the database the identification numbers of patients who have performed medical investigations outside the medical unit where they are registered and the list of server addresses of healthcare units where such medical examinations were performed. For each patient, the Supervisor Agent creates an Integration Agent, which receives, as parameters, his identification number and the list of non-HL7 servers corresponding to the medical units in question, along with the names of the DB Agents which they will communicate with for getting the necessary information. The Integration Agent sends REQUEST messages containing the patient's identification number to the DB agents of the partner medical units and then waits for answers from those agents. Each of these DB agents is familiar with the login details to the database from which information about the patient has to be retrieved (such as database type, address, user and password) and the database structure. Thus, based on the received identification number, the DB agent will extract data from the database tables containing the results of medical examinations undergone by the patient and will send them to the Integration Agent that requested it. The Integration Agent will mark in the database that it received the requested information from that server. In addition, it sends to Supervisor Agent the replies containing the requested information. The Integration Agent will end its execution when it has received responses to all performed requests or after a certain period of inactivity. With regard to getting necessary information from HL7- compliant servers, the Supervisor Agent will create one HL7 Agent for each HL7 server of the medical units of interest. An HL7 Agent receives as parameters the patient identification number along with details for connection to one of the considered servers. The HL7 agent initiates a communication channel with the appropriate server and attempts to obtain information from the patient's electronic medical record database through specific HL7 messages. The results received by the HL7 agent are also directed to the Supervisor Agent. As a result of the performed requests, the Supervisor Agent receives responses containing the results of patient's medical investigations from the Integration Agent or HL7 Agent. In this case, Supervisor Agent verifies that the information are not already stored in the system database and when there are no corresponding entries, adds them to the database and notifies the Physician Agent of the patient's family physician, with regard to newly received information. Moreover, when, for example, the family doctor/general practitioner recommended a specific medical investigation to a patient and got no answer, it can initiate the process of updating patient's electronic medical records, simply by selecting a command button in the user interface of Physician agent (*Refresh records* button in Figure 4). In this case, the Physician Agent will forward to the Supervisor Agent the request for updating medical records of the patient identified through identification number specified in the window.

Communications between agents comply with the FIPA interaction protocol. Interaction between agents is illustrated in Figure 6.

To develop the above-described multi-agent system, we selected the JADE platform. Jade is an open-source multi-agent platform that offers several advantages, such as the following: it is FIPA compliant (Foundation for Intelligent Physical Agents), allows the execution of agents on mobile devices (like PDA), provides a range of security services regarding the actions allowed for agents (via add-on module JADE-S) and provides intra and inter-platform mobility.

The SIMOPAC system also has a series of advantages. The integration of RFID technology provides the unique identification of patients, as well as fast retrieving of minimum patient health information, which is primordial in emergency cases. Moreover, given the fact that this system allows medical personnel to obtain information about the patient's medical history, it will increase the chances of accurate diagnoses and will decrease the number of medical errors.

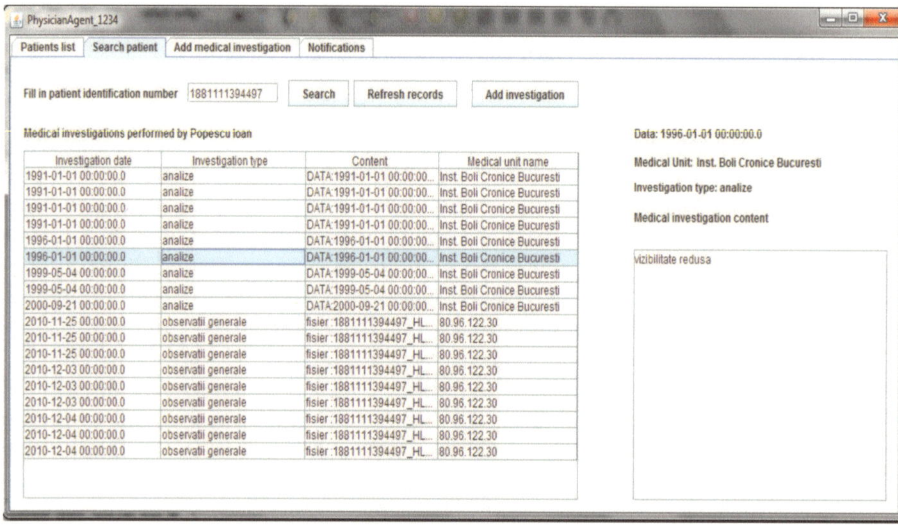

Fig. 5. The physician agent interface for displaying and updating patients' medical records

Regarding the information search performance, the eMAGS and MAMIS systems described above perform an exhaustive search for information related to a patient, in the first case on the servers that publish such services, and in the second case on servers from a particular community where medical units must register first. In SIMOPAC approach, it is only in the servers of healthcare facilities where the patient has performed medical examinations that the system runs a query, resulting in a general improvement of system efficiency.

By using dedicated agents, SIMOPAC proves to be an easy-to-use tool, which allows automation of some operations performed frequently in medical units.

6. Conclusions

A patient's medical history is very important for doctors in the process of diagnose and determination of the appropriate treatment for the patient. In emergency cases, when these operations must be carried out against the clock, fast retrieval of information related to patient's medical history may be of vital importance for the patient's life. RFID technology provides a solution for enabling the medical staff to access a patient's medical history, by using a device (RFID tag) that stores essential information about the patient, and acts as a gateway to the complete electronic healthcare records of the patient. Multi-agent systems provide, among others, the framework for collecting and integrating heterogeneous information distributed in various medical units specific systems in order to retrieve the patient's electronic healthcare records as comprehensively as possible.

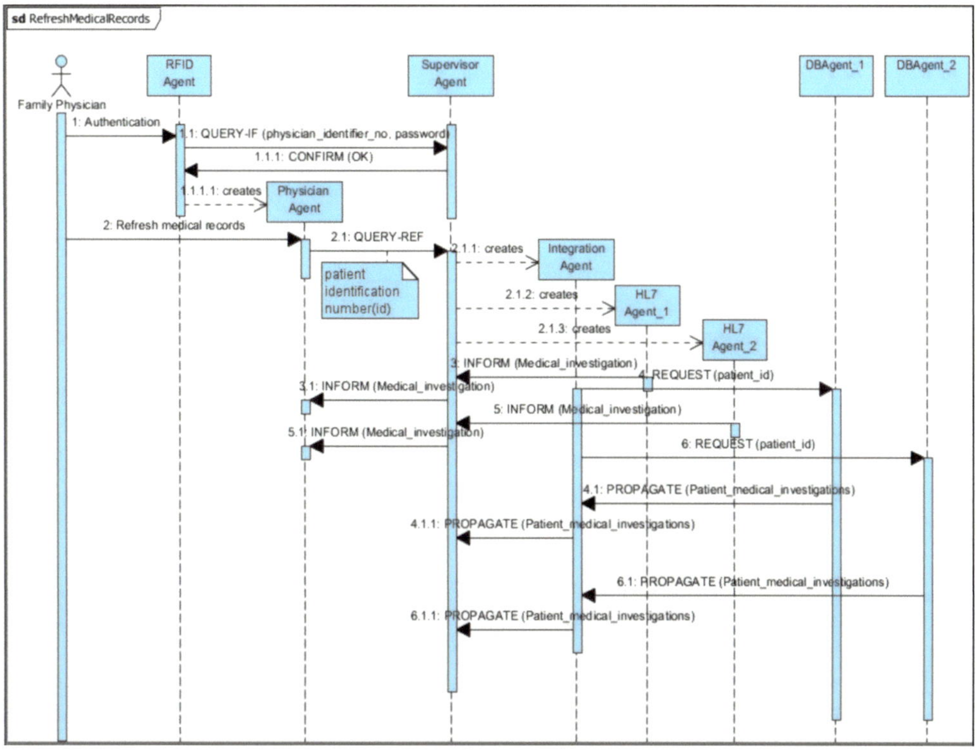

Fig. 6. Agent communication for updating electronic medical records for patients

The RFID-based multi-agent system, SMA-SIMOPAC, designed and implemented by our research team, facilitates the integration of data from heterogeneous sources (HL7-compliant or non-HL7 servers) in order to achieve a complete electronic medical record. The adoption of this system does not require major changes in terms of the software resources existing in the medical units. The proposed architecture is scalable, so that new sources of information can be added without amendment to the existing configuration. It also allows easy addition of new agents to provide other functionalities, without requiring changes of the existing agents. When a data source does not follow the HL7 standard, a new agent is developed to interface with this data source and to provide communication with the appropriate agent from the SIMOPAC system. The agents are independent of each other, and in order to retrieve information about patients, other agents are created to run the query again for sources of data. The agents previously created are disposed of when they accomplished the received task or after a preset time interval from the moment of receiving the task. The developed system is robust, each agent acting independently and autonomously. The failure of an agent does not cause overall system failure; other agents may take over the task of that agent. Last but not least, we should mention that the system is secure, as the access to the information about a patient is permitted based on an RFID tag specific to the patient or the doctor who wants to access the patient's electronic medical records.

7. Acknowledgments

The research results and technical solutions presented in this chapter have received the support of the Grant "SIMOPAC – Integrated System for the Identification and Monitoring of Patient" no. 11-011/2007, within the framework of the Romanian Ministry of Education and Research "PNCDI II, Partnerships".

8. References

Bajo, J., Corchado, J.M. & Rodriguez, S. (2008). GR-MAS: Multi-Agent System for Geriatric Residences. *ECAI 2008*

BioHealth and RFID (2007).www.gsf.de/imei/biohealth Vol. 3, April 2007. BioHealth is supported by the European Commissionunder the Europe INNOVA initiative. www.europe-innova.org

Bouzeghoub, A. & Elbyed, A. (2006). Ontology Mapping for Learning Objects Repositories Interoperability. *Intelligent Tutoring Systems'2006*. pp.794~797

BRIDGE Project, Logica CMG and GS1. European Passive RFID Market Sizing 2007-2022. February 2007. http://www.bridgeproject.eu/data/File/BRIDGE%20WP13%20European%20passive%20RFID%20Market%20Sizing%202007-2022.pdf (accessed on 31/07/08)

Cathleen F. Crowley and Eric Nalder (2009). Within health care hides massive, avoidable death toll, Aug. 10, Available at www.chron.com/disp/story.mpl/deadbymistake/6555095.html

Chen, M., González, S., Zhang, Q. & Leung, V. C.M. (2010). Code-Centric RFID System Based on Software Agent Intelligence. *IEEE INTELLIGENT SYSTEMS.*

Dias, J. C. Q., Calado, J. M. F., Osório, A. L. & Morgado, L. F. (2008). Intelligent Transport System Based on RFID and Multi-Agent Approaches. *IFIP International Federation for Information Processing*, Volume 283/2008, p.533-540

Fonseca, J.M., Mora, A.D. & Marques, A.C. (2005). A Multi-Agent Information System for Bioprofile Collection, *Proceedings of CIMED2005 - Second International Conference on Computacional Intelligence in Medicine and Healthcare.*

Hearst (2009). Dead by Mistake - Hearst Newspapers Report, August 2009

Iosep, C (2007). Standards save lives. *GS1 in Healthcare.* Healthcare Forum, Bucureşti, June 2007

Janz, .B, Pitts, M. & Otondo, R. (2005). Information Systems and Health Care II: Back to the Future With RFID: Lessons Learned - Some Old, Some New. *Communications of the Association for Information Systems*. Vol. 15, 2005:132-148

Laleci, G. B., Dogac, A., Olduz, M. , Tasyurt, I., Yuksel, M. & Okcan, A. (2008). SAPHIRE: A Multi-Agent System for Remote Healthcare Monitoring through Computerized Clinical Guidelines. *Whitestein Series in Software Agent Technologies and Autonomic Computing*, p.25-44

Lebrun, Y., Adam, E., Kubicki, S. & Mandiau, R. (2010). A Multi-Agent System Approach for Interactive Table Using RFID. *Advances in Practical Applications of Agents and Multiagent Systems.*Advances in Soft Computing, Volume 70/2010, 125-134

Nguyen, M. T., Fuhrer, P. & Pasquier, J. (2008). Enhancing Legacy Information Systems with Agent Technology. *Hindawi Publishing Corporation International Journal of Telemedicine and Applications.*

Orgun, B. & Vu, J. (2006). HL7 ontology and mobile agents for interoperability in heterogeneous medical information systems. *Computers in Biology and Medicine*, Volume 36, Issue 7, Pages 817-836 (July 2006).

Schweiger, A., Sunyaev, A., Leimeister, J.M. & Krcmar, H. (2007). Information Systems and Healthcare XX: Toward Seamless Healthcare with Software Agents. *Communications of the Association for Information Systems* (Volume 19, 2007) 692- 709

7

Farm Operation Monitoring System with Wearable Sensor Devices Including RFID

Tokihiro Fukatsu[1] and Teruaki Nanseki[2]
[1]*National Agricultural Research Center*
[2]*Kyushu University*
Japan

1. Introduction

To increase agricultural productivity and promote efficient management in modern agriculture, it is important to monitor the field environment, crop conditions, and farming operations instead of simply relying on farmers' experiences and senses. However, it is difficult to realize such monitoring automatically and precisely, because agricultural fields are widely spaced and have few infrastructures, monitoring targets vary according to crop selection and other variables, and many operations are performed flexibly by manual labor. One approach to monitoring in open fields under harsh conditions is to use a sensor network (Akyildiz et al., 2002; Delin & Jackson, 2000; Kahn et al., 1999) of many sensor nodes comprised of small sensor units with radio data links. In our previous study, we developed a sensor network for agricultural use called a Field Server (Fukatsu & Hirafuji, 2005, Fukatsu et al., 2006, Fukatsu et al., 2009a) that enables effective crop and environment monitoring by equipped sensors and autonomous management. Monitoring with Field Servers facilitates growth diagnosis and risk aversion by cooperating with some agricultural applications such as crop growing simulations, maturity evaluations, and pest occurrence predictions (Duthie, 1997; Iwaya & Yamamoto, 2005; Sugiura & Honjo, 1997; Zhang, et al., 2002). However, it is insufficient for obtaining detailed information about farming operations, because these operations are performed flexibly in every nook and cranny depending on crop and environment conditions.

Several approaches have been used to monitor farming operations, including writing notes manually, using agricultural equipment with an automatic recording function, and monitoring operations with information technology (IT)-based tools. Keeping a farming diary is a common method, but it is troublesome to farmers and inefficient to share or use their hand-lettered information. Some facilities and machinery can be appended to have an automatic recording function, but it requires considerable effort and cost to make these improvements. Moreover, it is difficult to obtain information about manual tasks, which are important in small-scale farming to realize precision farming and to perform delicate operations such as fruit picking.

Several researchers have developed data-input systems that involve farmers using cell-phones or PDAs while working to reduce farmers' effort of recording their operations (Bange et al., 2004; Otuka & Sugawara, 2003; Szilagyi et al., 2005; Yokoyama, 2005; Zazueta

& Vergot 2003). By using these tools, farmers can record their operations easily according to the input procedures of the systems, and the inputted data can be managed by support software and then shared with other farmers via the Internet. However, these systems cannot be easily applied for practical purposes because it is difficult to train farmers to use these tools, especially the elderly, and the implementation of these methods requires farmers to interrupt their field operations to input data.

Other systems equipped with a global positioning system (GPS) or voice entry have been developed to solve the problems of data input (Guan et al., 2006; Matsumoto & Machda, 2002; Stafford et al., 1996). These hands-free methods help farmers by inputting operation places or contents. However, the system that uses a GPS requires detailed field maps including planting information, the development of which requires significant costs and efforts, and with the system that uses cell phones, it is sometimes difficult for the device to recognize a voice entry because of loud background noises such as tractor sounds. Furthermore, for easy handling, these data-input systems only accept simple and general farming operations such as just spraying and harvesting. To allow flexible use and detailed monitoring, such as what farmers observe, which pesticide they choose, in what area they are operating and how much they spray, a more useful and effective support system is desired.

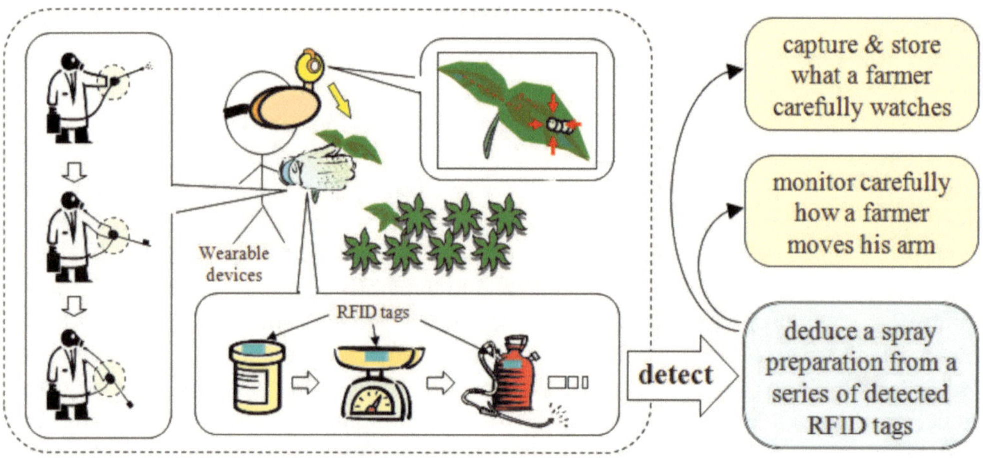

Fig. 1. Concept of farm operation monitoring system using wearable devices with RFID.

We propose a farm operation monitoring system using wearable sensor devices with radio frequency identification (RFID) readers and some sensing devices such as motion sensors, cameras, and a GPS (Fig. 1). This system recognizes detailed farming operations automatically under various situations by analyzing the data from sensors and detected RFID tags, which are attached to relevant objects such as farming materials, machinery, facilities, and so on. In this chapter, we describe the concept and features of the system, the results of several experiments using a prototype system, and the major applications and extensions of the current systems based on our research (Fukatsu & Nanseki 2009b; Nanseki et al., 2007; Nanseki 2010).

2. Farm operation monitoring system

Farmers want to record their farming operations in detail without interrupting their operations and without having to alter their farm equipment so that they can make effective decisions about future operations by utilizing the collected information with support applications. To meet such needs, we propose an innovative farm operation monitoring system with wearable sensor devices including RFID readers. In this section, we describe the concept, features, and architecture of our proposed system.

2.1 Concept

The concept of our farm operation monitoring system is to provide a versatile, expansible, practical, and user-friendly monitoring system that recognizes users' behavior in detail under various situations. To develop a useful monitoring system, we must consider the following requirements:

- The system should not encumber farmers' activities during farming operations.
- The system should be simple to use for non-experts without complicated processes.
- The system should be available without changing the facilities or equipment.
- The system should monitor detailed farming operations under various conditions.
- The system should be able to cooperate with various applications easily.

To meet these requirements, we propose a recognition method for farming operations by using RFID-reader-embedded wearable devices that are comfortable to wear, have unimpeded access to the farming situations they're supposed to monitor, and have sufficient sensitivity to RFID tags. Typical RFID systems, which can identify or track objects without contact, are used for individual recognition in some areas of logistics, security control, and traceability system (Finkenzeller, 2003; Rizzotto & Wolfram, 2002; Wang, et al., 2006; Whitaker, et al., 2007). For example, in the livestock industry, RFID tags are attached to or embedded in animal bodies, and some applications such as health control, fattening management, milking management, and tracking behavior are implemented by checking the detected RFID tags and using that data in combination with other measurement data (Gebhardt-Henrich, et al., 2008; Murray, et al., 2009; Trevarthen & Michael, 2008). In our system, however, we adapted an RFID system for use in the recognition of farming operations by analyzing patterns of the detected RFID tags. The procedure has the following steps:

1. RFID tags are attached to all relevant objects of farming operations such as farming materials, implements, machinery, facilities, plants, and fields.
2. A farmer performs farming operations with wearable devices that have RFID readers on them.
3. A sequence of RFID tags is detected throughout the farmer's activities.
4. The system deduces the farming operations by analyzing the pattern of the data.

In the conventional applications, RFID tags are attached to objects which themselves are important targets to be observed. In our system, however, a farmer puts on not an RFID tag but an RFID reader in order to apply this system to various operations easily. Also, in this system, not just single detected tags but series of detected tags are utilized to derive the desired information, unlike the conventional applications.

2.2 Features

The proposed system has some advantages and features. This method is flexible and available under various conditions without changing the facilities or equipment. All that is required is to attach RFID tags to existing objects and to perform farming operations while wearing the appropriately designed devices. For example, only by attaching RFID tags to many kinds of materials such as fertilizer and pesticide bottles, this method can automatically record which materials a farmer selects without interrupting his operations. With this system, we can easily collect an enormous amount of data about farming operations, and it helps to solve a shortage of case data for decision support systems (Cox, 1996). In the case of monitoring people who come and go at various facilities, in the conventional method the people carry RFID tags and RFID readers are set up at the gates to detect people's entrances and exits. In our proposed method, however, people wear RFID readers, and RFID tags, which are cheaper than the RFID readers, are attached to the gates. This will be effective in the situation in which a few people work in many facilities, such as in greenhouses. It can also be applied to monitoring operations with machinery at a low cost by attaching RFID tags to parts of operation panels such as buttons, keys, levers, and handles. The sequence of detected RFID tags tells us how a farmer operates agricultural implements.

By combining the data of RFID tags and other sensors, this system can monitor more detailed farming operations. For example, if an RFID tag is attached to a lever on a diffuser, we cannot distinguish between just holding the lever and actually spraying the pesticide. However, by using the data collected by wearable devices with finger pressure sensors, this system can distinguish between just holding the lever and actually spraying the pesticide accurately and specifically. Moreover, by connecting a GPS receiver to wearable devices, we can monitor when and where a farmer sprays the pesticide precisely. This information is now required to ensure the traceability of pesticides, and this system is expected to be an effective solution to the requirement of traceability, especially, when farmers manually perform the cultivation management (Opara & Mazaud, 2001). When attaching RFID tags to plants, trays, and partitions, we can also monitor the locations of farmers' operations in greenhouses where a GPS sometimes does not function well, and we can monitor even the time required for manual operations such as picking and checking of plants. The information about the progress and speed of farming operation can help in setting up efficient scheduling and labor management (Itoh et al., 2003). This system is effective for monitoring farming operations in detail, especially manual tasks that are difficult to record automatically in a conventional system.

2.3 Architecture

In our proposed system, a core wearable device is equipped with an RFID reader, an expansion unit for sensing devices, and a wireless communication unit (Fig. 2). The wireless communication unit enables the separation of heavy tasks such as data analysis and management processing from the wearable device. That is, the detected data can be analyzed at a remote site via a network instead of by an internal computer, so the wearable device becomes a simple, compact, and lightweight unit the farmer can easily wear. This distributed architecture allows for the implementation of a flexible management system and facilitates the easy mounting of various support applications that can provide useful information in response to recognized farming operations.

Thanks to the distributed architecture, the remote management system can be operated with high-performance processing. Therefore, the management system can recognize farming operations based on the patterns of detected RFID tags and sensing data with a complicated estimation algorithm. We can choose various types of algorithms such as pattern matching, Bayesian filtering, principal component analysis, and support vector machines by modifying the recognition function. A basic estimation algorithm is pattern matching in which a certain operation is defined by a series of data set with or without consideration of order and time interval. For example, an operation consisting of the preparation of a pesticide is recognized when the RFID tags attached on a pesticide bottle, a spray tank, and a faucet handle are detected within a few minutes in random order. Some estimation algorithms classify the data in groups of farming operations based on supervised learning, and they enable very accurate recognition, even though missed detection or false detection sometimes occurs.

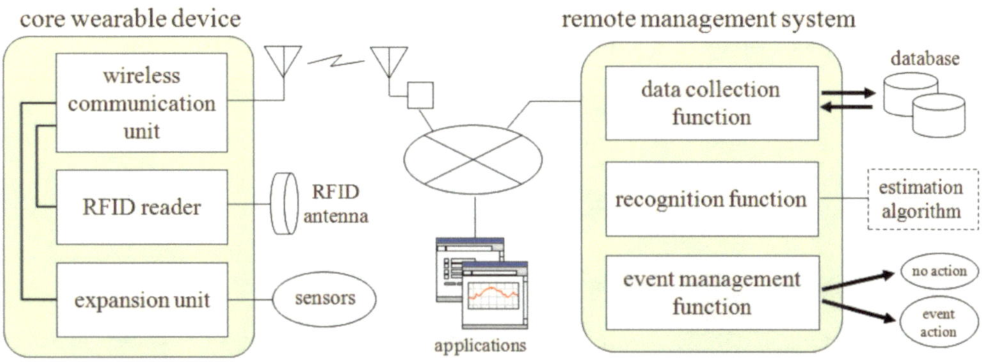

Fig. 2. Architecture of the farm operation monitoring system comprised of a core wearable device and a remote management system.

3. Prototype system

In our proposed system, farming operations are deduced by analyzing the patterns of detected RFID tags. To evaluate the possibility and effectiveness of this system, we developed a prototype system constructed of a glove-type wearable device, Field Servers for providing hotspot area, and a remote management system. With this prototype system, we conducted several experiments to demonstrate the system's functionality. In this section, we describe the architecture and performance of the prototype system and the results of the recognition experiments that involved a transplanting operation and greenhouse access.

3.1 System design

Figure 3 shows an overview of the prototype system and the wearable device which a farmer puts on his right arm. At a field site, we deployed several Field Servers that offer Internet access over a wireless local area network (LAN) so that the wearable device could be managed by a management system at a remote site. RFID tags were attached to some objects the farmer might come into contact with during certain operations. The information of the attached RFID tags and the objects including their category, was preliminarily registered in a database (DBMS: Microsoft Access 2003) named Defined DB in the management system. The remote management system constantly monitored the wearable device via the network, stored the data of detected RFID tags, and analyzed the farmer's operations.
The wearable device was equipped with a wireless LAN for communicating with the management system, an RFID reader for detecting relevant objects, and an analog-to-digital (A/D) converter with sensors for monitoring a farmer's motion. The RFID reader consisted of a micro reader (RI-STU-MRD1, Texas Instruments) and a modified antenna. The A/D converter consisted of an electric circuit including a microcomputer (PIC16F877, Microchip

Technology) with four input channels. A device server (WiPort, Lantronix), which served the function of a wireless LAN and enabled monitoring of the RFID reader and the A/D converter via the network, was also embedded. This wearable device worked for up to two hours when a set of four AA batteries was used. The battery life was able to be extended by using energy-saving units and modifying the always-on management. In some experiments, we added sensors such as pressure sensors to monitor the farmer's fingers and other wearable devices such as a network camera unit to collect user-viewed image data and a wearable computer display unit to provide useful information in real-time.

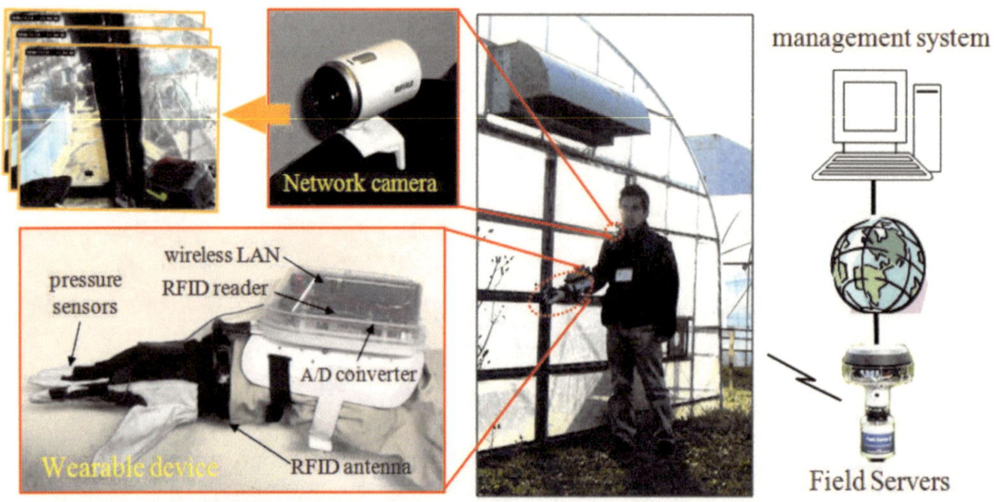

Fig. 3. Overview of the prototype system and the wearable device.

The type of RFID reader and the antenna shape are important factors for detecting RFID tags accurately without encumbering farmers' activities in various situations. There are RFID tags available with different frequencies (e.g., 2.45 GHz, 13.56 MHz, and 134.2 kHz) that differ in terms of communication distance, tag shape, antenna size, and broadcasting regulations (Khaw, et al., 2004). In this prototype system, the 134.2-kHz RFID was used because of the emphasis on the communication distance and the radio broadcasting laws in Japan. A bracelet-type antenna (85 mm in diameter) was developed with consideration of an easily wearable shape and adequate inductance of the antenna coil (47 uH for 134.2 kHz). The antenna had sufficient accessible distance (more than 100 mm) to detect RFID tags without any conscious actions.

Figure 4 shows a block diagram of the remote management system. It accessed the RFID reader and the A/D converter at high frequency (200 ms interval) and stored the data in a database (DBMS: Microsoft Access 2003) named Cache DB. In this system, we simply chose pattern matching as an estimation algorithm. The rules of expected farming operations were preliminarily defined into a pattern table with combinations or sequences of objects or categories that had already been registered in Defined DB. The management system checked the time-series data of Cache DB against the pattern table to detect defined farming operations. When the system recognized a certain farming operation, the information of the recognition result was recorded, and appropriate actions in response to the results were executed.

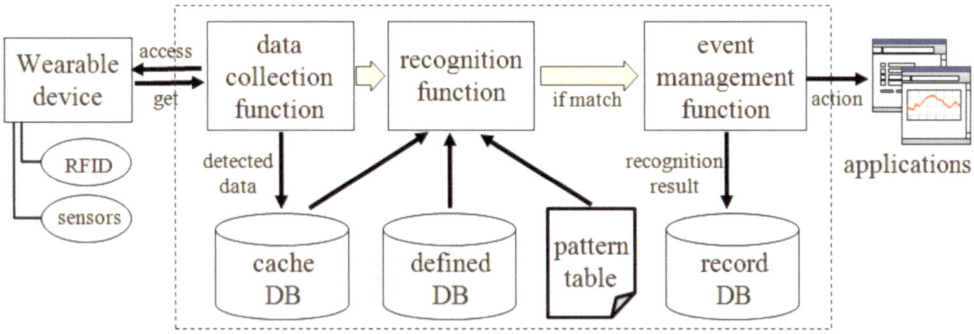

Fig. 4. Block diagram of the remote management system.

3.2 Recognition experiments

3.2.1 Transplanting operation

To evaluate the feasibility and the basic performance of this system, we performed a fundamental experiment to recognize transplanting operations in a field environment. In this experiment, a user took each potted seedling, checked the seedling's condition, and transplanted it to a large pot if it was growing well. RFID tags were attached to every pot including empty pots for transplanting, and a user performed the operation with the wearable device. Field Servers were deployed in the experimental area, and the remote management system accessed the wearable device via the Field Servers. We arranged twelve potted seedlings including two immature ones and tested whether the detailed information about this operation could be obtained by using our proposed system.

Figure 5 illustrates some results from this experiment. The white circle shows the detected RFID tags corresponding to each pot. The pots labeled pot-A to pot-E (categorized as small pots) were potted seedling, while the pots labeled pot-I to pot-IV (categorized as large pots) were empty pots for transplanting. The seedling in pot-B was an immature one that did not need to be transplanted. The transplanting operation was defined as occurring when a detected small pot was transplanted to a large pot detected within ten seconds, but only if the large pot was detected for over three seconds. The system was able to correctly identify every target pot that a user touched during the operation without any problem.

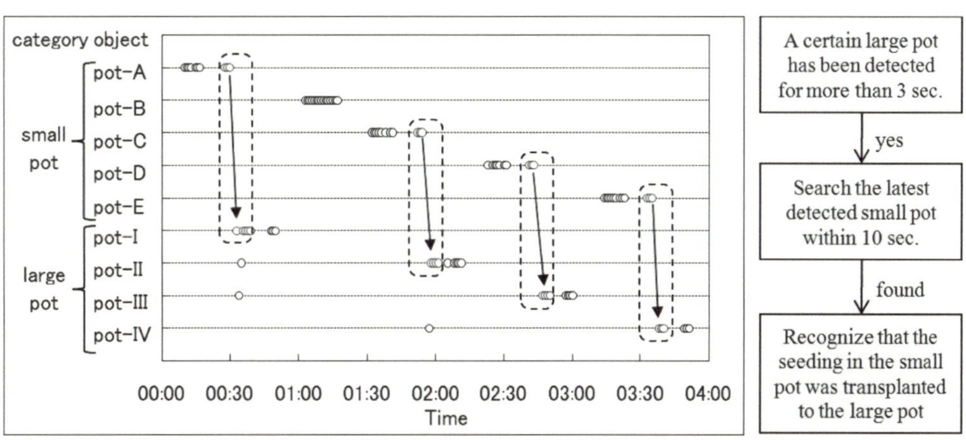

Fig. 5. Result of a recognition experiment about transplanting operation.

When a user took a large pot, an RFID tag of another large pot was mistakenly detected once in a while because these large pots were piled up. However, the defined rule was able to filter out the false detection, so this system was accurately able to recognize the operation. In this experiment, our proposed system was also able to recognize the correspondence relation of which large pot a seeding in a small pot was transplanted to. For example, the seedlings in pot-A, -C, -D, and -E were transplanted to pot-I, -II, -III, and –IV (the user didn't transplant pot-B, so that pot didn't have a corresponding large pot). In this system, not only the detected RFID tag identification number but also the detected time was stored in the database. By subtracting the first detected time of the small pot from the last detected time of the corresponding large pot, we were also able to obtain the process time of the transplanting operation as detailed information.

3.2.2 Greenhouse access

The next experiment was recognition of people entering and leaving greenhouses. In this experiment, RFID tags were attached to both sides of sliding doors (tag-A: outside; tag-B: inside) of greenhouses. A user equipped with the prototype wearable device entered and exited two different greenhouses eight times each to work inside and outside them. This system judged a greenhouse access by checking the sequence pattern of the detected RFID tags with pattern matching. The entering action was defined as occurring when the tag-B of either greenhouse had been detected for more than one second within ten seconds after the tag-A of the greenhouse was detected. The leaving action was defined as the opposite pattern of the entering action.
Figure 6 illustrates some results from the experiment. In this experiment, this system couldn't perfectly detect the entering and leaving actions; the percentage of accurate recognition in total was 87.5% for entering and 81.3% for leaving. The main reason for misrecognition was not missed detection due to inadequate antenna sensitivity but false detection caused by the excessive antenna range, which resulted in the antenna mistakenly detecting a far-side tag through the door once in a while. In this condition, the system was

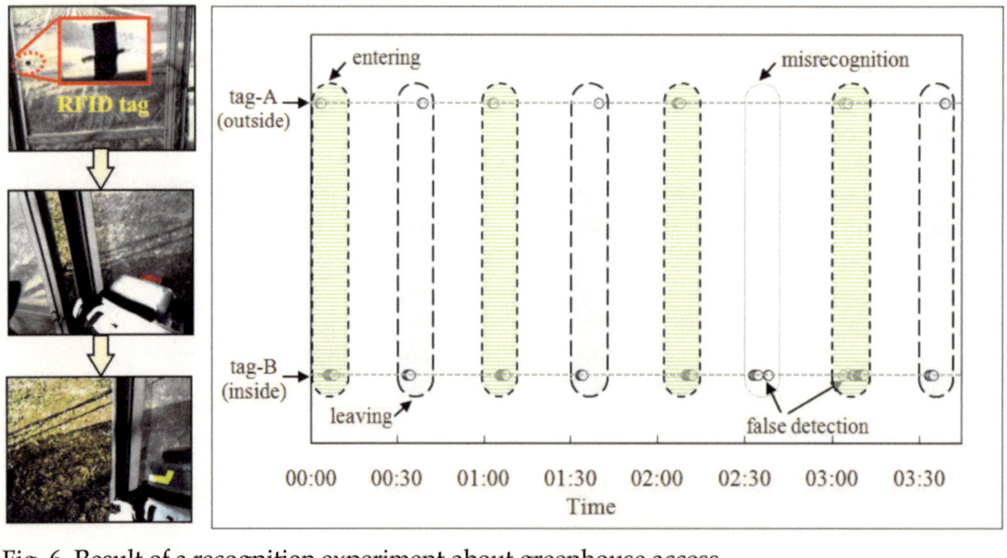

Fig. 6. Result of a recognition experiment about greenhouse access.

able to deduce the correct operations based on the detected patterns, even though false detections were included in them. At other times, the system was not able to deduce the correct operations that included false detections. To solve this problem, we must consider the allocation of the attached RFID tags so that the antennas can avoid false detections.

4. Applications

Our proposed system can recognize farming operations from the patterns of detected RFID tags. The farm operation monitoring system has the potential to be used effectively and to be implemented in a wide variety of applications. By using some sensor devices together, this system can recognize farming operation more accurately. By coordinating with Field Servers, we can also obtain more detailed information about farming operations. Moreover, this system enables us to provide useful information in response to the recognized operation by cooperating with agricultural support tools. In this section, we describe several applications of the system and the results of the experiments.

4.1 Recognition with RFID and sensing devices

Our prototype wearable device had an A/D converter with four input channels and an expansion port for RS232C. We used a pressure sensor to monitor the condition of the farmer's hand and a network camera unit to record user-viewed image data during farming operations. By using the enhanced wearable device, this system can recognize complicated farming operations and obtain useful information in detail. To evaluate the feasibility and effectiveness of the system, we conducted a recognition experiment of the snipping operation with a pair of scissors.

In this experiment, a user equipped with the enhanced wearable device took a plant tray, checked the condition of a plant in the tray, and snipped off unwanted leaves with scissors. RFID tags were attached to each plant tray and to the handle of the scissors. The system recognized the snipping operation when the RFID tag of the scissors was detected and simultaneously the value of the pressure sensor for the forefinger exceeded a certain threshold level that was set by preliminary test. By using the detected data of the RFID tag attached to the plant tray, this system deduced which plant was sniped off. The network camera unit on the user's shoulder captured several pictures of the operation after it was recognized.

Figure 7 illustrates some results from the experiment, which tested the snipping operation five times each in two kinds of plant tray. By using RFID tags and the pressure sensor together, this system was able to distinguish the status between just holding the scissors and actually using the scissors. In this experiment, the system had 80% accurate recognition of the snipping operation. The main reason for any misrecognition was that sometimes the value of the pressure sensor did not exceed the threshold level because the position of the sensor attached to the glove was not accurate for the user. The image data was adequately collected just when the user snipped a target leaf, and it enabled us to provide useful information about how the user performed the operation. In this experiment, the data of the pressure sensor was shown as an 8-bit raw data item with no calibration data. If we calibrated the sensor, we could get more detailed information about the user's technique with the scissors.

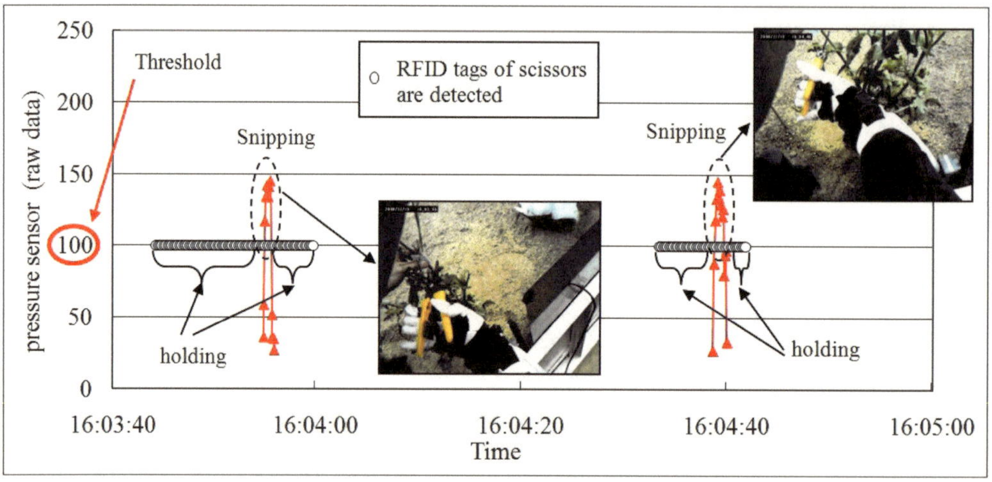

Fig. 7. Result of a recognition experiment about snipping operation.

4.2 Multi monitoring with field servers

Image data provides useful and helpful information for agricultural users to check crop conditions and to comprehend farming operations. Especially, recording operations of skilled farmers visually is very important for new farmers and agricultural researchers to understand practical techniques. We previously developed Field Servers with controllable cameras that can realize the distributed monitoring system. By using the Field Servers in cooperation with our proposed system, we can record the processes of farming operations carefully from a number of different directions in response to the results of recorded data. To evaluate the feasibility and effectiveness of the system in cooperation with Field Servers, we conducted an experiment in which the system collected pictures of recognized farming operation by controlling the camera of the surrounding Field Servers.

In this experiment, RFID tags were attached to a warehouse door, to some points on a rack in the warehouse, and to stored farming materials such as pesticide bottles. One Field Server equipped with a controllable camera was deployed near the warehouse. The Field Server periodically monitored field and crop conditions as part of a scheduled operation. The system recognized the preparing operation when a certain RFID tag of farming materials was detected after the RFID tag on the warehouse door was detected. We had previously registered the material places and preset camera positions and settings. When the system recognized that a certain material was being taken, it performed an event operation to record the target process by using the Field Server camera with a zoom function.

When two management systems share one controllable camera, there is a potential conflict between scheduled operations and event operations that require monitoring a different target. To solve this problem, we introduced a multi-management system (Fukatsu et al., 2007, Fukatsu et al., 2010). Figure 8 shows the operation status flow of the multi-management system and illustrates some results from the experiment designed to test the system. One management system (Agent-A) monitored the Field Server on the basis of its scheduled operation and the other system (Agent-B) periodically checked the RFID database. When a defined operation was recognized, Agent-B sent a stop signal to Agent-A to avoid access collision, and Agent-B preferentially directed the camera of the Field Server to the defined position.

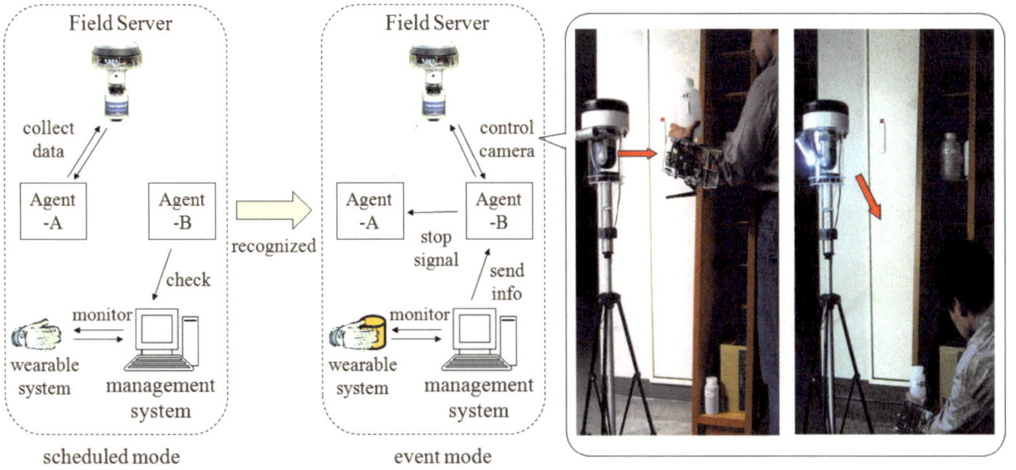

Fig. 8. Operation status flow of the multi-management system.

When a user with a wearable device tried to bring out the materials randomly, the system was able to record the target operation procedure as the image data. In some cases, it couldn't acquire desirable image data because the speed of the camera was not fast enough. To avoid the delay of the camera moving, we modified the camera control algorithm in which the camera was preliminarily directed to the expected position when the rack-attached RFID tag was detected. By introducing the modified algorithm, we were able to get more image data that included the scene of the operation.

4.3 Cooperation with support application

In agriculture, many support applications that provide useful information to farmers have been developed. Some support applications, such as a navigation system for appropriate pesticide use (Nanseki & Sugahara, 2006), are provided as Web application services, which are available for our proposed system. By combining our system and Web-based support applications, we can provide appropriate information in real-time in response to farming operations. For example, it is helpful for a farmer to get pointed advice regarding proper usage of a pesticide to avoid misuse of the pesticide.

To evaluate whether the system was able to cooperate with a Web-based support application easily, we conducted an experiment in which the system provided detailed information about the pesticide held by a farmer via a wearable computer display in real-time. In this experiment, we prepared a Web application service for pesticide management that outputted a target pesticide name, its detailed information including history of usage, and relevant links to information about the appropriate pesticide to use in response to an inputted query. By using the Web application service, we were also able to register and update target pesticide information via the Internet. When the system recognized that a certain pesticide bottle was taken, it sent the recognized pesticide ID to the Web application service and received detailed information about it with an HTML format. Then, the system outputted the information to the wearable computer display connected to the Internet via the Field Server.

Figure 9 illustrates some results from the experiment. RFID tags were attached to five kinds of pesticide bottle and a spray tank. A user with the prototype wearable device and a wearable computer display conducted the pesticide preparation. When the RFID tag attached to the pesticide bottle was detected, the system was able to provide the appropriate information to the user. When the RFID tag of the spray tank was detected after the recognition, the system judged that the pesticide was used, and it updated information about the pesticide's use history by accessing the Web application service. We confirmed that the target history information was automatically updated without problems when the user poured a certain pesticide into the spray tank.

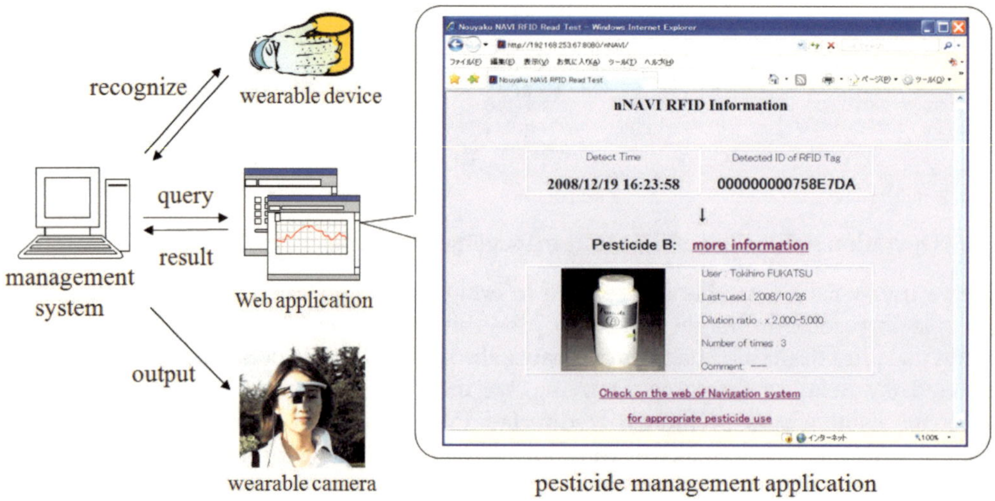

Fig. 9. Support application of providing useful information.

4.4 Extension of the system: the farming visualization system

Several types of the farm operation monitoring system have been developed according to the varied needs of farms. All of these systems are designed to record and replay all the information of farming operations based on combinations of data from several kinds of sensors, including RFID readers. Some farms need a low-end type of system with only a few sensors. This type of system is simple to use and has a low introduction cost. On the other hand, some farms need a high-end system with many sensors. This type of system can monitor many kinds of farming operations with high accuracy and frequency. Our proposed system can be modified to suit both kinds of farm.

Our system can also be extended in various directions within the field of agriculture, and one such extension is the farming visualization system (FVS) that has been developed based on our previous research (Fukatsu & Nanseki 2009b; Nanseki et al., 2007). One of the major application fields is to record precious and detailed farming history for good agricultural practice (GAP) and food traceability. Another major application field is the human development of young farming operators. These applications fields of the FVS are especially important in large farm cooperations, and the government has aided us in developing several types of FVS. The NoshoNavi project, begun in 2010 as a five-year period, is one such national research project (http://www.agr.kyushu-u.ac.jp/keiei/NoshoNavi/).

Figure 10 shows images of a high-end type of FVS. The wearable devices of the system include two wearable RFID readers (Wellcat), two cameras (Logicool), one differential GPS (Hemisphere), one mobile PC (Panasonic) and one head mount display (Mikomoto). The two RFID readers on both hands enable us to distinguish whether the right or left hand touches the RFID tags. One of the cameras captures a wide view of the farming environment, and the other camera captures a narrow view, focusing on the area immediately around the operator's hand. The differential GPS has 50-cm accuracy. The mobile PC controls each sensor and manages all the data, so this particular system does not need a network connection. The visualization software of the FVS can show an integrated view of all the data of these sensors.

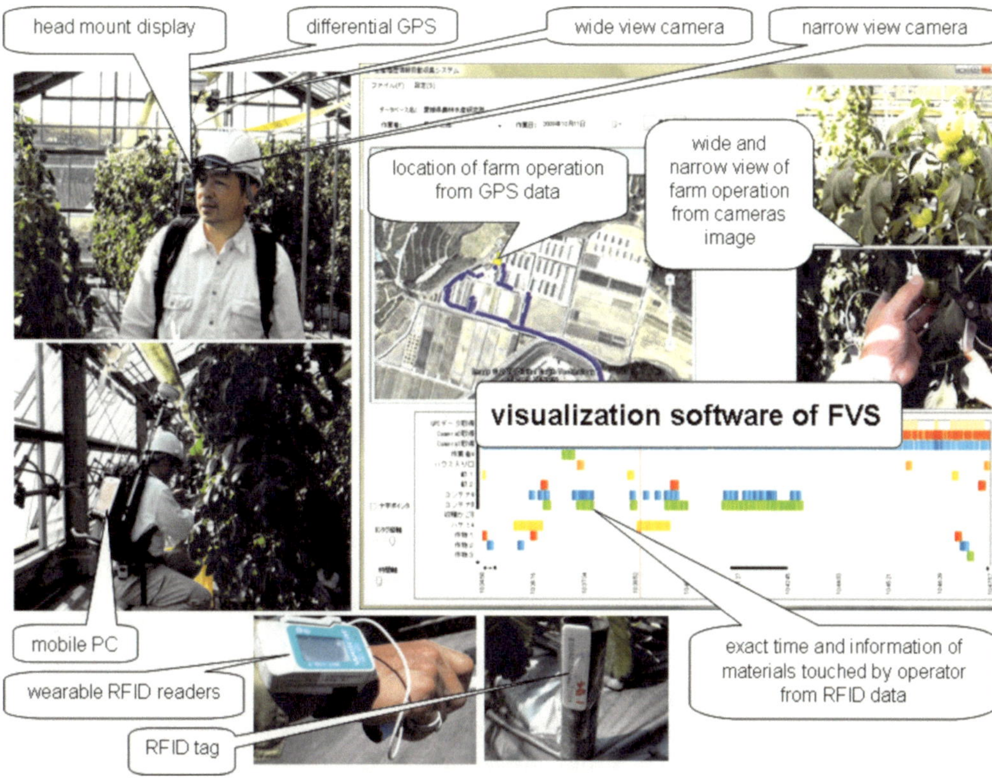

Fig. 10. High-end type of the farming visualization system.

The system enables a fully automatic recording of Five W's and one H information of the whole farming operation. With this system, non-skilled operators can learn farming skills based on recorded visual and audio data of a skilled operator, for example. The data of RFID readers gives exact information about the materials touched by a skilled operator. The data of the differential GPS gives the exact location where an operation is done. Images of farm operation from the viewpoint of the skilled operator give good guidance to non-skilled operators.

The low-end type of system has only one RFID reader with one GPS mobile phone. This type of system is suitable for automatic recording of the location of a farming operation and materials touched by operators. The system is now being tested on several farms, including one of the biggest rice farms in Japan. The farm grows many varieties of rice with several cultivation methods requested by the buyers, in 150 ha of paddy fields. There are more than 20 farming operators. One of the major issues of the farm management is the passing on of the farming skills of the skilled operators. The FVS is expected to be helpful in solving this problem.

5. Discussion and future work

We have proposed a farm operation monitoring system with wearable devices including RFID readers and conducted some experiments with a prototype system. These experiments show that the system can recognize farming operations appropriately and can provide useful information to users in response to the recognized operations. The feasibility and effectiveness of our proposed system has been evaluated experimentally, and we have discussed issues remaining to be solved future works, and the potential of the system for practical use.

One of the main issues of the system is recognition accuracy. In our experiments, false detection of RFID tags occurred once in a while because of excessive antenna sensitivity. To avoid false or missed detection, adequate design of an RFID antenna is required. For example, a ring-type or a fingertip-type antenna is capable of detecting only fingered objects selectively. By using another type of shaped antenna or combining several types of antennas, we can solve the problem. Adequate tag allocation is also important. Attaching many RFID tags while avoiding the mutual interference and false detection helps the system to increase the recognition accuracy, even when some RFID tags are not detected. We should also consider the position at which we attach RFID tags and the reading interval of RFID readers depending on the operation contents and farmers' activities, so that key RFID tags will not be missed.

To recognize farming operations more accurately, it is important not only to detect RFID tags adequately but also to estimate the farming operation effectively from detected data, including that from motion sensors. In our experiments, we appended a pressure sensor to the wearable device to recognize complicated operations and to interpolate the data of RFID tags. By using many kinds of motion sensors such as finger-bending sensors, acceleration sensors, capacitive sensors, and myogenic potential sensors depending on the situation, this system will be able to recognize farming operations with a high degree of accuracy. With regard to an estimation algorithm of farming operations, we use pattern matching in our experiments, because we attached a minimum amount of RFID tags for this testing. If we had many RFID tags attached to relevant objects, useful motion sensors, detailed rules with many steps for recognizing operations, and preliminary data for supervised learning, we could apply various kinds of estimation algorithms. These algorithms should be customized and adjusted on the basis of the performance of the wearable device, tag allocation, operation contents, and user requirements. It is also important to consider what farming operation should be recognized, how we should define the rules for the farming operation, and which tags and sensors we should use for the recognition. The preparation of registering many kinds of rules and tag information needs careful consideration and a lot of effort. To recognize various farming operations, some support tools for registering these data easily will be needed.

Another problem is the need to overcome fitting difficulties with the wearable devices. In our experiment, the glove with a pressure sensor didn't fit the target user and the sensor data sometimes indicated wrong values that caused mistaken recognition. In general, it is difficult to fit a wearable device to every user, so wearable devices should be designed with a target user in mind or have some key components such as the sensor position be adjustable. Where and how a wearable camera is set to record clear and desirable image data is also important. A wearable camera may be swung in response to the user's motion, and the position and direction of the camera may easily be changed. The desirable angle and direction of the camera also differs according to the operation and the user's request. For example, a head-mounted camera can record from the user's viewpoint, but the user's head will move frequently during some operations. In some situations, it is better to use a camera with a wide field of view mounted on the user's chest pocket or waist belt. It will be important to design a wearable device with ergonomics, operation contents, and each user's individual features.

Though the proposed system has some open issues, the system in its current form can record a farmer's operation easily and automatically. It is effective for realizing agricultural support applications such as labor control, precision management, and food traceability. To make improvements in farm management, it is important to know how long farmers perform each operation in detail. Our system enables farmers to record labor information easily. In some countries, large-scale farming is popular, so precision management can easily be conducted by using the automatic and mechanized operation system. On the other hand, manual operation tasks are still required in many countries and on many farms, so our proposed system helps to realize precision management in these situations. Especially in Japan, there are many small-scale farms on which it is difficult to perform mechanized farming. Moreover, to grow high-quality crops, practical farmers operate some implements manually, because each crop needs a different amount of fertilizer and chemicals. In food traceability, not only the supply chain but also farmers are required to record the processing of products (Smith & Furness, 2008). The record of cultivation management, especially pesticide use, has become increasingly important, but the task requires much effort. For a farmer to meet the legal requirements, this system is helpful to establish traceability and to provide detailed information such as image data.

This system also enables the production of advanced applications such as controlling equipment in a coordinated manner, useful databases of operation techniques, and navigation and attention systems for new farmers. Field Servers have a function to control peripheral equipment such as greenhouse heaters and sprays. By combining with Field Servers, our proposed system can control suitable machines automatically to reduce farmers' efforts in response to recognized farming operations. By combining the information of operation history and other monitoring information such as crop growth data, we can analyze the effects of operations on the crops. Practical farmers check various conditions with their senses based on their experience, and this system can record data of farming operations of skilled farmers. If we can obtain information not only on farming operations but also on the farmers' behavior, e.g., what they pay attention to and how they interact with crops and fields, the database will become an important tool for understanding their techniques and wisdom. Especially in Japan, the age of the farming population is increasing, and the number of farmers is rapidly decreasing. Therefore, practical techniques of skilled farmers are vanishing, and new generations of farmers lose an opportunity to learn from them. By storing a lot of information on farming operations in detail, this system can provide a useful digital archive of the agricultural system. By using our proposed system, we can realize an advanced decision support system such as the navigation and attention systems. Such a system can provide useful and suitable information such as a tutorial about the next operation in a sequence, the needed data for decision-making, and warnings about misuse to a farmer in real-time in response to recognized operations. Such a system will enhance the farmer's sensitivity, judgment, and activity.

We have proposed an innovative monitoring system to recognize farming operations easily, and have demonstrated the effectiveness and feasibility of a farm operation monitoring system with RFID readers and tags. Our proposed system can be applied to a wide variety of situations and purposes not only in agriculture but also in other fields. It is expected to be used as an effective tool for monitoring humans' behavior and experiences.

6. Conclusion

To monitor farming operations easily and automatically, we proposed a farm operation monitoring system with wearable sensor devices including RFID readers and tags. By attaching RFID tags to various objects such as farming equipment and performing farm operations with wearable devices including RFID readers, the system can recognize the operation by analyzing the pattern of detected RFID tags and sensor data. This proposed system can monitor farmers' operations flexibly without interrupting their activities or the necessity of changing their facilities or equipment. Moreover, this system can facilitate effective support applications that provide useful information to farmers in response to the recognized operations. To evaluate the feasibility of the system, we conducted several experiments with a prototype system. Through the experiments, we demonstrated the effectiveness and potentiality of our proposed system.

7. References

Akyildiz, I.F.; Su, W.; Sankarasubramaniam, Y. & Cayirci, E. (2002). Wireless sensor networks: a survey. *Computer Networks*, Vol.38, No.4, (March 2002), pp. 393-422, ISSN 1389-1286

Bange, M.P.; Deutscher, S.A.; Larsen, D.; Linsley, D. & Whiteside, S. (2004). A handheld decision support system to facilitate improved insect pest management in Australian cotton systems. *Computers and Electronics in Agriculture*, Vol.43, No.2, (May 2004), pp. 131-147, ISSN 0168-1699

Cox, P.G. (1996). Some issues in the design of agricultural decision support systems. *Agricultural Systems*, Vol.52, No.2, (October 1996), pp. 355-381, ISSN 0308-521X

Delin, K.A. & Jackson, S. P. (2000). Sensor Web for in situ exploration of gaseous biosignatures, *Proceedings of IEEE Aerospace Conference Vol.7*, pp. 465-472, ISBN 0-7803-5846-5, Big Sky, MT, USA, March 18-25, 2000

Duthie, J.A. (1997). Models of the response of foliar parasites to the combined effects of temperature and duration of wetness. *Phytopathology*, Vol.87, No.11, (November 1997), pp. 1088-1095, ISSN 0031-949X

Finkenzeller, K. (2003). *RFID handbook: fundamentals and applications in contactless smart cards and identification* (2nd Ed.), John Wiley and Sons, ISBN 978-0-470-84402-1, New York, NY, USA, 2003.

Fukatsu, T. & Hirafuji, M. (2005). Field monitoring using sensor-nodes with a Web server. *Journal of Robotics and Mechatronics,* Vol.17, No.2, (April 2005), pp. 164-172, ISSN 0915-3934

Fukatsu, T.; Hirafuji, M. & Kiura, T. (2006). An agent system for operating Web-based sensor nodes via the Internet. *Journal of Robotics and Mechatronics,* Vol.18, No.2, (April 2006), pp. 186-194, ISSN 0915-3934

Fukatsu, T.; Kiura, T.; Tanaka, K. & Hirafuji, M. (2007). Hierarchical agent system for Web-based sensor network, *Proceedings of the 2007 International Symposium on Applications and the Internet Workshops*, pp. 77, ISBN 0-7695-2757-4, Hiroshima, Japan, January 15-19, 2007

Fukatsu, T., Hirafuji, M., Saito, Y. & Suzuki, T. (2009a). Long-term field monitoring system with Field Servers at a grape farm. *ISHS Acta Horticulturae*, Vol.824, (April 2009), pp. 183-190, ISSN 0567-7572

Fukatsu, T. & Nanseki, T. (2009b). Monitoring system for farming operations with wearable devices utilized sensor networks. *Sensors*, Vol.9, No.8, (August 2009), pp. 6171-6184, ISSN 1424-8220

Fukatsu, T.; Hirafuji, M. & Kiura, T. (2010). Web-based sensor network with flexible management by agent system, In: *Advances in Practical Multi-Agent Systems. Studies in Computational Intelligence Vol.325*, Q. Bai and N. Fukuta (Eds.), pp. 415-424, Springrt-Verlag, ISBN 978-3-642-16097-4, Berlin Heidelberg

Gebhardt-Henrich, S.G.; Buchwalder, T.; Froehich, E. & Gantner, M. (2008). RFID identification system to monitor individual outdoor use by laying hens, *Proceedings of 42nd Congress of the International Society for Applied Ethology*, pp. 113, ISBN 978-90-8686-081-4, Dublin, Ireland, August 5-9, 2008

Guan, S.; Shikanai, T.; Minami, T.; Nakamura, M.; Ueno, M. & Setouchi, H. (2006). Development of a system for recording farming data by using a cellular phone equipped with GPS. *Agricultural Information Research,* Vol.15, No.3, (May 2006), pp. 241-254, ISSN 0916-9482

Itoh, T.; Ishii, H. & Nanseki, T. (2003). A model of crop planning under uncertainty in agricultural management. *International Journal of Production Economics*, Vol.81-82, (January 2003), pp. 555-558, ISSN 0925-5273

Iwaya, K. & Yamamoto, H. (2005) The Diagnosis of Optimal Harvesting Time of Rice Using Digital Imaging. *Journal of Agricultural Meteorology*, Vol.60, No.5, (2005), pp. 981-984, ISSN 0021-8588

Kahn, J.M.; Katz, R.H. & Pister, K.S.J. (1999). Next century challenges: mobile networking for "Smart Dust", *Proceedings of the 5th Annual ACM/IEEE International Conference on Mobile Computing and Networking*, pp. 271-278, ISBN 1-58113-142-9, Seattle, WA, USA, August 15-19, 1999

Khaw, M. K.; Mohd-Yasin, F. & Reaz, M. L. (2004). Recent advances in the integrated circuit design of RFID transponder, *Proceedings of the IEEE International Conference Semiconductor Electronics*, ISBN 0-7803-8658-2, Kuala Lumpur, Malaysia, December 7-9, 2004

Matsumoto, K. & Machida, T. (2002). Studies on development of work journal supporting systems with speech recognition, *Proceedings of the 3rd Asian Conference for Information Technology in Agriculture*, pp. 434-438, Beijing, China, October 26-28, 2002

Murray, B.B.; Rumbles, I. & Rodenburg, J. (2009). Application of RFID technology in herd management on dairy herds in Canada, *Proceedings of the Joint International Agricultural Conference 2009*, pp. 259-265, ISBN 978-90-8686-113-2,Wageningen, Netherlands, July 6-8, 2009

Nanseki, T. & Sugahara, K. (2006). A Navigation system for appropriate pesticide use: system development and application in Japan, *Proceedings of the 5th International*

Conference of the Asian Federation for Information Technology in Agriculture, pp. 293-300, ISBN 1403-93162-3, Bangalore, India, November 9-11, 2006
Nanseki, T.; Sugahara, K.; & Fukatsu, T. (2007). Farming Operation Automatic Recognition System with RFID (In Japanese with English summary). *Agricultural Information Research*, Vol.16, No.3, (October 2007), pp. 132-140, ISSN 0916-9482
Nanseki, T. (2010). Development of Decision Support Systems for Agricultural Management : An overview in Japan, *Proceedings of the 3rd Asian Federation for Information Technology in Agriculture*, pp. 434-438, Bogor, Indonesia, October 3-7, 2010
Opara, L.U. & Mazaud, F. (2001). Food traceability from field to plate. *Outlook on Agriculture*, Vol.30, No. 4, (December 2001), pp. 239-247, ISSN 0030-7270
Otuka, A. & Sugawara, K. (2003). A labor management application using handheld computers. *Agricultural Information Research*, Vol.12, No.2, (December 2003), pp. 95-104, ISSN 0916-9482
Rizzotto, P. & Wolfram, G. (2002). Intelligent tagging - getting supply chain smart!, *Proceedings of the Official ECR Europe Conference*, Barcelona, Spain, April 22-24, 2010
Smith, I. G. & Furness, A. (2008). *Food traceability around the world* (1st Ed.), Vicarage Publications Ltd., ISBN 978-0-9553707-1-7, York, United Kingdom, 2008.
Stafford, J.V.; Le Bars, J.M. & Ambler, B. (1996). A hand-held data logger with integral GPS for producing weed maps by field walking. *Computers and Electronics in Agriculture*, Vol.14, No.2, (February 1996), pp. 235-247, ISSN 0168-1699
Sugiura, T. & Honjo, H. (1997). A dynamic model for predicting the flowering date developed using an endodormancy break model and a flower bud development model in Japanese pear. *Journal of Agricultural Meteorology*, Vol.52, No.5, (1997), pp. 897-900, ISSN 0021-8588
Szilagyi, R.; Herdon, M. & Lengyel, P. (2005). Agricultural application development for mobile devices, *Proceedings of the International Conference on Agricultural Economics, Rural Development and Informatics*, Debrecen, Hungary, April 7-8, 2005
Trevarthen, A. & Michael, K. (2008). The RFID-enabled dairy farm: towards total farm management, *Proceedings of the 7th International Conference on Mobile Business*, pp. 241-250, ISBN 978-0-7695-3260-8, Barcelona, Spain, July 7-8, 2008
Wang, N.; Zhang, N. & Wang, M. (2006). Wireless sensors in agriculture and food industry - recent development and future perspective. *Computers and Electronics in Agriculture*, Vol.50, No.1, (January 2006), pp. 1-14, ISSN 0168-1699
Whitaker, J.; Mithas, S. & Krishnan, M.S. (2007). A field study of RFID deployment and return expectations. *Production and Operations Management*, Vol.16, No.5, (September 2007), pp. 599-612, ISSN 1059-1478
Yokoyama, K. (2005). Promoting the good agricultural practice movement through interactive and seamless communication based on user-friendly mobile information technology, *Proceedings of the International Seminar on Technology Development for Good Agricultural Practice in Asia and Oceania*, pp. 213-219, Tsukuba, Japan, October 24-28, 2005
Zazueta, F.S & Vergot, P.III. (2003). Use of handheld computers in agricultural extension programs, *Proceedings of the 4th International Conference of European Federation for Information Technology in Agriculture*, pp. 12-16, Debrecen, Hungary, July 5-9, 2003
Zhang, Y.; Li, C.; Zhou, X. & Moore, B. (2002). A simulation model linking crop growth and soil biogeochemistry for sustainable agriculture. *Ecological Modelling*, Vol. 151, No.1, (May 2002), pp. 75-108, ISSN 0304-3800

8

The Application of RFID in Automatic Feeding Machine for Single Daily Cow

Zhijiang Ni, Zhenjiang Gao and Hai Lin
China Agricultural University
China

1. Introduction

Chapter Objectives
In this chapter, you'll be able to do the following:

- You'll know why the identification of single daily cow is needed
- The RFID device used in this research
- The communication between RFID and PC, between RFID and MCU
- The good effect due to the technology (experiment)

2. Why the identification of single daily cow is needed

Daily cow is one kind of ruminant animal, whose rumen plays an important role in the digestive process. There are many kinds of microbes in the lumen. Actually it is these microbes that play a crucial part for the digestion. These microbes are sensitive to the pH value in the rumen environment. To keep these microbes be in active status, the pH value should be kept at stable (the pH range should be 6.4~6.8). The studies show that the pH value in the rumen is relative with the amount of the concentrated feed. So we need control the amount of the concentrated feed that each daily cow got. This process involves the feeding based on a single daily cow. To realize this process, we need to identify the daily cow, and then give it the amount of concentrated feed that it needs. This process could be realized by the application of RFID system.

Ni (2009) designed an intelligent moving precise feeding machine for single dairy cow. An RFID system was equipped on this machine, which can move and identify the single dairy cow, and then give it the amount of the concentrated feed needed. The schematic figure is showed in Fig.1.

Voulodimos (2010) established a complete farm management system based on animal identification using RFID. This system contains various kinds of workstations, such as desktop computers (servers, database), laptops, handheld mobile devices, and a number of different subsystems. Fig. 2 shows the main subsystems: the central database, the local database and the mobile—RFID subsystem.

The central database system (left down in Fig.2) is used to store all information related to the management of animal tracking and monitoring at central level.

The local database system (right-down in Fig.2) is based on an animal data management application, such as tracking of animal vaccination, tracking of animals' diet.

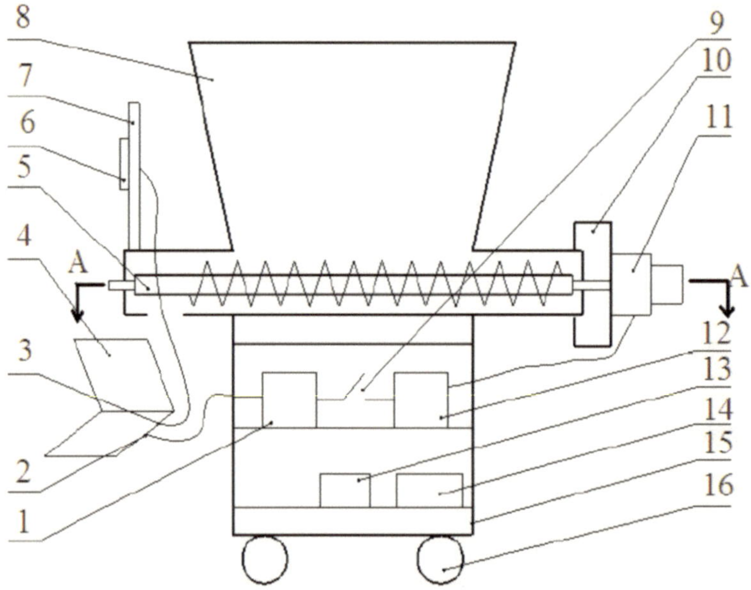

1-MCU, 2-First Serial Port, 3-Second Serial Port, 4-PC, 5-Auger, 6-RFID system, 7-Board for RFID, 8-Feed Bin, 9-Switch, 10-Board for Motor, 11-Motor, 12-Motor Actiyator, 13-Voltage Transfer Device, 14-Battery, 15-Frame, 16-Moving Device

Fig. 1. Schematic for Feeding Machine (Ni,2009)

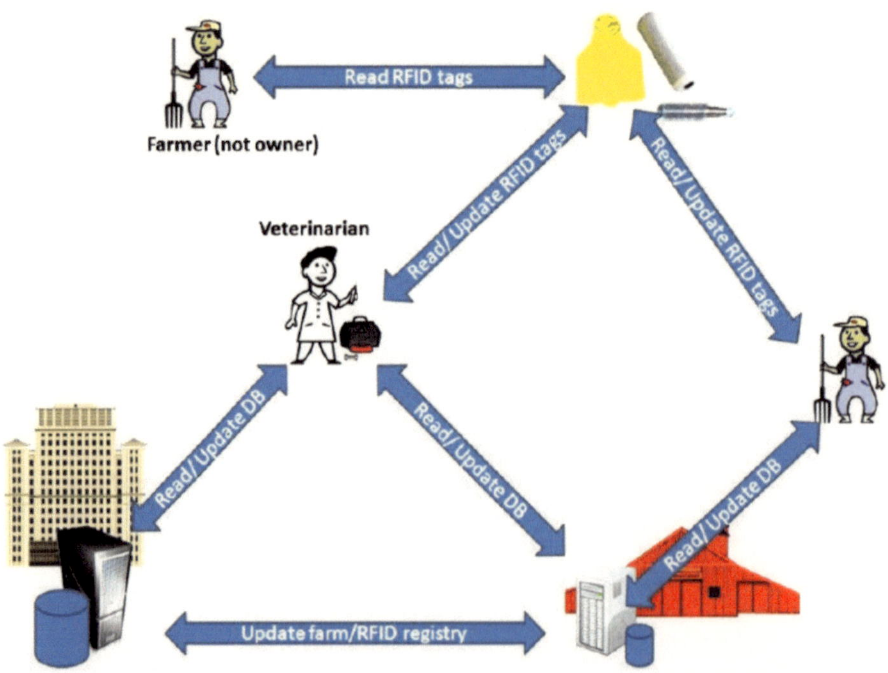

Fig. 2. Platform architecture (Voulodinos, 2010)

3. The RFID device used in this research

RFID is the abbreviation for Radio Frequency Identification, which is a technology that utilizes communication through electromagnetic waves to exchange data between an object and a terminal to realize the purpose of identification.

A RFID system (Fig.3) typically comprises following three parts (Roberts, 2005):

- An RFID device (tag);
- A tag reader with an antenna and transceiver;
- A host system or connection to an enterprise system.

Fig. 3. A typical RFID system (Roberts, 2005)

In the research of Ni (2009) and Li (2010), the reader used is SMC-R134 (Fig. 4), and the tag is SMC-E1334 (Fig. 5). Both the reader and the tag are the product of SMARTCHIP MOCROELECTRONIC CORP (SMC) in Taiwan.

Fig. 4. SMC-R134 Reader (Ni, 2009)

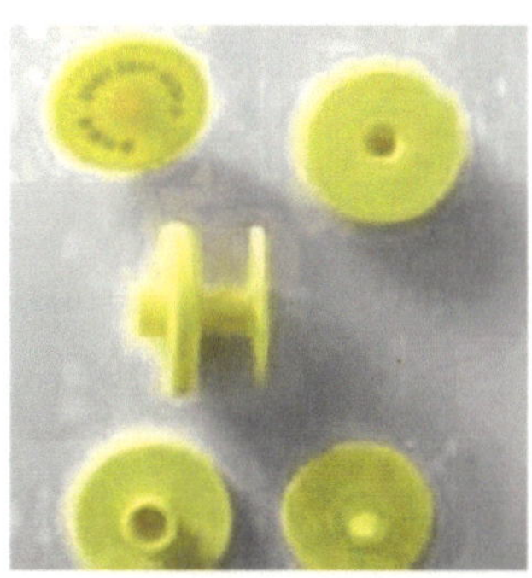

Fig. 5. SMC-E1334 Tag(Ni, 2009)

The maximum identify distance for this RFID system is 50cm ± 10%. The frequency is 134.2 kHz. The working voltage is DC 9V. The parameters are shown in table 1.

Name	Parameters
Type	SMC-R134
Frequency	AM 134.2 kHz
Voltage	Vcc = 9V
Current dissipation	Max: 200 mA (9V)
Induction distance	50cm ± 10%, working with SMC-E1334 tag
Weight	780g ± 2%
Length	264 mm
Width	264 mm
Height	30 mm

Table 1. Parameters for SMC-R134 Reader (Ni, 2009)

There are ten pins for the reading head of SMC-R134 reader. The colors for each pin (from left to right) are: red, black, yellow, purple, gray, green, brown, white, blue and orange, which is shown in fig. 6. The function for each pin is shown is table 2.

Pins	**Pin Color**	**Name**	**I/O**	**Sign**	**Min Value**	**Typical Value**	**Max Value**	**Instruction**
PIN1	red	V_{CC}	I	V_{CC}	8V	9V	10V	Power
PIN2	black	GND	I	V_{SS}	-	-	-	Power
PIN3	yellow	Program1	I	Vi-H Vi-L	V_{CC}- 0.2V-	V_{CC} GND	V_{CC}+0.2V V_{SS}+0.2V	Select mode
PIN4	purple	Program2	I	Vi-H Vi-L	V_{CC}- 0.2V-	V_{CC} GND	V_{CC}+0.2V V_{SS}+0.2V	Select mode
PIN5	gray		O			±5V	±8V	Used to select magnetic emulation
PIN6	green	DATA1	O	V_o-H V_o-L	V_{CC}- 0.2V-	V_{CC} GND	V_{CC}+0.2V V_{SS}+0.2V	Wiegand output
PIN7	brown	DATA0	O	V_o-H V_o-L	V_{CC}- 0.2V-	V_{CC} GND	V_{CC}+0.2V V_{SS}+0.2V	Wiegand output
PIN8	white							
PIN9	blue	RS232	O			±5V	±8V	RS232
PIN10	orange	For Customer	I	V_i-H V_i-L	V_{CC}- 0.2V-	V_{CC} GND	V_{CC}+0.2V V_{SS}+0.2V	Connected GND to Light orange LED

Table 2. The function for each I/O pin (Li, 2010)

Fig. 6. I/O pins for SMC-R134 (Ni, 2009)

There are two types of output format: RS232 and Wiegand. Ni (2009) used RS232 format to establish the communication between RFID and PC. Li (2010) used Wiegand format to establish the communication between RFID and MCU.

4. The communication between RFID and PC, between RFID and MCU

4.1 The communication between RFID and PC

Ni (2009) used visual basic 2005 (VB 2005) as the software to communicate RFID with PC. To realize this objective, RS232 output format was used. The function of ReadExisting() was used to read the data sent by RFID reader. Before doing this, we need to establish the serial port object in VB 2005. The block diagram of establishing serial port object is shown in Fig 7.

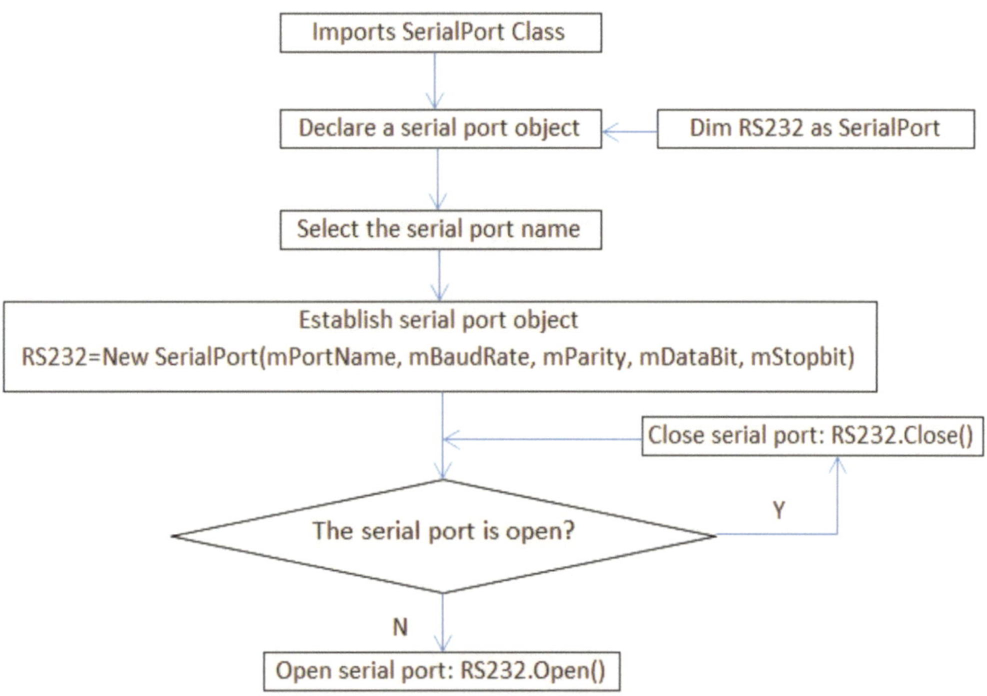

Fig. 7. The block diagram of establishing serial port object (Ni, 2009)

4.2 The communication between RFID and MCU

Li (2010) used Wiegand output format to establish the communication between RFID and MCU. This can realize the automatic control through only RFID and MCU without the help of computer. So the cost was minimized.

The software developed by Ni (2009) can be used as post processing tool to establish the dairy cow information system. The cow information, such as, ID number, weight, age, milk production, were recorded into the database. The ID number was gotten through the communication of RFID and PC. Then these ID numbers can be exported to the MCU to establish the communication between RFID and MCU.

Li (2010) imported two external interrupts to read the tag number. Once the tag number is matched with one of the ID stored in MCU, the single cow will be identified, and the cow's information will show in the LCD screen. The block diagram of communication between RFID and MCU is shown in Fig 8.

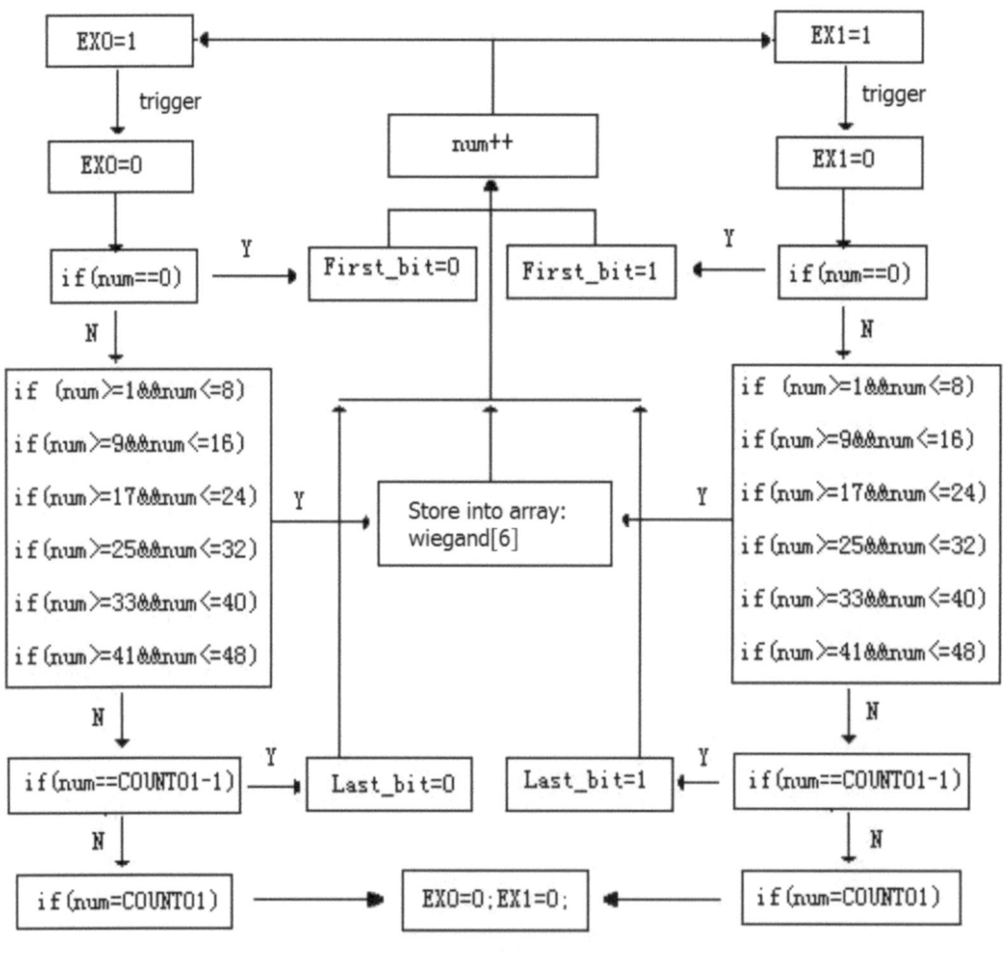

Fig. 8. The block diagram of communication between RFID and MCU (Li, 2010)

5. The good effect due to the technology (experiment)

Gao (2008) invented one kind of intelligent precise feeding machine. This machine was used in the research of Ni (2009) and Li (2010). This machine was equipped with the SMC-R134 reader. And the experimental cow was worn an ear tag (SMC-E1334). The feeding machine is shown in Fig 9.

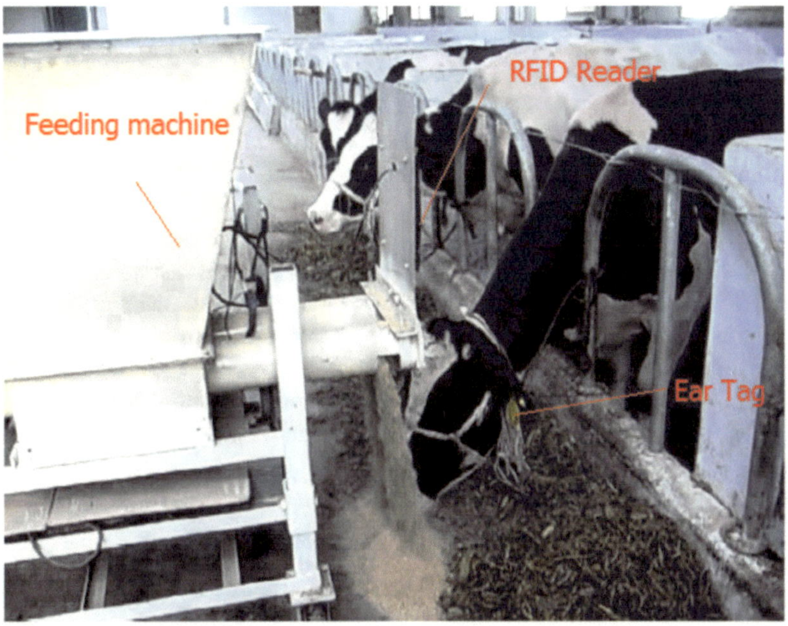

Fig. 9. Feeding Machine (Ni, 2009)

Ni (2009) did a basic experiment using this machine. Ten dairy cows were fed for one month. The concentrated feed was given by this machine based on the cow information (ID number, weight, age, milk production, etc). The result showed that the milk production can be added 4kg per day per cow.

Li (2010) did a deep experiment using this machine. 70 dairy cows were used. Besides the milk production, milk fat content and protein content were also be evaluated. The improved milk production is 3.9 kg, the average milk fat content is 3.74%, and the average protein content is 2.98%.

6. Summary

In this chapter, we introduced the application of RFID in daily cow industry. Firstly, we gave a brief introduction of why the identification of single daily cow is needed. By using RFID technology, the single daily cow information can be stored in database system. Through the tag ID, we can know the information about the cow. Later, we introduced one kind of RFID device used in the research. The communications between RFID and PC, RFID and MCU were established. Finally, two experiments based on the machine invented by Gao (2008) were introduced. The experimental results was good. The milk production were improved about 4kg per day for per cow.

7. Abbreviations and symbols

- RFID: Radio Frequency Identification
- PC: Personal Computer
- MCU: Micro Controller Unit

8. Reference

[1] Zhijiang Ni. 2009. *Design and Experiment on Intelligent Moving Precise Feeding Machine for Single Dairy Cow*. Master's dissertation. China Agricultural University, Beijing, China.

[2] Zhijiang Ni. 2009. *Intelligent Feeding Information System for Single Dairy Cow*. Software. China Agricultural University, Beijing, China. Software Register Number: 2009SRBJ0356.

[3] A. S. Voulodimos, C. Z. Patrikakis, A. B. Sideridis, V. A. Ntafis, E. M. Xylouri. 2010. *A complete farm management system based on animal identification using RFID*. Computers and Electronics in Agriculture, Volume 70, Issue 2, Pages 380-388

[4] C.M. Roberts. 2005. *Radio frequency identification* (RFID). Computers & Security, Volume 25, Issue 1, February 2006, Pages 18-26

[5] Jicheng Li. 2010. Design and Experiment on Auto-Control System of Dairy Cow Precising Feeding Equipment based on MCU. Master's dissertation. China Agricultural University, Beijing, China.

[6] Zhenjiang Gao, Zhijiang Ni, Za Kan. 2008. *One kind of intelligent precise feeding machine. China patent*. Patent id: 200810239151.

9

The U.S. National Animal Identification System (NAIS) & the U.S. Beef-Cattle Sector: A Post-Mortem Analysis of NAIS

Rhonda Skaggs
New Mexico State University
United States of America

1. Introduction

The appearance of bovine spongiform encephalopathy (BSE) in the United States in late 2003 resulted in severe economic impacts to the U.S. livestock sector. U.S. exports of beef and live cattle were immediately embargoed by importing countries as a result of BSE, and markets have not fully recovered eight years later. The trade status of the U.S. beef and cattle sectors was severely harmed when trading partners used BSE as justification for increased protectionism. The trade response to one BSE-infected cow and the desire to protect the U.S. livestock industry's economic interests enhanced concerns about intentional and accidental disease outbreaks. The first BSE-infected cow identified in the United States and ongoing fears that a virulent disease (foot and mouth disease, in particular) could cost billions and destroy the U.S. livestock sector led many people to conclude that a nationwide individual animal identification system was necessary. As a result, the National Animal Identification System (NAIS) was set forth in early 2004 by a working group including both industry and government officials. The NAIS built on the National Animal Identification Plan initiated in 2002. The goal of the NAIS was nationwide 48-hour traceback of all livestock and poultry in the event of a disease emergency.

The Animal Health Protection Act (AHPA) enacted with the 2002 Farm Bill set the legal stage for the federal government to be involved in the national animal identification effort. The 2002 AHPA includes language that indicates the federal government's intention to expand regulation of livestock due to interstate commerce and related movements of pest or disease threats (O'Brien, 2006). The AHPA was interpreted as giving the U.S. Secretary of Agriculture the ability to prohibit all movement of livestock unless producers participated in the NAIS. The NAIS entailed three components: Premises registration, animal identification, and animal tracking. Premises registration was the assignation of a unique premises number to all facilities where animals are managed or held. Animal identification assigned a unique number to individual animals or lots in the case of animals that stay with the same group their entire lives. Animal tracking involved the collection of data for animal movements and the recording of those data in a central recordkeeping system which could be quickly and comprehensively accessed in the event of an animal health emergency.

A 2005 USDA document indicated that the NAIS would begin as a voluntary program, but would become mandatory in 2009 (United States Department of Agriculture - Animal and

Plant Health Inspection Service [USDA-APHIS], 2005). The USDA stated in a 2006 document that while the agency had the authority to make the system mandatory, it had chosen to make every component of NAIS voluntary at the federal level (USDA-APHIS, 2006a). In a 2008 report, the USDA designated cattle as the highest priority species with respect to NAIS implementation and presented revised timelines and benchmarks for NAIS progress by species (USDA-APHIS, 2008a). Implementation benchmarks for cattle were scaled down from previous NAIS documents, and the cattle implementation timeline was also extended. NAIS benchmarks were scaled back for other species, although not as much as for cattle.

In June 2006 the USDA published a document intended to provide guidance for "non-commercial" livestock producers and their position within the NAIS. This guide attempted to alleviate small-scale livestock producers' concerns about the system, stating that NAIS participation was voluntary and that the NAIS would "largely focus on commercial operations and animals" (USDA-APHIS, 2006b). Critics of NAIS quickly pointed out that many statements in the report were inconsistent with other NAIS documents regarding the government's plan to extend NAIS coverage to all livestock and livestock movements within the United States.

The federal government issued numerous grants and cost-shares to states and tribes as inducements for premises registration and spent more than $120 million in the process; however, at the end of 2009, only 36% of premises were registered nationwide (USDA-APHIS, 2010). Some states achieved higher levels of premises registration by tying it to other state-level licenses or programs. In September 2008, the USDA issued a memorandum which stated that premises registration would be mandatory for emergency disease management or for state or federal activities involving diseases regulated through the Code of Federal Regulations. Although this memorandum was cancelled in December 2008, the USDA maintained that the federal government has broad authority to assign premises identification numbers as part of their normal animal health program activities. Recent livestock disease outbreaks in some states thus have resulted in mandatory NAIS participation for affected producers.

In June 2009, federal funding for NAIS in its current form was dropped from the fiscal 2010 spending bill by the House Agriculture Appropriations Subcommittee, with House leaders indicating that no future funds would be available for the program unless USDA developed and implemented a mandatory NAIS. The USDA conducted numerous NAIS "listening sessions" throughout the country in 2009 and received many more comments on NAIS at the Regulations.gov website.

Since the inception of NAIS, the federal government has asserted that the future economic viability of the U.S. livestock industry rests on improved disease management through nationwide animal identification and traceability. However, over the last several years, many U.S. livestock producers raised concerns about the security and confidentiality of premises and animal data provided to the national system, increased liability on the part of producers as a result of traceback to the farm level, the costs of NAIS participation, and the overall feasibility of the system. Opponents of NAIS claimed it was unconstitutional, a violation of their property rights, inconsistent with religious beliefs, an invasion of their privacy, and a loss of freedom. They did not believe USDA's assurances that NAIS information would not be subject to Freedom of Information Act requests or that use of the information would be restricted to animal health emergencies. The 2009 "listening sessions" were dominated by NAIS opponents, with a small minority of session participants speaking out in favor of the system. The comments posted at Regulations.gov were nearly unanimous against NAIS.

In February 2010, the USDA announced that it was abandoning the NAIS (USDA-APHIS, 2010). The agency indicated that it was going to "revise prior animal identification policy and offer a new approach to achieving animal disease traceability" (USDA-APHIS, 2010). The new approach will apparently only apply to animals moving interstate, although the operational details of the approach have yet to be developed. The agency's February 2010 Factsheet also stated that the new approach intends to "help overcome some of the mistrust caused by NAIS."

For almost a decade, proponents maintained that NAIS would protect producers' animals, investments and neighbors, and that "as producers become increasingly aware of the benefits of the NAIS and the level of voluntary participation grows, there will only be less need to make the program mandatory" (USDA-APHIS, 2006a). The USDA stated that NAIS would help protect U.S. livestock and poultry from disease spread, maintain consumer confidence in the food supply, and retain access to domestic and foreign markets (USDA-APHIS, 2007). In 2010, the federal government was forced to admit that arguments in favor of NAIS had fallen flat with a large segment of U.S. livestock producers.

The cattle industry was designated by the USDA as having the highest priority for full NAIS implementation; however, the cow-calf portion of the beef cattle sector was very resistant to NAIS (evidenced by continuously extended timelines and increasingly modest benchmarks for implementation). The economic, structural, and socio-cultural reasons for cow-calf producer resistance are the subject of the rest of this paper. If future livestock disease traceability efforts in the United States are to be successful (and disease catastrophes are to be avoided), it is absolutely essential that the context of cow-calf producer resistance to NAIS be fully understood. The objective of this paper is to describe the context and implications for the post-NAIS traceability framework.

2. Overview of U.S. agriculture and the beef-cattle sector

The history of U.S. agriculture is dominated by a relentless march toward increased concentration. Ever fewer numbers of farms are producing an ever larger percentage of total agricultural output. Of the 2.2 million farms enumerated in the 2007 Census of Agriculture, 10% generate almost 85% of the value of all agricultural sales (United States Department of Agriculture - National Agricultural Statistics Service [USDA-NASS], 2009). The remaining 90% of farms are responsible for 15% of output value. U.S. agriculture wasn't always this concentrated and much of the history of U.S. settlement and economic development is one of smallholders supporting their household through agricultural production, while generating a small marketable surplus. Technological changes occurring throughout the 19th and 20th centuries worked to increase productivity and drive down per unit production costs; new lands and resources were brought into production, and real prices for agricultural commodities plunged. As the relative purchasing power of raw agricultural commodities decreased, so did farm household incomes. Extreme structural upheaval occurred, many farms failed and millions of farm families exited agriculture. Their land was subsequently absorbed by survivor farms which grew larger. The remaining farms were successful as long as they managed to stay on the technology treadmill or otherwise survive decreasing real prices for their products. Consequently, many farm households now achieve acceptable income levels as a result of non-farm income sources. One-third of all U.S. farms have consistently negative net farm incomes and nearly 83% of total national farm household income in 2004 originated from off-farm sources (Hoppe et al., 2007). At first glance, it would seem that negative net farm incomes should prompt continued outmigration of people and resources from agriculture. But, it isn't happening.

U.S. farm-level commodity production is very diverse although 98% of U.S. farms are family farms, organized as proprietorships, partnerships, or family corporations that do not have hired managers (Hoppe et al., 2007). U.S. family farms range from small limited resource operations, to the extremely large industrialized farms that account for the majority of farm-level production. The USDA estimated that in 2004 57% of U.S. farms were retirement or residential/lifestyle farms, and that these farms' off-farm income as a share of total household income was 98% (Hoppe et al., 2007). According to the USDA, rural-residential farms account for only 7% of the value of production and include 35% of farm assets (including land). Small farms of all types, defined as having annual sales of less than $250,000, are 90% of farms, generate 25% of production value, and hold 68% of farm assets. Small farms, and especially retirement and residential/lifestyle farms, tend to specialize in the production of beef cattle, primarily cow-calf enterprises (Hoppe et al., 2007). There are several economic reasons for this specialization, including lower labor and management intensity (desirable to operators who are retired or who hold full-time non-farm jobs), relatively low cash costs of beef cattle production, and favorable tax treatment.

Productivity gains in U.S. agriculture over the last century have been astounding. However, the beef cow-calf industry is a notable exception to the productivity increases which characterize agriculture overall. This is due to the biological limitations of bovine reproduction. The rate of reproduction in cattle continues to be stable and low, with one cow rarely producing more than one calf. Natural twin production continues to be an unusual occurrence in beef cattle herds, and often results in extra production costs and/or sterile female offspring. By comparison, the U.S. hog industry has been characterized by steady increases in piglets/litter and litters/sow/year. Genetic advances and the adoption of industrialized confinement production by the hog industry in the post-World War II era led to dramatic increases in productivity, decreases in real hog prices, and industry concentration. The lack of equivalent productivity gains in beef cattle production are reflected in the much less drastic decrease in the real purchasing power of the calf commodity over the last half century, and an unconcentrated cow-calf sector.

The nature of the bovine digestive system also has contributed to relatively low productivity gains and limited adoption of capital and management intensive technologies in U.S. cow-calf production. Land-extensive calf production processes continue to be used in much of the cow-calf sector because the beef animal functions as a scavenger, using and transforming low value forages produced on marginal lands into a higher-valued product. Land-extensive production processes are generally not compatible with management intensive technologies, adoption of which is driven by the need and opportunity to increase returns per unit of capital and management input.

Most of the advances in technology and increases in efficiency in the beef industry have occurred beyond the farm gate at the feeding and packing levels. The feedlot and meat packing sectors have dramatically increased in size and concentration to achieve economies of scale. The beef feeding sector is increasingly dominated by a small number of extremely large operations, while the four largest beef packers controlled 84% of the market in 2007 (Hendrickson and Heffernan, 2007).

The beef cow-calf sector is the foundation of the beef cattle industry. Cow-calf production is not concentrated, dispersed nationwide, and occurs in every state, with an estimated 33 million national beef cow inventory living on almost 765,000 farms and ranches (USDA-NASS, 2009). Cow-calf operations produce the calves (or the animal frames - including skeleton, internal organs, and hide) upon which the cattle feeding sector accumulates meat using higher energy feed resources (usually under confinement conditions).

The USDA's National Animal Health Monitoring System (NAHMS) divides cow-calf producers into three groups: Those who have cow-calf herds primarily for income objectives (14% of producers), those whose beef cow-calf operation is a supplemental source of family income (72%), and those who keep cattle for some reason other than for providing family income (e.g., pleasure) (14%) (USDA-APHIS, 2008b). Differences in management practices for calving, animal health, feeding, marketing, and record keeping for different types of cow-calf operations are statistically significant and strikingly obvious in the NAHMS survey results (USDA-APHIS, 1998). Management of non-primary income herds is consistently less intensive, and productivity indicators for the herds are less favorable.

The technologies used in cow-calf production have not changed greatly over the last century, although some advances in cow-calf productivity have been made through selective breeding, use of veterinary pharmaceuticals, and improved forage management. Cow-calf production in the United States continues to be characterized by low entry costs, low cash production costs, low technology requirements, and low management intensity. Cow-calf operations also have lower exit probabilities than other farm enterprises because of their compatibility with off-farm work (Hoppe & Korb, 2006).

The technological stability of the U.S. cow-calf industry is evidenced by the small change in the average size of a U.S. beef cow herd over the last ~30 years (it went from 40 in 1974 to 43 in 2007) (USDA-NASS, 2009). By comparison, the average size of a U.S. milk cow herd went from 26 in 1974 to 133 in 2007. Nationally, almost 80% of U.S. beef cow-calf operations have fewer than 50 cows with these farms accounting for 29% of the country's beef cow herd.

Most research exploring U.S. cow-calf producers' motivations has been conducted in the West by investigators interested in rangeland management and public land policy issues. For example, the desire to have a rural lifestyle was found to inflate the value of farms and ranches in the West (Gosnell & Travis, 2005) while a relatively small percentage of ranchland value can be explained by livestock income in the Southwest (Torell et al., 2005). Gentner & Tanaka (2002) found that half of western public land ranchers earn less than 22% of their total income from ranching, that a ranch business "profit motivation" is a relatively low-ranked objective for all types of ranchers, and that public land ranchers are strongly motivated to be in ranching for tradition, family, and lifestyle reasons (i.e., consumptive objectives). Similarly, Cash (2002) noted that most U.S. beef cattle producers are not actually in the business of farming.

The multiple roles of livestock in traditional societies have long been recognized by anthropologists, human ecologists, and other social scientists. In traditional societies, livestock are mobile stores of wealth and status. And even though the United States has a very advanced economy, cattle continue to be viewed as "banks-on-the-hoof" by cow-calf producers (Eastman et al., 2000), who say that when they "need the money" is a key factor in determining when they market their cattle (Lacy et al., 2003). For many cow-calf producers, cattle and the land used to produce them are investments, savings, and financial safe-havens. Cattle provide emergency funds, and are also a stable supply of high quality meat for family consumption. Similar to their counterparts in traditional societies, cattle are also a source of identity and a cultural touchstone for many U.S. cow-calf producers. Pope (1987) concluded that "romance, recreation, the achievement of a desired social status, or simply the maintenance of a family tradition" are the primary motives for many western U.S. cattle producers. Identity objectives are financially feasible, compatible with other lifestyle and household objectives, and are encouraged by the nation's tax system. Lifestyle goals, particularly the desire to live in the country, were the most highly ranked strategic ranch goals among small-acreage livestock producers interviewed by Rowan (1994).

Technological advances, structural adjustment in response to technology, economies of size, and the wringing out of cultural identity objectives have not occurred at the cow-calf producer level as they occurred throughout much of U.S. agriculture in the 20th century. As a result, household-level cow-calf production has maintained more of its traditional economic, social, and cultural character than any other geographically dispersed agricultural commodity sector in the United States today.

3. The NAIS pushback

The trend of fewer numbers of ever-larger beef feeding and packing operations throughout the United States has led many cow-calf producers to be concerned about the structure of the overall beef industry, the negative effects of downstream concentration, and their belief that they are at the losing end of the structural change. Many believe that prices received by cow-calf producers are depressed as a result of non-competitive market behavior by feeders and packers. Domestic cow-calf producers feel threatened by the market impacts of imported feeder cattle from Mexico and imported fed cattle from Canada. Live cattle imports are viewed favorably by a majority of feeders and packers, who generally welcome the flow of the animals into the U.S. market. Many in the cow-calf sector vigorously promoted country of origin labeling (COOL) for U.S. beef. COOL was opposed by feeders and packers as a result of their integration with the rest of the North American as well as the global cattle-beef markets.

The schism between the cow-calf sector and the feeding and packing sectors led to the creation of a new industry lobbying group, the Ranchers-Cattlemen Action Legal Fund, United Stockgrowers of America (R-CALF USA). R-CALF consistently appeals to cow-calf industry fears about trade liberalization and global market integration, property rights erosion, loss of freedoms, and invasions of privacy. R-CALF was opposed to the NAIS. The National Cattlemen's Beef Association (NCBA) represents cow-calf producers, as well as feeders and packers. In the view of R-CALF, the NCBA and the United States Department of Agriculture do not represent the interests of "independent cattlemen." The NCBA publishes *Beef Magazine,* was very supportive of the NAIS, and was a key player in the effort to establish a centralized, NCBA-affiliated, privately held database for animal tracking information. In 2005 *Beef Magazine* reported that 76% of survey respondents said a national system of individual animal ID and traceback was needed for health monitoring purposes, and 63% indicated such a system should be mandatory. According to the magazine, 83% of cattle producers who responded to their survey individually identify their cattle and 12% use electronic ID tags. These results are very different from USDA NAHMS 2007-08 survey results, which found that 53% of U.S. cow-calf producers use no form of individual calf identification and less than 1% of producers use electronic ID technology (USDA-APHIS, 2009a). In 2006, the Cattle Industry Work Group (established by the USDA to develop NAIS guidelines and standards for the cattle industry) declared electronic ID technology (specifically, radio frequency identification (RFID)) as the technology to be used to individually identify cattle under NAIS (USDA-APHIS, 2006c).

Although originally conceived as a means to deal with animal health emergencies (zoonotic and otherwise), NAIS proponents and technology vendors consistently emphasized the valuable management benefits to producers from individual animal identification and performance record keeping (particularly in their RFID and electronic forms). NAIS proponents and technology vendors have assumed that management intensification and the tools to accomplish it are desired by producers. However, cow-calf production is an intrinsically low-management intensity activity. It is a land-extensive activity and one where it is often not desirable, necessary, or feasible for producers to increase management intensity or capital investments. NAIS proponents touted individual animal identification's role in maintaining international market access and cattle and meat trade flows. This justification has not been well received by cow-calf producers who believe international trade is a threat to their industry. In their opinion, shutting off beef exports would be a small price to pay for shutting off the live cattle imports with which they directly compete.

For the cow-calf sector, NAIS became an attempt to impose a technology mandate and modernization on an industry where cow reproductive limitations, producer household and personal objectives, and cattle's efficient use of low-value forage have limited and will continue to limit technology adoption and modernization. Much of cow-calf producer opposition to NAIS was founded on fears that they would pay for the NAIS while the feeding and packing sectors would benefit from animal tracking and performance information derived from the electronic data.

Cow-calf producers' fears about the costs of NAIS were confirmed in a 2009 USDA benefit-cost analysis of the system (USDA-APHIS, 2009b, 2009c). The analysis concluded that beef cow-calf operations would incur 79% of the total annual beef cattle industry cost of a fully implemented NAIS. Given existing economies of size, the cost of an individual cow-calf animal ID system with full traceability ranged from a low of $2.48 per head for the largest operations to a high of $7.17 per head for the smallest operations. These data supported NAIS opponents' long-running contention that NAIS would benefit large agribusiness at the expense of the smallest farming and ranching operations in the country.

4. Conclusion

A few years ago, the author of this paper was forcefully told by a USDA official that anyone who wanted to "produce or market cattle in the United States" would have to comply with NAIS. This official clearly did not recognize what a critical wedge issue NAIS would become within the U.S. beef-cattle industry. He and the broad complex of government animal health personnel, large agribusiness interests (particularly feeders and packers), and established industry associations failed to appreciate the deep distrust many cattle producers have of them. The proponents of NAIS also seem to have been unaware or dismissive of the deeply ingrained socio-cultural aspects of cow–calf production and traditional small-scale lifestyle agriculture in the United States. Although this paper focuses on the cow-calf sector, many traditional small-scale producers of other species objected to the NAIS using arguments similar to those of cow-calf producers.

Serious miscalculations by government officials about livestock producers and owners fed and strengthened grassroots-level resistance to increased animal health regulations. NAIS proponents in government and the private sector sent too many conflicting messages to NAIS skeptics. Official NAIS reports and documents that appeared on and disappeared from the USDA's website following criticism added to confusion, suspicion, and hostility regarding NAIS. As a consequence, new disease management risks have been created and the ability of the nation to effectively deal with real animal health emergencies has been compromised. The level of suspicion created by NAIS among traditional livestock producers led to an environment where, should a disease such as FMD arise in the United States, many producers will not respond as they should in a true emergency. Rather, they will suspect that a false emergency is being used to expand government control of their activities. Efforts to implement livestock movement control, quarantine, condemnation, and depopulation will be hampered and defied by some producers. Under these circumstances, disease outbreaks could be catastrophic for the entire nation.

The USDA appears to have recognized the suspicions and potential for civil disobedience within the livestock sector which resulted from the NAIS experience, as evidenced by official statement that the new animal disease traceability framework has trust issues to overcome (USDA-APHIS, 2010). However, memories of NAIS will negatively affect whatever form a federally-promoted traceability framework takes in the future. Cow-calf producers' distrust of federal regulation and their suspicions about relationships between large agribusiness NAIS supporters and the federal government are unlikely to moderate under any new federal traceability program. NAIS became part of the paranoia smaller (and many larger) producers feel about industry structure and market power relationships within the U.S. beef-cattle sector. The USDA's recent statements that the new traceability framework will apply only to animals moving interstate will not mollify many cow-calf producers, as the vast majority of beef calves produced in the United States cross state lines at some point in their lives (even if they are first sold "locally"). Specifically, the February 2010 statement from USDA-APHIS that small producers who sell animals "to local markets" will not be a part of the new disease traceability framework has yet to be operationally defined.

Unfortunately, much federal and state credibility has been lost in the rush to mandate a culturally insensitive, high technology, management-beneficial, and trade-oriented animal identification program. NAIS represented an enormous leap in government involvement in the beef cow-calf sector. From the beginning of NAIS, government was under the impression that it was dealing with an "industry"; however, much of U.S. livestock production is deeply grounded in culture and lifestyle. Expanded regulation of culture and lifestyle choices was an uphill battle for NAIS, and will continue to be so in the future. USDA's unsuccessful efforts to promote NAIS as a management tool and as a means for supporting trade carried little weight with the large percentage of non-management intensive, non-trade oriented cow-calf producers. These producers' concerns about competition from U.S. imports of feeder and fed cattle aren't going away simply because federal animal disease traceability efforts are being renamed.

Successful animal disease management in the future will require significant rebuilding of trust between state and federal animal health officials and grassroots-level producers. This will require that animal health officials credibly demonstrate their independence from large-scale agribusiness and from identification technology vendors.

Previous disease management and eradication programs (e.g., scrapie, brucellosis) haven't required producer investments in electronic eartags and other equipment. Furthermore, a comprehensive, nationwide, 48-hour traceback objective probably is infeasible under *any* existing and future technology and management assumptions, regardless of what technology vendors say.

The USDA-APHIS announcement that future federal animal disease traceability efforts will apply to animals moving interstate means that any new program is likely to have much in common with NAIS. A future federally-influenced traceability program will thus encounter resistance and disease management will be compromised because of the NAIS experience. The loss of federal credibility and increased mistrust of government which resulted from NAIS has made the United States beef industry vulnerable to trade barriers and protectionism. The U.S. beef industry needs international trade, and post-NAIS, also needs programs that assure the quality and safety of U.S. beef products to overseas buyers. The demise of NAIS and potential cow-calf producer resistance to future government-mandated traceability systems have created a vacuum that industry-driven quality assurance or process verification programs can fill. In the wake of NAIS, an industry-driven system that covers willing buyers and sellers and financially rewards specific attributes or processes will be more successful than government regulation at holding and growing international markets for U.S.-produced beef.

Even though NAIS was not implemented, animal disease hazards haven't disappeared. In their recent factsheet, the USDA indicated that post-NAIS animal disease management and traceability efforts will be led by the states and tribal nations (USDA-APHIS, 2010). NAIS-related damage control needs to be high on the agenda for state and tribal agencies responsible for animal disease management. Whatever reservoirs of trust grassroots livestock producers have for state- or tribal-level animal health agencies desperately need to be refilled before new or well-known pathogens emerge to threaten livestock or human health throughout the United States.

5. Acknowledgement

This research was supported by the New Mexico Agricultural Experiment Station, New Mexico State University, Las Cruces, New Mexico, USA.

6. References

Beef Magazine. (2005). What You Told Us. Vol.41, No.11(July 1). 11.03.2011, Available from http://beefmagazine.com/mag/beef_told_us/

Cash, J.A. (2002). Where's the Beef? Small Farms Produce Majority of Cattle. *Agricultural Outlook*. USDA Economic Research Service, pp. 21–24. 11.03.2011, Available from http://www.ers.usda.gov/publications/agoutlook/Dec2002/ao297g.pdf

Eastman, C., Raish, C. & McSweeney, A. (2000). Small Livestock Operations in Northern New Mexico. In: *Livestock Management in the American Southwest: Ecology, Society, and Economics,* R. Jemison & Raish, C., (Eds.), 523-554, ISBN 0-444-50313-7, Elsevier Science, Amsterdam, Netherlands

Gentner, B.G. & Tanaka, J.A. (2002). Classifying Federal Public Land Grazing Permittees. *Journal of Range Management,* Vol.55, pp. 2-11, ISSN 0022-409X

Gosnell, H. & Travis, W.R. (2005). Ranchland Ownership Dynamics in the Rocky Mountain West. *Rangeland Ecology and Management,* Vol.58, No.2, pp. 191-198, ISSN 1550-7424

Hendrickson, M. & Heffernan, W. (2007). *Concentration of Agricultural Markets April 2007.* 11.3.2011, Available from http://civileats.com/wp-content/uploads/2009/05/2007-heffernanreport.pdf

Hoppe, R.A. & Korb, P. (2006). *Understanding U.S. Farm Exits.* United States Department of Agriculture Economic Research Service Report #21. 11.3.2011, Available from http://www.ers.usda.gov/publications/err21/err21.pdf

Hoppe, R.A., Korb, P., O'Donoghue, E.J., & Banker, D.E. (2007). *Structure and Finances of U.S. Farms: Family Farm Report, 2007 Edition.* United States Department of Agriculture, Economic Research Service, Economic Information Bulletin #24. 11.3.2011, Available from http://www.ers.usda.gov/publications/eib24/eib24.pdf

Lacy, R.C., Little, R.D., Forrest, C.S., & Gregory, T.L. (2003). *Attitudes of Small Beef Producers toward Selected Production and Marketing Practices.* Mississippi Agricultural and Forestry Experiment Station Bulletin 1126. Mississippi State University. 11.3.2011, Available from http://msucares.com/pubs/bulletins/b1126.pdf

O'Brien, D. (2006). Animal Identification and the Next Farm Bill. University of Arkansas School of Law, National Agricultural Law Center. 11.3.2011, Available from http://www.nationalaglawcenter.org/assets/articles/obrien_animalid-newfarmbill.pdf

Pope, C.A. (1987). More Than Economics Influences the Allocation of Rangeland Resources. *Choices* 4th Quarter, pp. 24-25

Rowan, R.C. (1994). Are Small-Acreage Livestock Producers Real Ranchers? *Rangelands,* Vol.16, No.4, pp. 161-166, ISSN 0190-0528

Torell, L.A., Rimbey, N.R., Ramirez, O.A., & McCollum, D.W. (2005). Income Earning Potential versus Consumptive Amenities in Determining Ranchland Values. *Journal of Agricultural and Resource Economics,* Vol.30, No.3, pp. 537-560, ISSN 0162-1912

United States Department of Agriculture, Animal and Plant Health Inspection Service, Veterinary Services. (1998). *Importance of Income in Cow-Calf Management and Productivity.* 11.3.2011, Available from http://ageconsearch.umn.edu/bitstream/32792/1/info03.pdf

United States Department of Agriculture Animal Plant Health Inspection Service. (2005). National Animal Identification System Draft Strategic Plan 2005-2009. 11.3.2011, Available from http://xstatic99645.tripod.com/sitebuildercontent/sitebuilderfiles/nais_draft_strategic_plan_42505.pdf

United States Department of Agriculture Animal and Plant Health Inspection Service. National Animal Identification System. (2006a). *A User Guide and Additional Information Resources.* Draft Version. 11.3.2011, Available from http://www.aphis.usda.gov/traceability/downloads/NAIS-UserGuide.pdf

United States Department of Agriculture Animal and Plant Health Inspection Service. (2006b). The National Animal Identification System: A Guide for Small-Scale or Non-Commercial Producers. 11.3.2011, Available from http://www.nh.gov/agric/divisions/animal_industry/documents/Guide_Smal_NonCommercial_6_2_06.pdf

United States Department of Agriculture Animal and Plant Health Inspection Services, (2006c). National Animal Identification System Cattle Industry Work Group Report "Executive Summary." 11.3.2011, Available from http://xstatic99645.tripod.com/sitebuildercontent/sitebuilderfiles/cwg_executive_summary_9_5_06.pdf

United States Department of Agriculture Animal and Plant Health Inspection Service. (2007). *National Animal Identification System- A User Guide and Additional Information Resources*. 11.3.2011, Available from http://www.cdfa.ca.gov/Ahfss/Animal_Health/pdfs/NAIS/NAIS-UserGuide12-07.pdf

United States Department of Agriculture Animal and Plant Health Inspection Service. (2008a). A Business Plan to Advance Animal Disease Traceability Through the Harmonization of State, Federal, and Industry Programs and Convergence with the National Animal Identification System, Version 1.0. 11.3.2011, Available from http://www.aphis.usda.gov/traceability/downloads/TraceabilityBusinessPlanSept2008.pdf

United States Department of Agriculture Animal and Plant Health Inspection Service, Veterinary Services. (2008b). *Beef 2007-08 Part I: Reference of Beef Cow-Calf Management Practices in the United States, 2007-2008*. 11.3.2011, Available from http://www.aphis.usda.gov/animal_health/nahms/beefcowcalf/downloads/beef0708/Beef0708_dr_PartI_rev.pdf

United States Department of Agriculture Animal and Plant Health Inspection Service, Veterinary Services. (2009a). *Beef 2007-08 Part III: Changes in the U.S. Beef Cow-Calf Industry, 1993-2008*. 11.3.2011, Available from http://www.aphis.usda.gov/animal_health/nahms/beefcowcalf/downloads/beef0708/Beef0708_dr_PartIII.pdf

United States Department of Agriculture Animal and Plant Health Inspection Service. (2009b). Benefit-Cost Analysis of the National Animal Identification System, January 2009, 11.3.2011, Available from http://www.aphis.usda.gov/traceability/downloads/Benefit_Cost_Analysis.pdf

United States Department of Agriculture Animal and Plant Health Inspection Service. (2009c). Overview Report of the Benefit-Cost Analysis of the National Animal Identification System, April 2009 (slightly revised), 11.3.2011, Available from http://www.aphis.usda.gov/traceability/downloads/NAIS_overview_report.pdf

United States Department of Agriculture Animal and Plant Health Inspection Service Veterinary Services. (February 2010). *Questions and Answers: New Animal Disease Traceability Framework*. Factsheet. 11.3.2011, Available from

http://www.aphis.usda.gov/publications/animal_health/content/printable_version/faq_traceability.pdf
United States Department of Agriculture National Agricultural Statistics Service. (2009). *2007 Census of Agriculture.* 11.3.2011, Available from http://www.agcensus.usda.gov/Publications/2007/Full_Report/index.asp

10

Mine Planning Using RFID

Vladimir Konyukh
Novosibirsk State Technical University
Russia

1. Introduction

A mine is considered to be a plant that produces useful mineral with a given percentage of ore and given quantity, whereas the cost of mining is expected to be minimum. The places of extraction (faces) move in space and time in accordance with the extraction of a rock mass. Mine planning is used to plot variables such as, from which places and how much rock mass to extract, where a miner is working, and what the utilization of any machine is, what the cost of mining is. Geological conditions for mining were determined by nature. They are unpredictable. The environment in a mine is especially harsh: dirty, dusty, and damp. Conditions of mining change randomly all the time.

Many technologies are used in the mining. However, any technology needs a real time data. The data are necessary for a decision-maker to get information about extraction out of each part of a deposit, utilization of each machine, working time of each miner, etc. The information will be used to keep cost to a minimum. For example, a diagram of an underground ore mining is shown on the Fig.1.

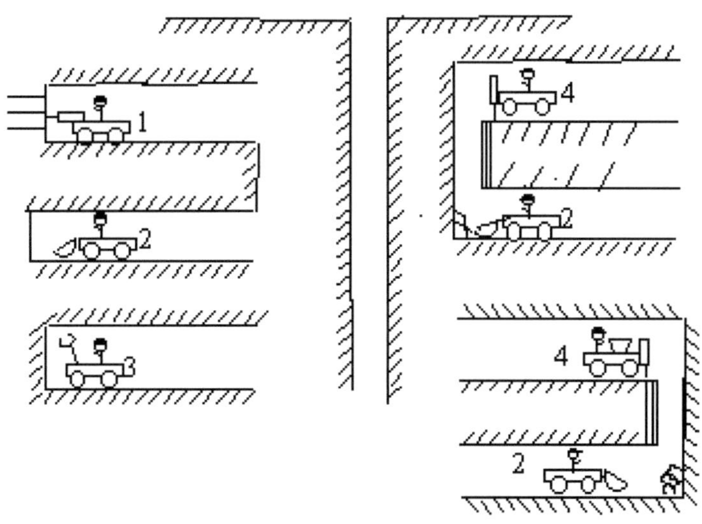

Fig. 1. An example of underground ore mining: 1-drilling machine; 2- loading-haulage-dumping machine; 3- concreting machine; 4- charging machine

The roadways are developed by drilling (1) and charging (4) machines to get access to deposits of ore. After blasting, a concreting machine (3) prepares a roadway. Other drilling and charging machines prepare an extraction chamber. After blasting the rock mass is transported by mobile loading- haulage-dumping machines with a bucket (2) on the distance 30-100 meters to a dumping place.
Many questions for management are not clear enough, such as:

- how much rock mass was delivered from each face;
- what the state of each face is ;
- what condition mine machines are ;
- how long each machine is working;
- who is the driver of each machine;
- what is the utilization of each machine in the mine ;
- where is each miner at present;
- what is the distribution of mine machines in the mine.

To answer such questions, reliable sources of information are necessary.

2. Problems of mine planning

2.1 Surface mining

The shovels, which extract a rock mass, are distributed in a space. Extracted rock mass is transported by trucks to refinery, storage or waste. For example, the layout of an open pit mine is shown in Fig.2.

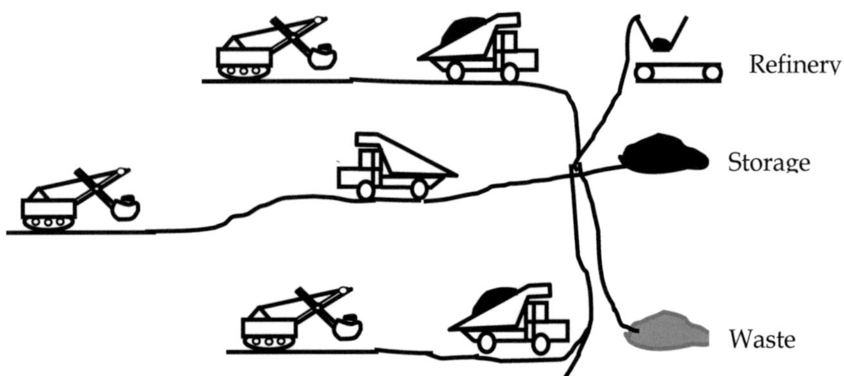

Fig. 2. Mining at an open pit mine

Both shovels and trucks are historically various and have a high cost. That is why the full utilization of shovels and trucks in the system " *N* shovels - *M* trucks" is actual for surface mining equipment. It is actuality for management to get current information about trucks and shovels to improve the work of the open pit mine. There were many efforts to determine placement of the trucks and shovels by using a Global Positioning System (GPS), to measure the load of a truck by vibration of body, and to identify a truck by bar-coding. Unfortunately, these ways are difficult and have many limitations for mine planning.
A dispatcher of an open pit mine would like to get the following information:

- ID of i-th ($i=1,M$) truck;
- ID of j-th ($j=1, M$) driver of the truck;
- ID of k-th ($k=1,N$) shovel;
- ID of k-th ($k=1, M$) of the shovel's driver;
- ID of the dumping place;
- current load of i-th truck;
- starting point of i-th truck;
- finish point of i-th truck;
- time of i-th truck's arrival from a known starting point;
- time of i-th truck's departure to a known finishing point;
- quantity of fuel in a i-th truck;
- what a condition of k-th shovel (waiting, loading of rock mass into a back, breakage) is;
- what a condition of i-th truck (movement with rock mass, movement without rock mass, refueling, breakage, loading, dumping, waiting) is;
- what number of trips has the i-th truck taken;
- what number of k-th shovel's buckets were carried by i-th truck.

If the truck situated in loading zone, the shovel loads the truck by several scooping. The number of scooping depends on many factors.
A total time of working cycle T_s for the each machine consists of two parts:

$$T_s = T_w + T_d,$$

where T_w=working time of k-th shovel;
T_d = idle time of k-th shovel.
Using this information, management could get the following indices to improve the activity of the open pit mine:
-the utilization of the i-th truck

$$K_i = \frac{\sum_{l=1}^{Li} T_{lw}}{\sum_{l=1}^{Li} (T_w + T_d)_l},$$

where L_i =number of trips for i-th truck;
T_w =working time of l-th trip for the i-th truck;
T_s= total time (including idle time) of l-th trip for the i-th truck;
-accumulated working time of i-th truck

$$A_i = \sum_{l=1}^{Li} T_l ;$$

-the utilization of the k-th shovel

$$K_k = \frac{\sum_{g=1}^{G}\sum_{v=l}^{V} T_{gv}}{\sum_{g=1}^{G}(T_w + T_s)_g},$$

where $g = (1,G)$-number of the shovel's working cycles;
$v = (1, V)$ = average number of shovel's buckets to load the i-th truck;
T_{gv} =working time for v-th loading cycle;
-accumulated working time of the k-th shovel

$$B_k = \sum_{g=1}^{G}\sum_{v=l}^{V} T_{gv};$$

-need for fuel for the i-th truck

$$Q_i = \sum_{l=1}^{L} q_l,$$

where q_l = consumption for fuel for the l-th trip;
-need for energy for the k-th shovel

$$E_k = \sum_{g=1}^{G} q_g,$$

where q_g = consumption for energy of the k-th shovel;
-distribution of energy between machines;
-cost of mining;
-quantity of rock mass that was extracted from various places in a deposit;
-placement of each person at mine;
-working time of each person at mine.
Thus, an on-board medium source must keep the following information about a truck (table 1).

Information	*Regularity*	*Use for mine planning*
ID of a truck	Shift	All trucks on the open pit mine
ID of a truck's driver	Shift	Assigning of a driver to a truck
State of a truck	Every hour	Utilization of a truck
Quantity of fuel inside a truck's tank	Every hour	Need for fuel for a truck
Working time of a truck	Current	Accumulated working time for maintenance
Load of a truck's body	Every trip	Accumulated quantity of extracted rock mass
Place of dumping	Every trip	Distribution of rock mass between a refinery, storage, and wastes
Number of a trip	Every trip	Comparison of a trucks' utilization

Table 1. Information about a truck

Other mobile objects, such as a drilling machine, must store and transfer various information for a dispatcher (at least the ID of a machine, its placement, its condition, and duration of its work). Current information about the placement of each working person is necessary for efficient management of the open pit mine.

2.2 Underground coal mining

The most widespread technology of coal mining is shown on the Figure 3.

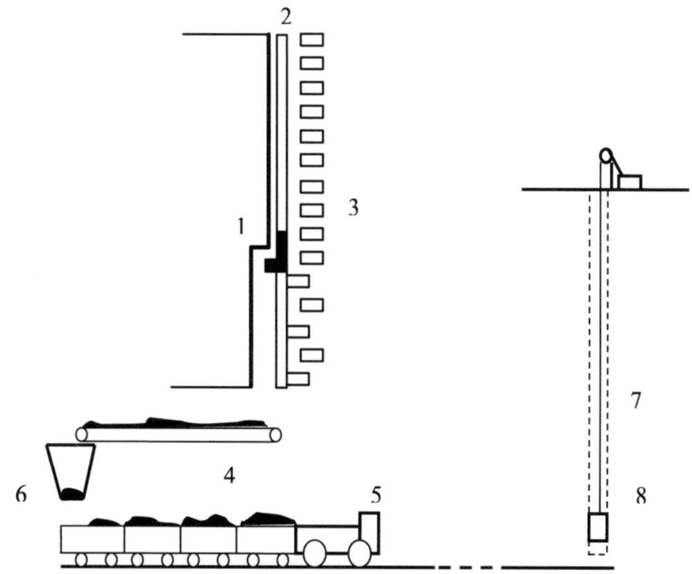

Fig. 3. Technology of coal mining at a fully - mechanized face

A shearer (1) extracts of a strip of coal, moving on a metal conveyor (2). At the same time, support units (3) are drawn up a face. The metal conveyor delivers extracted coal to a conveyor network (4). An underground train (5) is loaded under a bin (6).After loading an underground train transports the coal to a shaft (7). Then the coal is lifted by a skip (8) to the surface.

Many underground roadways for ventilation and transportation are inside an underground mine. Some of them are abandoned, some roadways are in development. All roadways form an underground network.

At present, some information about current work is transferred by a team-leader by telephone. Objective information in real time will improve mine planning (table 2).

Planning at the mine will be more effective because new information can be acquired on the basis of the initial data:

-utilization of j-th mobile machine at i-th face

$$K_j = \frac{\sum_{g=1}^{G} T_{wg}}{\sum_{g=1}^{G} (T_w + T_s)_g},$$

Information	*Regularity*	*Use for mine planning*
ID of the i-th (i=1; n) face	All the time	State of the mine
State of the i-th (i=1; n) face (work, stoppage)	Every hour	Re-distribution of faces
Quantity of coal, that was extracted out of the i-th face	Every shift	Comparison of faces, output of the mine
State of k-th machine (work, stoppage or breakage)	Every hour	Utilization of j-th mobile machine, timely repair
State work, stoppage, breakage of k-th stationary machine	Every hour	Utilization of k-th stationary machine, timely repair
Grade of the bin's filling	All the time	Utilization of the bin
Placement of the underground train: on the way to the dumping place, on the way to the loading place, in front of the dumping place, in front of the loading place	All the time	Control of transportation
Condition of the underground train: under a loading, under a dumping, movement, waiting, breakage	All the time	Control of transportation
ID of each miner during the working shift for the i-th face	Twice in the shift	Calculation of working time, identification in cases of accidents
ID of each miner in the various places of the mine	Twice in the shift	Calculation of working time, identification in cases of accidents
Placement of miners in the mine	Every hour	Calculation of working time, identification in cases of accidents

Table 2. The initial data about underground coal mining

where $g= (1,G)$-number of the machine's working cycles;
T_w =working time in g-th working cycle;
T_s= total time (including idle times) of the g-th working cycle;
-utilization of k-th stationary machine at i-th face

$$K_k = \frac{\sum_{g=1}^{G} T_{wg}}{\sum_{g=1}^{G} (T_w + T_s)_g},$$

where $g= (1, G)$-number of the working cycles;
T_w=working time for the g-th working cycle;
T_s=total time (including idle times) of the g-th working cycle;

-accumulated working time of the *j*-th mobile machine

$$A_j = \sum_{g=1}^{G} T_{wg};$$

-accumulated working time of the *k*-th stationary machine

$$A_k = \sum_{g=1}^{G} T_{wg};$$

-need for energy for the *i*-th face

$$Q_i = \sum_{k=1}^{K} q_k + \sum_{j=1}^{J} q_j ,$$

where K= number of stationary machines in the *i*-th face;
J=number of mobile machines in the *i*-th face;
-distribution for energy between faces;
-time-table of trains' movement,
-load of a train,
-number of trains for each placement of mining
-cost of mining;
-quantity of rock mass, that was extracted out of various places of a deposit;
-working time of each person at the mine.
Managers of the mine will be able to organize mining at a minimum cost.

2.3 Underground ore mining

The widespread technology of ore mining by extraction chambers is shown in the Figure 4.

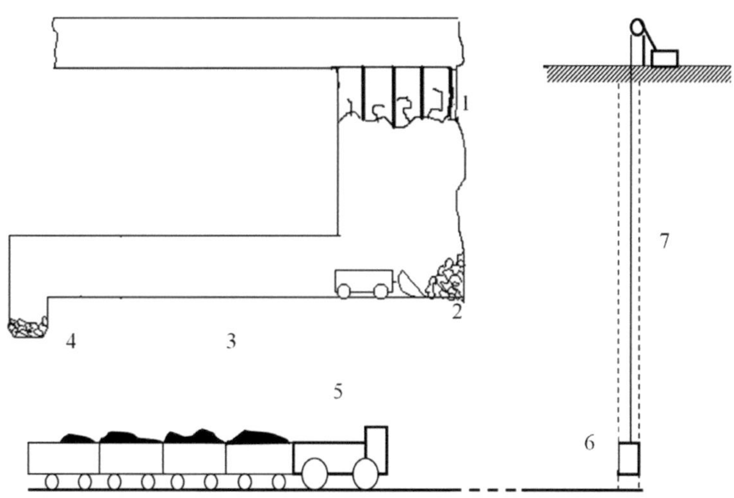

Fig. 4. Underground mining at an ore mine: 1- a roadway for ventilation and drilling machine; 2- a loading-haulage-dumping machine (LHD); 3- a roadway for transportation by LHD; 4- a dumping place; 5- an underground train; 6- a skip for lifting of rock mass; 7- shaft.

First, many vertical boreholes (40-60 meters long) are drilled from a drilling roadway (1) (Fig. 4). Then the boreholes are charged by explosive partially. After the blasting the ore mass drops to the bottom of a chamber. After that, a diesel Loading-Haulage-Dumping machine (LHD) (2) scoops the rock mass and transports it via roadway (3) for a distance 50-100 m to a dumping place (4). Finally, an underground train (5) transports the ore mass to the shaft (7). The rock mass is lifted by skip (6) to surface for refinery. A percentage of useful mineral for a chamber is variable. As a rule, one is known before.

The problem is- how to distribute extraction between the chambers to ensure the given percentage for ore mass on output of the mine. All of mobile machines are attached to the chamber. A behavior of a chamber without extraction is unpredictable.

The following data can be extracted during current work of the mine (table 3).

After preparation of the initial data, a manager of the mine can determine:

-utilization of *i*-th LHD-machine for the *j*-th extraction chamber

$$K_i = \frac{\sum_{g=1}^{G} T_{wg}}{\sum_{g=1}^{G} (T_w + T_s)_g},$$

where g= *(1,G)*-number of the machine's working cycles;

T_w =working time in g-th working cycle;

T_s= total time (including idle times) of the g-th working cycle;

-utilization of *k*-th train

$$K_k = \frac{\sum_{n=1}^{N} T_{wn}}{\sum_{n=1}^{N} (T_w + T_s)_g},$$

where n= *(1,N)*-number of trips for *k*-th train;

T_v =working time for the *n*-th trip;

T_f=time of the *n*-th trip;

-accumulated working time of the *i*-th LHD-machine

$$A_i = \sum_{g=1}^{G} T_{wg}$$

-accumulated working time of the *k*-th train

$$A_k = \sum_{n=1}^{N} T_{wn};$$

-need for fuel for the *i*-th LHD-machine

$$Q_i = \sum_{g=1}^{G} q_g,$$

where q_g=fuel consumption for g-th trip of *i*-th LHD-machine;

Information	*Regularity*	*Use for mine planning*
ID of the *i*-th (*i=1; n*) mobile machine	Start of the shift	Consideration of machines,
ID of the driver on *i*-th (*i=1; n*) mobile machine	Start of the shift	Consideration of drivers, permission for driving
Placement of the *i*-th (*i=1; n*) mobile machine	All the time	Control of mining
State of *j*-th mobile machine: work, stoppage or breakage	All the time	Utilization of the machine
State (work, stoppage, breakage) of *k*-th stationary machine	All the time	Utilization of the machine
Quantity of rock mass that was extracted out of a *j*-th chamber for the trip of the *i*-th (*i=1; n*) mobile machine	Each trip	Output of the *i*-th (*i=1; n*) mobile machine
Quantity of rock mass, that was delivered to a dumping place after a trip of LHD	Each trip	Calculation of rock mass
Total quantity of rock mass that was extracted out of a *j*-th extraction chamber	Finish of the shift	Output of the mine
Grade of the bin's filling	All the time	Utilization of the bin
State of the underground train : loading, movement to the shift , movement to the bin , dumping, breakage	All the time	Utilization of the train
ID of a miners in the various places of the mine	Start and finish of the shift	Calculation of working time, identification in case of accident
Placement of miners in the mine	Every hour	Calculation of working time, identification in case of accident
ID of a miners in the *j*-th face	Start and finish of the shift	Calculation of working time, identification in case of accident
Quantity of explosive that was expended for the j-th extraction chamber	After blasting	Need for materials
Percentage of dangerous gases inside the j-th extraction chamber	All the time	Danger warning
Placement of the underground train: on the way to the dumping place, on the way to the loading place, in front of the dumping place, in front of the loading place	All the time	Control of transporting
ID of the train on the mine	Start of the shift	Utilization of the train
State of the underground train: under a loading, under a dumping, movement, waiting, breakage	All the time	Control of transporting
Volume of fuel for *i*-th (*i=1; n*) mobile machine	All the time	Need for fuel for the *i*-th (*i=1; n*) mobile machine

Table 3. The initial data about underground ore mining

-distribution for fuel for LHD-machine between places of extraction;
-time-table of train movement,
-load of a train,
-dynamics of rock mass extraction out of the j-th extraction chamber;
-number of trains needed for each place of mining;
-cost;
-quantity of rock mass extracted from various places in a mine;
-working time of each person at mine.
As a result, the more effective work, e.g. a redistribution of LHD-routes can be organized according to the concentration of desirable minerals in various places of the deposit.

2.4 Delivery of supplies to distributed underground faces

The distributed faces need the various supplies from the surface to continue mining. At present, a number of supplies loads, such as support units, are delivered from surface storage to a face with the aid of a shaft dropping, transporting by underground rail train, and transporting by winches (Fig.5).

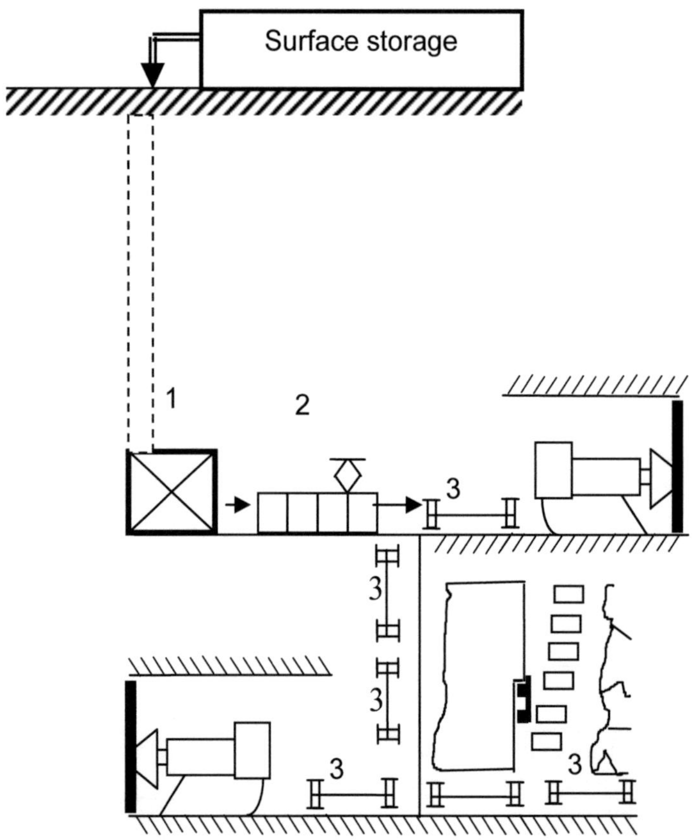

Fig. 5. Existing delivery of loads to distributed faces: 1- underground storage; 2-rail train; 3-winches

The disadvantage of such delivery is the long delay in delivery of supplies to a face. Besides, subjective mistakes for distribution of supplies between faces take place. Underground faces move all the time. A limited space and movement of a face do not permit to have an own storage for a face. Equipping of supplies by medium sources makes it possible, to ensure a face by necessary supplies in "Just-In-Time"- mode (Krieg, 2005).

3. Mobile objects in mining

Moving mine faces require mobile objects. Let's describe some peculiarities of such objects from the point-of-view of mine planning.
It is necessary to know where each miner is in real-time. Consequently, each miner must have an ID. Many ID-readers must be distributed on the pathways miners and connected to the surface via an information network. Mine machines and dangerous places must determine the ID of a miner. Every miner must wear a helmet (hard hat) (Fig. 6).

Fig. 6. A miner as a mobile object for mine planning

A helmet is the most suitable place for a medium source that must be cheap, light, and stable to harsh environment. Its capacity may be small. A mine has up to 1000 miners. In underground mining, every miner has a lamp with a battery pack in addition to a helmet.
A truck moves on the fixed road during surface mining (Fig.7). To make a decision about a distribution of trucks, a dispatcher would like to know ID of a machine, ID of a driver, a fuel need, time of loading, movement, and dumping, start and finish points of movement, current load, and placement on a pit mine, and state of a truck. A medium source will work in harsh conditions, with natural temperature, under metal environment. An open pit mine has until 50 trucks.

(a) (b)

Fig. 7. A truck for surface mining: a- without rock mass; b- with rock mass

There are many mobile machines for underground mining, such as a Loading-Haulage-Dumping Machine (Fig. 8).

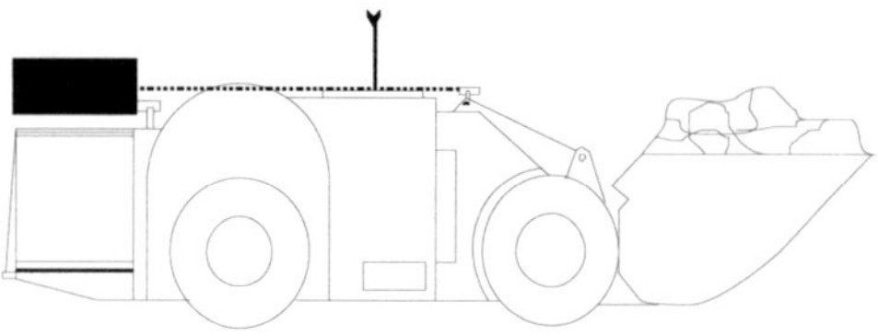

Fig. 8. Loading-Haulage-Dumping Machine as an example of mine machine for underground ore mining

An underground mine has up to 50 such machines. As a rule, a Loading-Haulage-Dumping machine has a diesel drive and rotating bucket.
A surface dispatcher would like to know ID of the machine, ID of the driver, current fullness of the bucket, the fuel need of each truck, current placement and state of each machine, time for each trip. An on-board medium source should work in metal environment and harsh conditions.

4. Identification of mobile objects

Like identification of mobile objects in industry, such decision is the obvious application of miners' identification (Wilma's, 2009). A miner has an own transponder, that is placed on a miner's helmet or on a battery pack (Fig. 9).

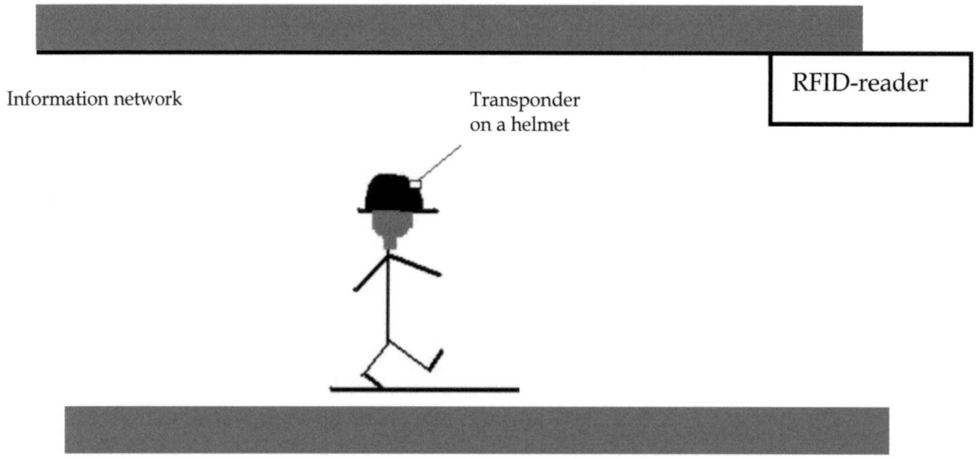

Fig. 9. Identification of a miner

A transponder can be used for identification of other mobile objects. ID of an object is written into a transponder. Stationary RFID-readers with RFID- antennas are connected with the information network and placed in various points of a working zone. If a mobile object moves near an RFID-antenna, the data about his ID and placement have been introduced into the information network (Fig. 10).

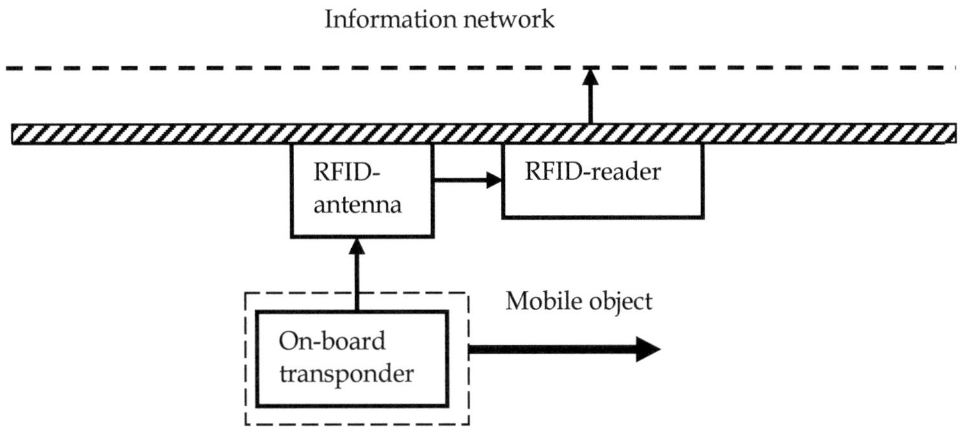

Fig. 10. Information accompaniment of a mobile object

It makes it possible to determine the time of arrival to working place; time of work's finish, placement of a miner at present; give permission for control of a machine.
Additional information can be derived on the basis of the data:

- how long did each miner work?
- where is a miner after his shift?
- how long was each machine used?
- by which miner was driven each machine?
- was access to the machine permitted for the miner?
- who is left in an emergency zone at present?

This information makes it possible to discover the placement of any miner, calculate his working time, and identify a miner in case of accident. The decision could be applicable also for other mobile objects in mining (Spadavecchia, 2007). An RFID-reader can read a vehicle's ID and switch a color-light signal in front of a crossroads. Many RFID-readers on the way of a vehicle can form its route.

5. Requirements to medium sources for mine planning

Many peculiarities of mining make special demands on medium sources.
A transponder for surface mining will be able to work in a natural temperature ranging from -50 ° C up to +50° C. A transponder for underground mining will be able to work in a temperature ranging from 0° C up to + 50° C.
Mostly an underground roadway is up to 4 meters wide. That is why the distance between a mobile object and an antenna is up to 3. 5 meters. The same distance is required for surface mining.
Mine environment is especially damp, dusty, and dirty. An on-board medium source will be able to work in metal environment with electromagnetic violence.
Mostly mobile machines work individually. That is why anti-collision prevention is necessary for miners mainly. However, sometimes it is necessary to determine, how many machines are situated together, e.g. inside a repair shop.
A transponder to identify a mobile object must keep information about the ID of the object only. It is enough to have a capacity app. 1 Kb.

A storage capacity to add information about current mining must be up to 200 Kb. Such information will be written up to 100 times per shift and will be kept at least for one month. The speed of mobile objects during underground mining is up to 1 meter per second. The data for mine planning are mostly constant. The distance between distributed stationary readers must be app.50 meters. The information network must be applicable for many readers, which could be work at the same time.

There are app. 150 mobile machines at a mine. A mine machine has the high cost. That is why; an on-board medium source can be expensive also. The information in a transponder must be protected against non-permitted access and non-permitted rewriting to avoid intentional falsification. A transponder in front of an emergency zone must be intended not only for a miner, but also for a mobile machine.

The acquired information must be complete to reduce cost, close down exhausted areas of deposit and mine new areas of a deposit, stabilize the percentage of useful mineral for a customer, plan the maintenance of mine machines, determine requirements of machines for development of mining, and etc.

6. A mobile mine machine as the data medium

If a mobile machine has an on-board RFID-writer and other mobile objects have on-board transponders, one can determine the way to continue a movement at underground crossroads, determine a miner or another machine in front of the location the mobile machine. It is difficult to get data out of moving working places. A mobile machine with an on-board sensor, transponder, and RFID-writer is suggested by the data on current mining. There are no limitations for the transponder's size.

We suggest using a mobile machine with RFID-writer that writes current information from an on-board sensor to an on-board transponder (Fig. 11).

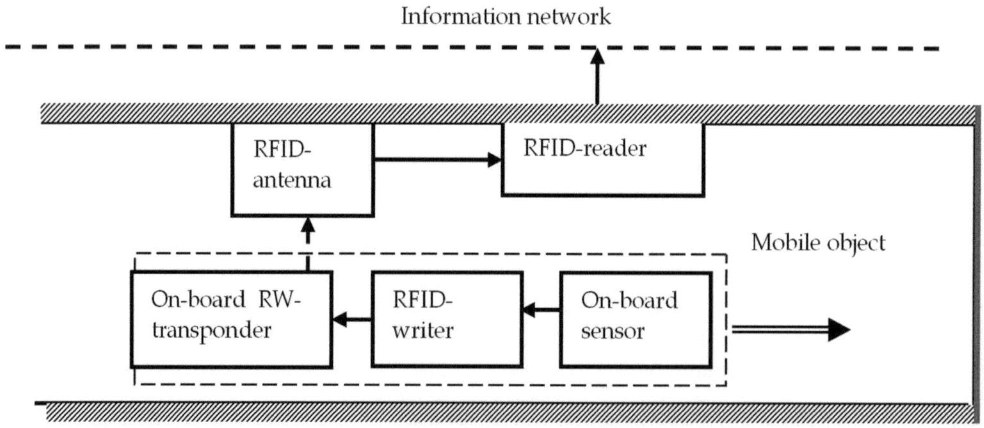

Fig. 11. Delivery of underground information from a mobile object to RFID-reader

A stationary RFID-antenna is placed on a wall of the roadway. The information from a mobile object is transferred to the RFID-reader, and then is introduced to the information network, which connects the distributed moving medium sources with an information centre on the surface (Fig. 12).

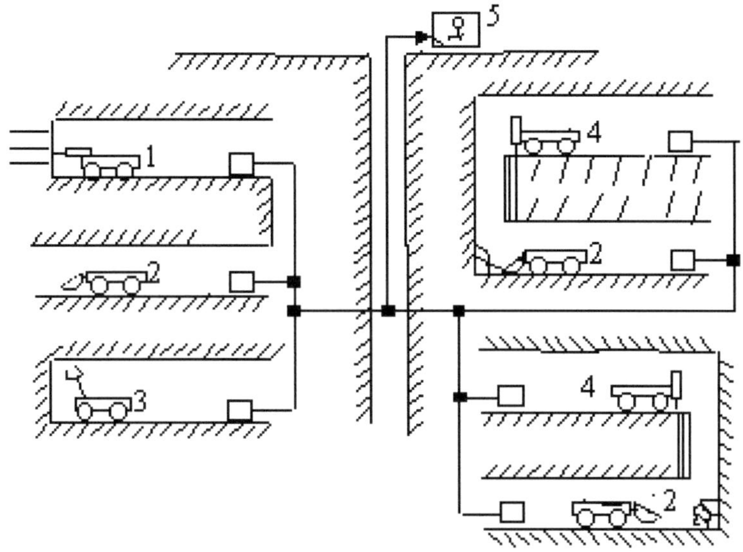

Fig. 12. Connection of underground RFID-readers with an information centre 5 on surface: 1-drilling machine; 2- loading-haulage-dumping machine; 3- shotcreting machine; 4-charging machine

The on-board RW-transponder can collect the data of various sensors to read not only current information but also track that accumulate a quantity of extracted rock mass. Besides, one must keep track of the ID of the various machines.

7. Reliability of primary information

To transfer primary information about current mining, a reliable information network must be introduced. Methodology of the volume balance can be used to have reliable information for mine planning.

$$\sum_{i=1}^{m} a_i = \sum_{j=1}^{n} b_j ,$$

where $\sum_{i=1}^{m} a_i$ = quantity of rock mass, that was extracted out of m distributed faces; $\sum_{j=1}^{n} b_j$ = quantity of rock mass, that was delivered to storages of mine.

This is estimated on the condition, that extracted rock mass has no addition or loss on the way to a dump.

Besides, the time from between the start of work at a face and appearance of each load at a dump must be calculated and compared with real time.

8. Total planning of mining for the mine

The structure of mine planning can be compared to a manufacturing planning. "Computer Integrated Manufacturing" is a plant that fulfills some orders of customers. One consists of standards "Manufacturing Resources Planning," "Manufacturing Execution System," "Supervisory Control And Data Acquisition," "Control," and "Input/Output" (Knuth, 2005). All standards are exchanged by information in real-time.

"Computer Integrated Mining" consists of "Mine's Resources Planning," "Mining Execution System", "Supervisory Control And Data Acquisition". Two lower standards "Input/Output" and "Control" are necessary for control only.

At first, orders for useful minerals are analyzed by management of mine on the standard "Mine's Resources Planning"
(Fig.13).

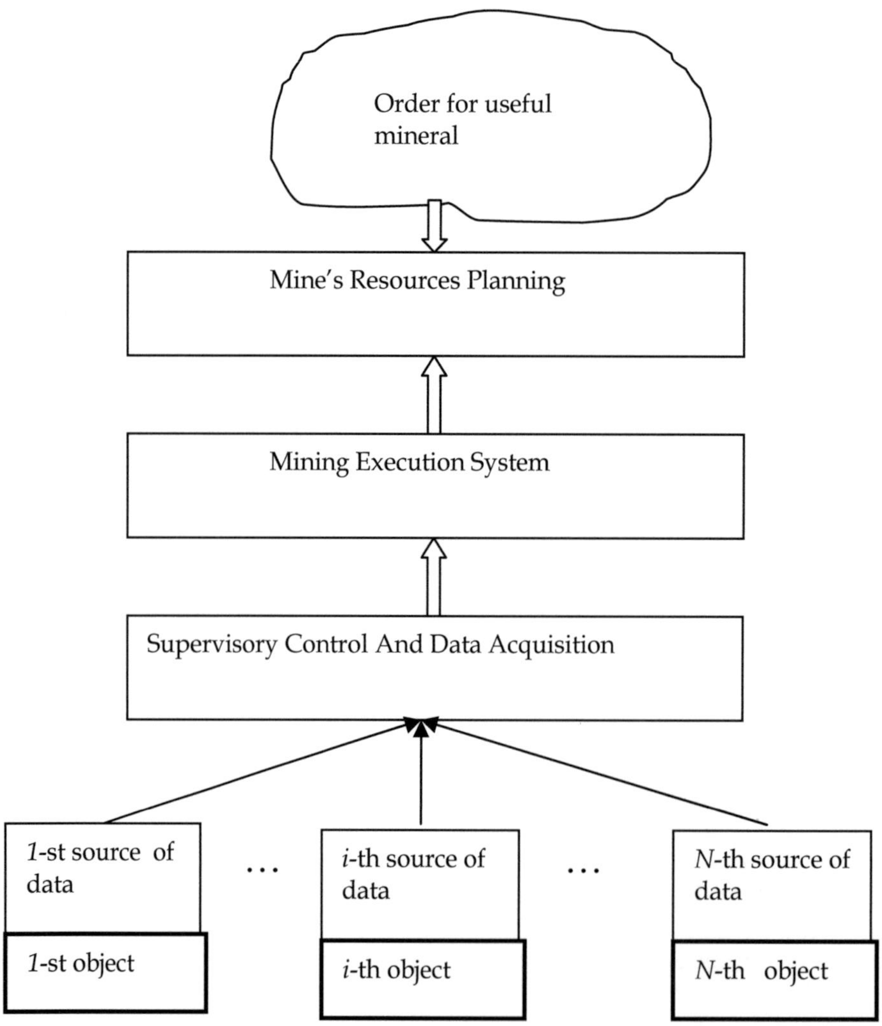

Fig. 13. Standards of mine planning

Managers on the standard "Mine's Resources Planning" evaluate the potentiality of the mine to fulfill a received order. They plan the development of the mine: existing places, abandoned places, and new places for mining. In addition, managers plan a changing of a machine, maintenance of mine machines, need for energy, need for materials, need for labor, etc. Decision-makers need various information about current mining (table 4).

Information	*Regularity*	*Effect for mine planning*
Current places of mining	A month	Evaluation of current state of mining
Abandoned places of mining	A month	Control of abandoned places
Perspective places of mining	A month	Development of the mine
Percentage of ore in extracted rock mass	A shift	Preparation of a refinery, discharge of a customer's order
Energy consumption for distributed places	A shift	Planning of need for energy sources
Distribution of materials consumption	A shift	Order of materials for distributed working places
Cost of mining	A month	Comparison with the current cost
Accumulated working time of a machine	Time of a working cycle	Planning of a energy consumption, maintenance, and replacement of a machine
Accumulated working time of a miner	A shift	Calculation of earnings
Orders of distributed places for materials	An hour	Distribution of materials

Table 4. Information for the standard "Mine's Resources Planning "

Energy consumption for *j*-th place of mining is calculated automatically as

$$E_j(T) = \sum_{i=1}^{n} E_i(T),$$

where $E_i(T)$ = energy consumption by *i*-th machine for the period *T*; n= number of machines for j-th place of mining.
Need for *k*-th kind of materials (explosive, cables, pipes, etc.) for the mine is calculated as

$$S_K(T) = \sum_{j=1}^{m}\sum_{k=1}^{K} S_{jk}\,.$$

Plans for the mine should be corrected according to current orders out of distributed faces.
Cost of mining is summarized as the share expenditures on working powers, energy, materials, amortization, etc.
Accumulated working time of a machine is necessary to replace machines workout:

$$T_i = \sum_{p=0}^{w} t_p\,,$$

where W= summary working time of *i*-th machine.

Accumulated working time of α-th miner is necessary to determine the earnings of α-th miner:

$$P_\alpha = \sum_{i=0}^{F} (p_i \times n),$$

where p_i= hour's earnings of α-th miner; n= (0;F) = number of working hours for α-th miner. Analyzing current information about the activity of the mine, a manager is able to re-distribute the mobile machines between places of mining, re-distribute teams of miners for a working shift, change the faces according to a current order, re-distribute the working shifts, etc.

9. Reflection of current mining

It is necessary for a dispatcher, to observe the current activity in a mine. To realize it, a Supervisory Control And Data Acquisition (SCADA system) is used. The existing SCADA-packages, such as In Touch®, can be adapted to the mine. The self-designed SCADA-system can be developed for the mine also. The development of the own SCADA-system is cheaper, but depends on developers of the mine.
There are static and dynamic information about current mining. Static information consists of the plan of the mine; the placement of stationary equipment, network of communications, placement of faces, some inscriptions. Dynamic information consists of current data about the mine: the state of any face (work or idle time), output of any face, state of any mobile object, current placement of any mobile machines; fullness of any bin. For example, the open pit mine is presented on a screen (Fig. 14).

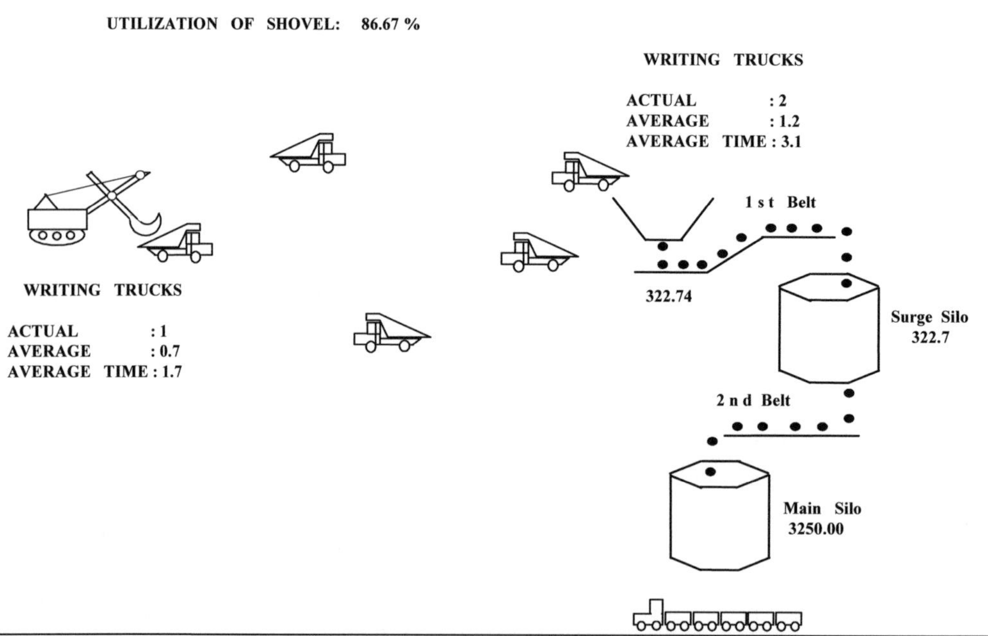

Fig. 14. Reflection of the open pit mine in USA (Sturgul, 1995)

Static information is the placement of shovel, silos, belts, railway, and inscriptions. Dynamic information is placement of trucks, state of the shovel, number of empty and loaded trucks, utilization of the shovel, time of the trip, filling of the silos, and load of the belts.
At first, static information must be constructed on a dispatcher's screen (table 5).

Information	*Details*	*Accuracy*
Scheme of the mine	no required	± 10 m
Places of loading	no required	± 10 m
Places of unloading	no required	± 10 m
Network of existing faces	no required	± 10 m
Network of abandoned faces	no required	± 10 m
Network of communications	no required	± 10 m
Transport network	no required	± 10 m
Placement of the stationary machines	no required	± 10 m
Various tables	standard	standard
Various inscriptions	standard	standard

Table 5. Static information for a dispatcher's screen

Then dynamic information about current time, output of the face, current plan's execution, pre-recognition of future accidents, and support of operative decisions in case of accidents is presented on a screen in real-time mode (table 6).

Information	*Regularity*	*Reflection*
State of a face	Every hour	Color of a face
Distribution of mobile objects	Every 15 minutes	Placement on the network
Output of a face	Every hour	Current data
Time of a working cycle	Each working cycle	Data
Output of the part of the mine	Each shift	Data
Fullness of every bin	Every 15 minutes	Full part of bin
State of the transport machine	Each trip	Color of a machine

Table 6. Dynamic information for visualization of current mining

Information is changed on a dispatcher's screen by introduction of global variables (by tags). Connection of medium sources with virtual reflection of mining is realized using OLE for Process Control (OPC).
The main rule for visualization is that the information must be enough to make a decision about improvement of current mining. For example, a decision-maker can compare the activity in various places of the mine.
Watching current mining information, a dispatcher can step and call the concrete persons, such as a team's leader to clear the matter up. The SCADA-system recognizes pre-accident situations in good time and notifies about beginning violations in normal work of the mine. If a random accident takes place, the SCADA-system produces recommendations to a dispatcher, who can prevent a deterioration of the situation , e.g. localize a random fire in various places of the mine.

As well as current information, the SCADA-system keeps detailed information about past mining, such as utilization of a mine machine. Comparison of current information with former information can improve the current mining.
Using this system, the information about total working time, expenses of energy, total output, utilization of mobile objects, and utilization of bins can be acquired for managers of the mine.

10. Mining execution system

The system is geared to control execution of shift planning and prepare information for the standard "Mine's Resources Planning ".
Sometimes mine equipment units have failures. Breakages lead to random refusals of a total technological chain.
Mining Execution System (MES) redistributes the faces and mine machines to ensure the same output of mine. The standard needs current information about mining (table 7).

Information	*Regularity*	*Effect for mine planning*
Output of a face	All the time	Contribution of a face to the mine's output
State of a face	All the time	Re-distribution of mining's places
Working time of a face	All the time	Fulfillment of a face's plan
State of a machine	All the time	Control of mining
Working time of a machine	All the time	Planning of maintenance
Placement of a machine	All the time	Planning of mining
Placement of miners	All the time	Planning of miners' distribution
Working time of a miner	All the time	Evaluation of miner's use
Fulfillment of a mine's plan	All the time	Evaluation of plan's fulfillment
Real time	All the time	Evaluation of the shift's time

Table 7. Information for "Mining Execution System"

Using this information, a mine dispatcher can determine how to maintain output during of unpredictable situations.

11. Suitability of RFID for mine planning

Optical character recognition needs comparison with a model. Random forms of objects, such as surge pile of rock mass make this impossible for mining. Infrared identification is not applicable for mining, because there is limited potential for a changing environment, requires the line of sight between a transmitter and receiver of information, needs comparison with a pattern. Bar coding has no protection to soiling and can not be attached by new information.
As a rule, voice sources of information are in use for mine planning. Voice sources are non-exact and non-reliable for mine planning.
Mobile data mediums on the basis of RFID produce many opportunities for mine planning. RFID- system can work under the harsh mine environment and does not require the light-of -sight between a transponder and a writer. Active transponders can be read at great distances. It is an obvious use of an RFID- system for identification and positioning of mobile objects.

Some mines introduce RFID to identify miners (RFID for Mining, 2008), like identification of goods in commerce. Many transponders can be read at once. Nobody can avoid being identifies before work. RFID-systems present the data in real time. It is impossible to forge information inside a transponder.
The possibility exists to add information and use machines to deliver data about working places in real time. Active transponders for mine applications may be smart. RFID- systems have no moving parts and do not require regular maintenance.
However, all miners must be informed in case of an accident. RFID may not be used to transfer accident information. The special design of RFID- system for a metal, dirty, and dusty environment is necessary. A mine must be equipped with an information network. Underground mines for coal mining require special permission to use RFID-system in an explosion-dangerous environment.

12. Towards intellectual mining

Deposits of useful minerals that were easily accessible for traditional mining are exhausted already. Historically, an underground mine is dangerous and unpleasant for miners. At present, the average depth of mines is 1200 meters. The deeper a mine is, the worse and more dangerous miners' work is and the more expensive miners' work is. The high temperature of the Earth's centre raises the temperature of the underground mine and it will be impossible to work.
It is too hard to co-ordinate underground mining actions in space and time. There are idle times of underground equipment owing to inadequate information about current mining. Employers waste a lot of money transporting miners for underground work.
The long-term dream of mining engineers is to be able to mine without underground miners. The main idea is - the control of underground machines from the surface (Fig. 15).

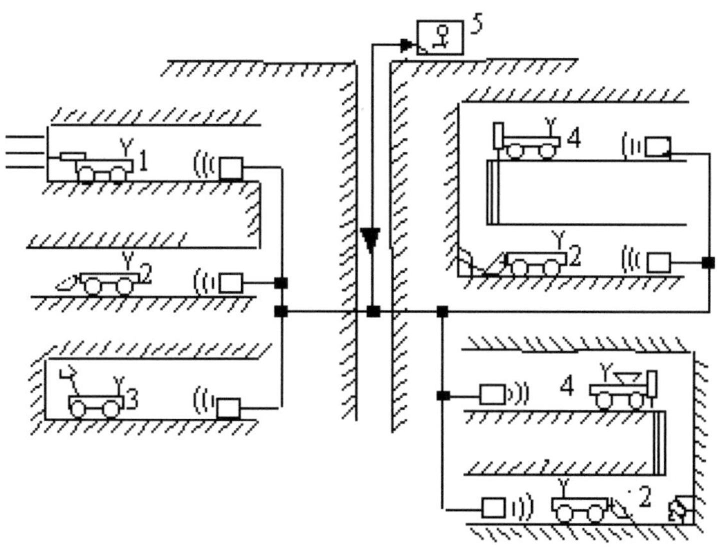

Fig. 15. Underground mining without underground drivers: 1-drilling machine; 2- loading– haulage-dumping machine; 3- shotcreting machine; 4- charging machine; 5- drivers' box

A console for remote control is situated in front of a working place. One is connected via an underground information network with the driver's box on surface. Mobile mine machines move along a guideline, which is placed in roadways. A driver observes a working place as if he is on a machine and transfers control commands to the machine. Each of the mine machines is equipped with an on-board receiver.

A broadband information network is the backbone of future mining. Such a network must transfer video, audio, and data information from distributed working places to the surface and back.

A machine in intellectual mine can adapt itself to changing working conditions: to change positions of working heads, direction of movement, step size of a roof support, and speed of a roof support. Such opportunities will make it possible to avoid some geological hazards, avoid dangerous rock pressure manifestations, stabilize the quality of mining, and increase the utilization of machinery. Existing information networks for voice exchange is not available for intellectual mining because the control of an autonomous machine in real-time needs a broad transmission band for video information.

Information network for a future mine could be used not only for remote control of underground machines, but also for mine planning using RFID.

As the long-term, an RFID-system for mining on other planets without direct visibility of a working place can be created.

13. System approach to use RFID for mine planning

The main idea of system approach consists of the creation of elements for the future system using step-by-step development. Each element will be included in a future system later without changes.

An RFID-system will be included in future mining that is based on control without direct visibility. How to transfer current information about mining to management of the mine? Many distributed working places are moving all the time during mining.

The existing information network in a mine was created for telephonic communication only which has a narrow communication band. Probably, transmission of data information via such a network will be incorrect for future mine planning.

A distributed information network for a future mine must transfer video information in real time mode to a remote driver. That is why one must connect moving transmitters with stationary receivers and be broad- band. Later, the network for future mining will be used for transferring information from on-board transponders without additional expense.

14. Need for research on the way to mine planning using RFID

It is necessary to test the RFID- system for the harsh mine environment that is metal, dirty, dusty, and damp.

An on-board RFID-writer for a suitable mine machine must be selected. One should have input for a sensor and output for the transponder. Existing telephonic network must be tested for suitability to transfer data information from the transponder.

The influence of random electromagnetic interference on RFID-system must be evaluated.

Placement of RFID-writer and RFID-transponder on a mine machine must be carefully chosen. The packages must be developed for each stage of mine planning. A human-machine interface must be developed for the visualization of current mining.

15. Conclusion

Mining has many peculiarities to get reliable information for mine planning. Environment for a data medium is humid, dirty, and dusty. Mine machines are metal. Working places are distributed in a space and move all the time. At present, RFID is used for identification of miners only, like identification of moving goods using EPC.
The connection of a sensor on a mobile object allows an RFID-writer to develop new potential for RFID-applications in mine planning.
Such a mobile data medium allows the gathering of various information: current reports about an extraction in various places of a deposit, placement of mobile objects during mining in real time, avoidance of non-permitted access to control, acquisition of full information about current mining, warning about emergency situations, and etc. An RFID-system can be used to visualize the placement of machines along roadways; to monitor miners with personal transponders; to prevent non-permitted control of machines; to give priority control of machines; to evaluate productivity of both machines and mining areas; to evaluate fuel consumption and machine resources. This information can be used for management of the mine.

16. Acknowledgment

This work is supported by the Russian Foundation of Basic Researches, grant № 10-08-01211-a "Modeling of mining on deep mines" and the State Program "Joining of Science and High Education in Russia for 2002-2006", grant № U0043/995 "Preparation of experts in information technologies for Kuzbass region". Many thanks to my old friends Prof. J.Sturgul and his wife Alison (Australia) for the thorough correction of English text.

17. References

Konyukh, V.; Tchaikovsky, E.& Rubtzova, E. (1988). Ways for the measurement of a LHD- bucket filling during extraction of ore out of dangerous places. *Physics-technical problems of mining,* No.2 (March-April1988), pp.67-73, ISSN 0015-3273 (in Russ.).

Konyukh, V. (2005). Achievements in industrial automation and their possible applications for underground mining, *Proceedings of 14-th Int. Symp. on Mine Planning and Equipment Selection (MPES2005),* pp. 645-661, ISBN 093-0-9968-835-9, Canada, Calgary, Sept. 16-20, 2005

Konyukh, V. (2010). Simulation of mining in the future, *Proceedings of IASTED International Conference on Control, Diagnostics, and Automation (ACIT 2010),* pp.1-6, ISBN 078-0-88986-842-7, Novosibirsk, Russia, June 15-18, 2010

Krieg, G. (2005). *Kanban-Controlled Manufacturing Systems,* Springer-Verlag, ISBN 3-540-22999-X, Berlin Heidelberg

Wilma's, C. (2009). Applying active RFID in mining, In: *Instrumentation and Control,* 1 Jan. 2009, Available from *www.instrumentation.co.za/papers/C9205.pdf*

Spadavecchia, O. (2007). RFID technology searching for more mining applications, In: *Mining weekly,* 13th April 2007, Available from www.miningweekly.com

RFID for Mining (2008). Available from www.falkensecurenetworks.com

Sturgul, J. (1995). Simulation and animation: come of age in mining , In: *Engineering and Mining Journal*, October 1995, pp.17-19.

11

The Applicability of RFID for Indoor Localization

Apostolia Papapostolou, Hakima Chaouchi
Telecom & Management Sudparis
France

1. Introduction

Although RFID has a relatively long history of more than 50 years in the field of wireless communications, only the last decade it has received a considerable attention for becoming a useful general purpose technology. Actually, RFID was initially developed as an automatic identification system consisting of two basic component types, a reader and a tag (Want, 2006). The reader is able to read the IDs of tags in its vicinity by running a simple link-layer protocol over the wireless channel. RFID tags can be either active or passive depending on whether they are powered by battery or not, respectively. Passive tags are prevalent in supply chain management as they do not need a battery to operate. This makes their lifetime large and cost negligible. The low cost of passive tags, the non-LOS requirement, the simultaneous reading of multiple tags and the reduced sensitivity regarding user orientation motivated the academia and industry for exploring its potentials in more intelligent applications Baudin & Rao (2005).

This chapter studies whether an RFID deployment can be applied for the purpose of indoor localization. It is widely accepted that location awareness is an indispensable component of the future ubiquitous and mobile networks and therefore efficient location systems are mandatory for the success of the upcoming era of pervasive computing. However, while determining the location of objects in outdoor environments has been extensively studied and addressed with technologies such as the Global Positioning System (GPS) (Wellenhoff et al., 1997), the localization problem for indoor radio propagation environments is recognized to be very challenging, mainly due to the presence of severe multi-path and shadow fading. The key properties of RFID motivated the research over RFID-based positioning schemes. Correlating tag IDs with their location coordinates is the principle concept for their realization.

Though RFID offers promising benefits for accurate and fast tracking, there are some technology challenges that need to be addressed and overcome in order to fully exploit its potential. Indeed, the main shortcoming of RFID is considered the interference problem among its components, mainly due to the limited capabilities of the passive tags and the inability of communication between readers (GP & SW, 2008). There are three main types of RFID interference. The first one is due to the responses of multiple tags to a single reader's query, the second is related to the queries of multiple readers to a single tag and finally, the third is due to the low signal power of weak tag responses compared to the stronger neighbor readers' transmissions. The first type affects the time response of the system, whereas the other two reduce the positioning accuracy. In addition, interference from non-conductive materials such as metal or glass imposes one more concern regarding the appropriateness of RFID for widespread deployment.

In this chapter, deploying cheap RFID passive tags within an indoor environment in order to determine the location of users with reader-enabled mobile terminals is proposed. The rationale behind selecting such configuration is mainly due to the low cost of passive tags, making their massive deployment a cost-effective solution. Moreover, next generation mobile terminals are anticipated to support RFID reading capabilities for accessing innovative tag-identifiable services through the RFID network. Three popular positioning algorithms are compared. The reason of their selection is because they can be all easily implemented on either the mobile or a central engine but they differ in their processing requirements. This chapter also studies the impact of several system design parameters such as the positioning algorithm, the tag deployment and the read range, on the accuracy and time efficiency objectives. Finally, mechanisms for dealing with these problems are also discussed.
The rest of this chapter is organized as follows: section 2 provides essential background for indoor localization and popular RFID positioning systems. In section 3 we explain the main shortcomings of RFID regarding localization which was our main motivation for conducting this study. In section 4 the conceptual framework of a RFID-based positioning system is described and section 5 provides simulation-based analysis results. Finally, in section 6 we give our main conclusions.

2. Background and related work

This section provides an overview of the indoor localization problem and a literature review in RFID indoor positioning systems.

2.1 Indoor localization

The localization problem is defined as the process of determining the current position of a user or an object within a specific region, indoor or outdoor. Position can be expressed in several ways depending on the application requirements or the positioning system specifications.
Localization using radio signals has attracted considerable attention in the fields of telecommunication and navigation. The most well known positioning system is the Global Positioning System (GPS) (Wellenhoff et al., 1997), which is satellite-based and very successful for tracking users in outdoor environments. However, the inability of satellite signals to penetrate buildings causes the complete failure of GPS in indoor environments. The indoor radio propagation channel is characterized as site specific, exhibiting severe multi-path effects and low probability of line-of-sight (LOS) signal propagation between the transmitter and the receiver (Pahlavan & Levesque, 2005), making accurate indoor positioning very challenging. For indoor location sensing a number of wireless technologies have been proposed, such as infrared (Want et al., 1992), ultrasound (Priyantha et al., 2000), WiFi (Bahl & Padmanabhan, 2000), (Youssef & Agrawala, 2005), (King et al., 2006), (Papapostolou & Chaouchi, 2009a), (Ubisense, n.d.), UltraWideBand (UWB) (Ingram et al., 2004), and more recently RFID (Hightower et al., 2000), LANDMARC, (Ni et al., 2004), (Wang et al., 2007), (Papapostolou & Chaouchi, 2009b).
Localization techniques, in general, utilize metrics of the Received Radio Signals (RRSs). The most traditional received signal metrics are based on angle of arrival (AOA), time of arrival (TOA), time difference of arrival (TDOA) measurements or received signal strength (RSS) measurements from several Reference Points (RPs). The reported signal metrics are then processed by the positioning algorithm for estimating the unknown location of the receiver, which is finally utilized by the application. The accuracy of the signal metrics and the complexity of the positioning algorithm define the accuracy of the estimated location.

Depending on how the signal metrics are utilized by the positioning algorithm, we can identify three major families of localization techniques (Hightower & Borriello, 2001), namely *triangulation*, *scene analysis* and *proximity*.

2.1.1 Triangulation

Triangulation methods are based on the geometric properties of a triangle to estimate the receiver's location. Depending on the type of radio signal measurements, triangulation can be further subdivided into *multi-lateration* and *angulation* method. In *multi-lateration* techniques, TOA, TDOA or RSS measurements from multiple RPs are converted to distance estimations with the help of a radio propagation model. Examples of such positioning systems include GPS (Wellenhoff et al., 1997), the Cricket Location System (Priyantha et al., 2000), and the SpotON Ad Hoc Location (Hightower et al., 2000). However, models for indoor localization applications must account for the effects of harsh indoor wireless channel behavior on the characteristics of the metrics at the receiving side, characteristics that affect indoor localization applications in ways that are very different from how they affect indoor telecommunication applications. In *angulation* techniques, AOA measurements with the help of specific antenna designs or hardware equipment are used for inferring the receiver's position. TheUbisense (Ubisense, n.d.) is an example of AOA-based location sensing system. The increased complexity and the hardware requirement are the main hindrances for the wide success of such systems.

2.1.2 Scene analysis/fingerprinting

Scene analysis or *fingerprinting* methods require an offline phase for learning the RRS behavior within a specific area under study. This signal information is then stored in a database called *Radio Map*. During the real-time localization phase, the receiver's unknown location is inferred based on the similarity between the Radio Map entries and the real-time RSS measurements. RADAR (Bahl & Padmanabhan, 2000), HORUS (Youssef & Agrawala, 2005), COMPASS (King et al., 2006) and WIFE (Papapostolou & Chaouchi, 2009b) follow this approach. The main shortcoming of scene analysis methods is that they are susceptible to uncontrollable and frequent environmental changes which may cause inconsistency of the signal behavior between the training phase and the time of the actual location determination phase.

2.1.3 Proximity

Finally, *proximity* methods are based on the detection of objects with known location. This can be done with the aid of sensors such as in Touch MOUSE (Hinckley & Sinclair, 1999), or based on topology and connectivity information such as in the Active Badge Location System (Want et al., 1992), or finally with the aid of an automatic identification system, such as credit card point of cell terminals. Such techniques are simple but usually suffer from limited accuracy.

2.2 RFID positioning systems

RFID positioning systems can be broadly divided into two classes: *tag* and *reader localization*, depending on the RFID component type of the target.

In *tag localization* schemes, readers and possibly tags are deployed as reference points within the area of interest and a positioning technique is applied for estimating the location of a tag. SpotON (Hightower et al., 2000) uses RSS measurements to estimate the distance between a target tag and at least three readers and then applies trilateration on the estimated

System	Target	Deployment	Approach	Accuracy
Hightower et al. (2000)	Tag	Readers	RSS trilateration	3 m
Ni et al. (2004)	Tag	Readers & Tags	RSS Scene Analysis	1 - 2 m
Wang et al. (2007)	Tag	Readers & Tags	RSS proximity and optimization	0.3 - 3 ft
Stelzer et al. (2004)	Tag	Readers & Tags	TDoA weighted mean squares	-
Bekkali et al. (2007)	Tag	Readers & Tags	RSS mean squares and Kalman filtering	0.5 - 5 m
Lee & Lee (2006)	Reader	Tags (dense)	RSS Proximity	0.026 m
Han et al. (2007)	Reader	Tags (dense)	Training and RSS Proximity	0.016 m
Yamano et al. (2004)	Reader	Tags	RSS Scene Analysis	80%
Xu & Gang (2006)	Reader	Tags	Proximity and Bayesian Inference	1.5 m
Wang et al. (2007)	Reader	Tags	RSS proximity and optimization	0.2 - 0.5 ft

Table 1. RFID Localization systems.

distances. LANDMARC (Ni et al., 2004) follows a scene analysis approach by using readers with different power levels and reference tags placed at fixed, known locations as landmarks. Readers vary their read range to perform RSS measurements for all reference tags and for the target tag. The *k* nearest reference tags are then selected and their positions are averaged to estimate the location of the target tag. Wang et al. (Wang et al., 2007) propose a 3-D positioning scheme which relies on a deployment of readers with different power levels on the floor and the ceiling of an indoor space and uses the Simplex optimization algorithm for estimating the location of multiple tags. LPM (Stelzer et al., 2004) uses reference tags to synchronize the readers. Then, TDoA principles and ToA measurements relative to the reference tags and the target tag are used to estimate the location of the target tag. In (Bekkali et al., 2007) RSS measurements from reference tags are collected to build a probabilistic radio map of the area and then, the Kalman filtering technique is iteratively applied to estimate the target's location. If the target is a RFID reader, usually passive or active tags with known coordinates are deployed as reference points and their IDs are associated with their location information. In (Lee & Lee, 2006) passive tags are arranged on the floor at known locations in square pattern. The reader acquires all readable tag locations and estimates its location and orientation by using weighted average method and Hough transform, respectively. Han et al. (Han et al., 2007) arrange tags in triangular pattern so that the distance in x-direction is reduced. They show that the maximum estimation error is reduced about 18% from the error in the square pattern. Yanano et al. (Yamano et al., 2004) utilize the received signal strength to determine the reader position by using machine learning technique. In the training phase, the reader acquires the RSS from every tag in various locations in order to build a Support Vector Machine (SVM). Since it is not possible to obtain the signal intensity from every location, they also propose a method to synthesize the RSS data from real RSS data acquired in the training phase. When the reader enters the area, it will pass the received signal intensity vector to the SVM to determine its position. A Bayesian approach is also proposed to predict the position of a moving object (Xu & Gang, 2006). Having the posterior movement probability and the detected tags' locations, the reader location is determined by maximizing the posterior probability. Then, the reader position is calculated by averaging the inferred position from all tags. However, the accuracy of the algorithm depends on the movement probability model. Finally, (Wang et al., 2007) proposes also a reader localization scheme by employing the Simplex optimization method. Table 1 summarizes the main characteristics of the above systems.

Apparently, selecting a best scheme is not trivial since it depends on several factors such as deployment cost, processing requirements, time and power constraints, scalability issues etc. The second type of positioning schemes attracted our attention because they are easier to be implemented since low cost passive tags can be deployed in a large extent in most indoor environments. Additionally, it is anticipated that future mobile terminals will have a reader extension capability for gaining access at a wide range of innovative applications and services supported by RFID systems. However, there is lack in the literature of a research study regarding the impact of the interference problem, persisting in RFID, on the localization performance. To that end, we have selected three positioning algorithms differing in their complexity level in order to investigate their behavior when multiple reader-enabled mobile nodes need to be localized simultaneously. We believe that examining this parameter is crucial for verifying the efficiency of employing RFID in general location sensing applications.

3. RFID shortcomings

The communication link between the main RFID components is half duplex, reader to tag and then tag to reader. In the forward link, the reader's transmitting antenna (transmitter) sends a modulated carrier to tags to power them up. In the return link, each tag receives the carrier for power supply and backscatters by changing the reflection coefficients of the antenna. In such a way, its ID is sent to the reader's receiving antenna (receiver). The path loss of this two way link may be expressed as:

$$PL(d) = PL_o + 10N \log\left(\frac{d}{d_o}\right) + X_\sigma, \tag{1}$$

where d the distance between the reader and a tag, PL_o the path loss at reference distance d_o given by $PL_o = G_t G_r (g_t \Gamma g_r)\left(\frac{\lambda}{4\pi d_o}\right)^4$ and G_t, g_t, and G_r, g_r are the gains of the reader and tag transmit and receive antennas, respectively. Γ is a reflection coefficient of the tag and λ the wavelength. $N = 2n$, where n the path loss component of the one way link. The path loss model defines the received power $RSS(d)$ at the receiver given the transmit power P_t of the transmitter, i.e.:

$$RSS(d) = P_t - PL(d). \tag{2}$$

In the absence of interference, the maximum read range a reader receiver can decode the backscattered signal is such that:

$$R_{max} = \arg\max_{d \geq 0} RSS(d) \geq TH, \tag{3}$$

where TH represents a threshold value for successful decoding.
Even though RFID technology has promising key characteristics for location sensing, it has also some limitations which become more intense in the case of simultaneous tracking in a multi-user environment and thus should be taken into account before employing an RFID system for localization.
Since RFID technology uses electromagnetic waves for information exchange between tags and readers, how radio waves behave under various conditions in the RFID interrogation zone (IZ) affects the performance of the RFID system. Radio waves propagate from their source and reach the receiver. During their travel, they pass through different materials, encounter interference from their own reflection and from other signals, and may be absorbed or blocked by various objects in their path. The material of the object to which the tag is attached may change the property of the tag, even to the point it is not detected by its reader.

However, the most harmful type of interference is the one among its components which is known as the RFID collision problem. Three are its main types: tag collision, multiple reader-to-tag collision and reader-to-reader collision.

3.1 Multiple tags-to-reader interference

When multiple tags are simultaneously energized by the same reader, they reflect simultaneously their respective signals back to the reader. Due to a mixture of scattered waves, the reader cannot differentiate individual IDs from the tags. This type of interference is known as multiple tags-to-reader interference or tag identification problem.

3.1.1 Anti-collision algorithms

For resolving multiple tag responses an anti-collision mechanism is essential. Reviewing the literature, several anti-collision protocols have been proposed, such as time-division multiple or binary tree-based schemes (GP & SW, 2008). For instance, the EPCglobal (EPCglobal, n.d.), an organization that recognized the potential of RFID early, proposed bit-based Binary Tree algorithm (deterministic) and Aloha-based algorithm (probabilistic). The International Standards Organization (ISO) as part of the ISO 18000 family proposed the Adaptive Protocol which is similar to the Aloha-based algorithm proposed by EPCglobal, and binary tree search algorithm. These protocols mainly differ in the number of tags that can be read per second, their power and processing requirements.

In this work, we selected the Pure and Slotted Aloha schemes (Klair et al., 2009) as basis for our analysis. Let $\mathcal{D}_u$ the set of tags simultaneously energized by the reader r_u. When reading starts, each tag transmits its ID irrespectively of the rest $|\mathcal{D}_u| - 1$ tags. The communications from a tag to the reader is modeled as a Poisson process (Schwartz, 1986). Each tag responds on average λ times per second. The model requires independence among tag transmissions, which is supported by the lack of tag-to-tag communication capabilities. Since each tag's transmission is Poisson distributed, there is a mean delay of $1/\lambda$ between consecutive transmissions. This is referred to as the arrival delay (Schwartz, 1986). Thus, on average each tag takes $\frac{1}{|\mathcal{D}_u|\lambda}$ time to transmit its ID for the first time. This is referred as arrival delay (Schwartz, 1986). During collisions, colliding tags retransmits after a random time. In Aloha-based schemes, the retransmission time is divided into K time slots of equal duration s and each tag transmits its ID at random during one of the next time slots with probability $1/K$. This means tags will retransmit within a period of $K \times s$ after experiencing a collision. On average, a tag will retransmit after a duration of $\frac{K+1}{2} \times s = a$ slots. The number of collisions before a tag successfully responds is $e^{xG_A} - 1$, where e^{xG_A} denotes the average number of retransmission attempts made before a successful identification, where $G_A = |\mathcal{D}_u|\lambda s$ is the offered load and $x = 1$ for Pure Aloha (PA) and $x = 2$ for Slotted Aloha (SA). Since each collision is followed by a retransmission, the average delay before a successful response is $(e^{xG_A} - 1)a$, followed by a single successful transmission of duration s. In total, the average delay a tag takes to transmit its ID successfully is $\mathrm{t}_{TR} = (e^{xG_A} - 1)as + s + \frac{1}{|\mathcal{D}_u|\lambda}$. For non-saturated case, i.e. tags to be detected are less than the maximum number of tags that can be read per inventory round, the total time needed for reading successfully $|\mathcal{D}_u|$ tags follows the linear model

$$\mathrm{T}_{TR} = |\mathcal{D}_u| \times \mathrm{t}_{TR} = |\mathcal{D}_u| \times \left\{ s\left[1 + (e^{xG_A} - 1)a\right] + \frac{1}{|\mathcal{D}_u|\lambda} \right\}. \quad (4)$$

3.2 Multiple readers-to-tag interference

Multiple readers-to-tag interference occurs when a tag is located at the intersection of two or more readers' interrogation range and the readers attempt to communicate with this tag simultaneously. Let R_i and R_j denote the read ranges of readers r_i and r_j and d_{ij} their distance. Apparently, if

$$R_i + R_j > d_{ij} \tag{5}$$

and r_i and r_j communicate at the same time, they will collide and the tags in the common area will not be detected.

Figure 1(a) depicts two readers r_1 and r_2 which transmit simultaneously query messages to a tag t_1 situated within their overlapping region. t_1 might not be able to read the query messages from neither r_1 nor r_2 due to interference.

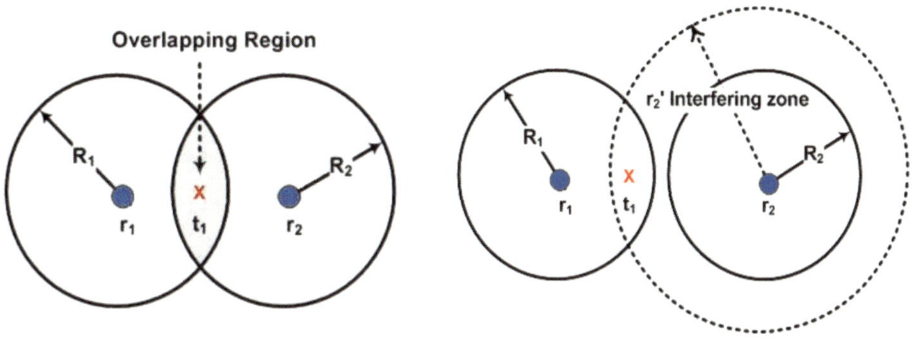

(a) Many Readers-to-Tag Interference. (b) Reader-to-Reader Interference.

Fig. 1. Two types of interference in RFID.

3.2.1 Reader collision probability

The probability P_{ij}^C of such collision type between readers r_i and r_j, if equation (5) is satisfied, depends on the probabilities r_i and r_j are simultaneously trying to communicate with their common tag. For characterizing the probability of simultaneous reader communication, we assume that each reader is in a scanning mode with probability p^{scan}. Thus, P_{ij}^C depends on the probabilities r_i and r_j are in a scanning mode, p_i^{scan} and p_j^{scan}, respectively, i.e.

$$P_{ij}^C = p_i^{scan} \times p_j^{scan}. \tag{6}$$

A mechanism coordinating reader transmissions as the one proposed in (Papapostolou & Chaouchi, 2009a) can compensate this type of interference.

3.3 Reader-to-reader interference

Reader-to-reader interference is induced when a signal from one reader reaches other readers. This can happen even if there is no intersection among reader interrogation ranges ($R_i + R_j < d_{ij}$) but because a neighbor reader's strong signal interferes with the weak reflected signal from a tag. Figure 1(b) demonstrates an example of collision from reader r_2 to reader r_1 when the latter tries to retrieve data from tag t_1. Generally, signal strength of a reader is superior to that of a tag and therefore if the frequency channel occupied by r_2 is the same as that between t_1 and r_1, r_1 is no longer able to listen to t_1's response.

3.3.1 Read range reduction

Reader-to-reader interference affects the read range parameter. In equation (3) this factor had been neglected. However, when interfering readers exist, the actual interrogation range of the desired reader decreases to a circular region with radius R^I_{max}, which can be represented by

$$R^I_{max} = \arg\max_{d \in [0, R_{max}]} SIR(d) \geq TH, \tag{7}$$

where

$$SIR(d) = \frac{P_s(d)}{\sum_i I_i} \tag{8}$$

and I_i the interference from reader r_i.

The Class 1 Gen 2 Ultra High Frequency (UHF) standard ratified by EPCGlobal (EPCglobal, n.d.), separates the readers' from tags' transmissions spectrally such that tags collide only with tags and readers collide only with readers.

4. RFID Positioning system framework

From architectural point of view, a location determination scheme can be either user-based or network-based. In the first case, each user is responsible for collecting and processing information necessary for determining his location, whereas, in the second case, a dedicated server is responsible for gathering all required data and finally providing the location estimates for all users. Processing capabilities, privacy and scalability issues, link quality are usually the main factors for selecting the appropriate approach. Since a RFID system includes tags, readers and servers, we propose a hybrid architecture as a compromise between them, i.e. both user and a dedicated location server participate in the location decision process.

Figure 2 depicts the proposed architecture. The reader embedded at each user device queries for reference tags within its coverage in order to retrieve their IDs. Then, the list of the retrieved tag IDs with the corresponding RSS levels is forwarded to the *Location Server* within a TAGLIST message. Based on the received TAGLIST messages and a repository which correlates the IDs of the *reference* tag with their location coordinates, the *Location Server* estimates the location for all users by employing a RFID-based positioning (see subsection 4.1) algorithm and finally returns the estimated locations back to the corresponding users in LOCATIONESTIMATE messages.

The communication between the reader and the tags is done through the RF interface of the reader, whereas the communication between the reader and the server is possible through the communication interface of the reader, such as IEEE 802.11. Alternatively, assuming multi-mode devices, the TAGLIST and location estimation messages can be exchanged by the wireless interface of the user device.

It is worthy mentioning that the proposed architecture may not be always the optimal choice. For example, if the wireless medium between users and the *Location Server* is not robust enough for exchanging messages successfully, a user-based approach would be more efficient. In this case, when a new user enters the indoor area it can receive information regarding the tag deployment automatically or after having subscribed to a relevant service. Then, by following a positioning algorithm, it can estimate its own location. However, in such approach, greater attention should be given regarding the complexity of the positioning algorithm since mobile terminals have limited resources compared to servers.

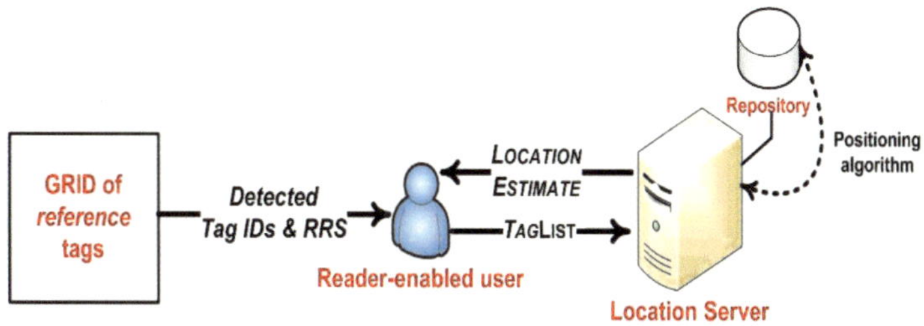

Fig. 2. Proposed RFID-based Positioning Architecture.

4.1 Positioning algorithms

A positioning algorithm defines the method of processing the available information in order to estimate the target's location. The main metrics for evaluating its performance are its accuracy, memory requirements and complexity. In this paper, we study three positioning algorithms which can be easily implemented in the sense that they do not require any special hardware, but differ in their complexity and memory requirements.

Let $\mathcal{D}_u$ denote the set of *reference* tags successfully detected from a user's reader r_u and $\mathbf{SS}_u$ a vector of the corresponding RSS measurements such that the entry RSS_t is the RSS from the tag $t \in \mathcal{D}_u$ to r_u.

4.1.1 Simple Average (SA)

This algorithm is based on the assumption that the reader radiation pattern forms a perfect circle. Thus, the user's location is estimated as the simple average of the coordinates (x_t, y_t) of all tags $t \in \mathcal{D}_u$, i.e.:

$$(\widehat{x}_u, \widehat{y}_u) = \left(\frac{\sum_{t \in \mathcal{D}_u} x_t}{|\mathcal{D}_u|}, \frac{\sum_{t \in \mathcal{D}_u} y_t}{|\mathcal{D}_u|} \right) \tag{9}$$

This scheme has the minimum memory requirements since only the ID information from the detected *reference* tags is used for estimating the unknown location. Regarding its processing requirements, it involves $2 \times |\mathcal{D}_u|$ additions of the coordinates of the detected tags and 2 divisions. Therefore, it has linear complexity $O(|\mathcal{D}_u|)$.

4.1.2 Weighted Average (WA)

Since some of the detected tags may be closer than others, biasing the simple averaging method is proposed as an alternative approach. This can be achieved by assigning a weight w_t to the coordinates of each tag $t \in \mathcal{D}_u$. These weights are based on their RRS from the reader. Thus, (9) becomes:

$$(\widehat{x}_u, \widehat{y}_u) = \left(\frac{\sum_{t \in \mathcal{D}_u} w_t \cdot x_t}{\sum_{t \in \mathcal{D}_u} w_t}, \frac{\sum_{t \in \mathcal{D}_u} w_t \cdot y_t}{\sum_{t \in \mathcal{D}_u} w_t} \right) \tag{10}$$

where $w_t = 1/|RSS_t|$ and RSS_t the measured RSS value from tag t.

This scheme requires more memory than the SA, since RSS information is used in addition to tags' IDs for estimating the unknown location. Regarding its processing requirements, it involves $4 \times |\mathcal{D}_u|$ addition, $2 \times |\mathcal{D}_u|$ multiplication and 2 division operations. Thus, its complexity remains linear, i.e. $O(|\mathcal{D}_u|)$.

4.1.3 Multi-Lateration (ML)

Finally, we investigate a multi-lateration based approach which tries to take into account the imperfection of the readers' radiation pattern. The distances from all detected tags $\mathcal{D}_u$ are first estimated and then (x_u, y_u) can be obtained by solving the following system of $|\mathcal{D}_u|$ equations:

$$\begin{aligned} (x_1 - x_u)^2 + (y_1 - y_u)^2 &= \widehat{d_1^2} \\ &\vdots \\ (x_{|\mathcal{D}_u|} - x_u)^2 + (y_{|\mathcal{D}_u|} - y_u)^2 &= \widehat{d_{|\mathcal{D}_u|}^2} \end{aligned} \tag{11}$$

The above system of equations is not linear. According to (Caffery, n.d.) it can be linearized by subtracting the last equation from the first $|\mathcal{D}_u| - 1$ equations. The resulting system of linear equations is given then given by the following matrix form:

$$\mathbf{A}[x_u, y_u]^T = \mathbf{b}, \tag{12}$$

where

$$\mathbf{A} := \begin{pmatrix} 2(x_t - x_1) & 2(y_t - y_1) \\ \vdots & \vdots \\ 2(x_t - x_{|\mathcal{D}_u|}) & 2(y_t - y_{|\mathcal{D}_u|}) \end{pmatrix},$$

$$\mathbf{b} := \begin{pmatrix} x_1^2 - x_{|\mathcal{D}_u|}^2 + y_1^2 - y_{|\mathcal{D}_u|}^2 + \widehat{d_1^2} - \widehat{d_{|\mathcal{D}_u|}^2} \\ \vdots \\ x_{|\mathcal{D}_u|-1}^2 - x_{|\mathcal{D}_u|}^2 + y_{|\mathcal{D}_u|-1}^2 - y_{|\mathcal{D}_u|}^2 + \widehat{d_{|\mathcal{D}_u|-1}^2} - \widehat{d_{|\mathcal{D}_u|}^2} \end{pmatrix}. \tag{13}$$

Since $\widehat{d_t}$ are not accurate, the above system of equations can be solved by a standard LS approach (Caffery, n.d.) as:

$$[\widehat{x}_u, \widehat{y}_u]^T = (\mathbf{A}^T\mathbf{A})^{-1}\mathbf{A}^T\mathbf{b} \tag{14}$$

with the assumption that $\mathbf{A}^T\mathbf{A}$ is nonsingular and $|\mathcal{D}_u| \geq 3$, i.e. at least three tags are detected. This scheme has similar memory requirements with the WA. However, it has polynomial complexity $O(|\mathcal{D}_u|^3)$ and it involves complex matrix operations such as creating an inverse matrix.

5. Performance analysis

In this section we evaluate the performance of our approach through simulations, using Matlab, (Matlab, n.d.), as our simulation tool. As performance metric we use the Mean Location Error (MLE) and Mean Localization Time (MLT). MLE is defined as the Euclidean distance between the actual and the estimated position of a user. The MLT includes the time T_{TR} needed for retrieving successfully all $|\mathcal{D}_u|$ tags' IDs within range, given by eq. (4), the processing time of the positioning algorithm, which depends on its complexity and the time needed for sending successfully the TAG LIST message from the reader (or user terminal) to the server and the time needed for sending successfully the location estimation from the server to the reader (or user terminal).

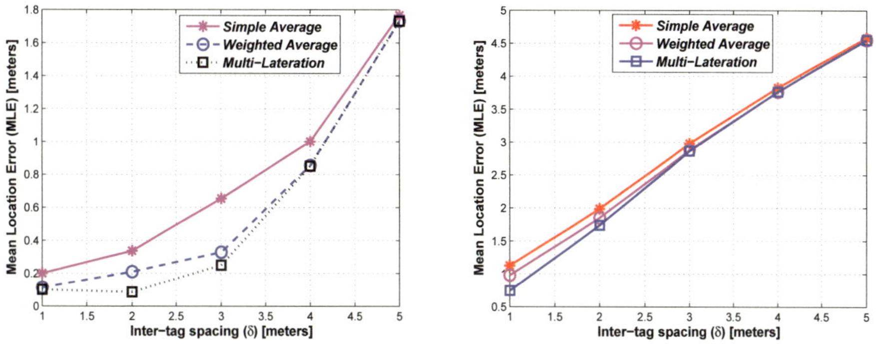

(a) Multi-user environment with $\beta = 0$ (b) Multi-user environment with $\beta = 1$

Fig. 3. Impact of tag density (δ)

We provide and interpret results of the simulations we conducted for evaluating the impact of the parameters δ, β, and R_{max} on the system performance. In order to illustrate the performance degradation due to the collision problem and the essentiality of an anti-collision mechanism, we considered two multi-user environmental cases which differ in the level of the collision problem. Assuming that the probability user readers query their tags follows uniform distribution $U(\beta, 1)$, we set $\beta = 0$ for the first case and $\beta = 1$ for the second case. Apparently, for the second environment all users readers scan simultaneously for their tags and thus its performance is anticipated to be worse due to the collision problem among them.

5.1 Localization accuarcy

Figures 3(a) and 3(b) illustrate the dependency of the MLE on the tag density, δ, when $\beta = 0$ and $\beta = 1$, respectively, and for the three RFID-based positioning methods described in subsection 4.1, i.e. AVG, W-AVG and ML. For all cases, increasing the inter-tag spacing reduces the accuracy. However, when the collision problem is severe, the achieved accuracy and performance reduction are worse and thus a dense tag deployment is required for providing robustness. Finally, comparing the behavior of the three positioning schemes, we note that there is a benefit from the added complexity but in highly colliding environments the achieved benefit is not significant.

In figures 4(a) and 4(b) the influence of the maximum read range, R_{max}, is depicted when $\delta = 2$. For both scenarios we observe that when $R_{max} = 1$, the MLE is increased and this is because tags are not detected. When $\beta = 0$, $R_{max} = 2$ gives the optimum performance for two main reasons; further than this collisions are more probable but also location information from far-away tags is included. For the second case, the optimum performance is achieved when $R_{max} = 3$ meters because of the collisions which prevents tags from being detected.

5.2 Time response

In Figure 5 we study the time-response performance of the positioning system, focusing on the time needed for retrieving the ID information from detected tags, i.e. T_{TR}. From equation (4) we see that T_{TR} depends on the total number of detected tags $|\mathcal{D}_u|$ and the PA or SA anti-collision algorithm which affects parameter x. $|\mathcal{D}_u|$ depends on the reference tag density δ and the read range R_{max}. Obviously, as δ increases $|\mathcal{D}_u|$ decreases, whereas when R_{max} is higher more tags are detected. The MLT versus the inter-tag spacing δ for both anti-collision

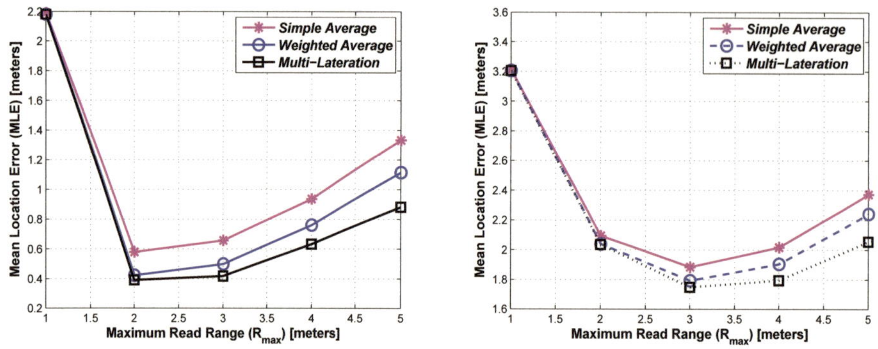

(a) Multi-user environment with $\beta = 0$ (b) Multi-user environment with $\beta = 1$

Fig. 4. Impact of maximum read range (R_max)

algorithms when $R_{max} = 3m$ and $R_{max} = 5m$ is depicted in Figure 5(a) and Figure 5(b), respectively. First of all, we observe that Slotted Aloha has better performance than Pure Aloha, due to the reduction of the vulnerability period 2*s* (Burdet, 2004). In both figures, when the grid deployment is dense, the tag reading time is very high due to the big number of responding tags. Comparing the two cases of R_{max} values, when $R_{max} = 3m$ less tags are within a reader's interrogation zone and thus, less reading time is required. Finally, recalling Figure, we conclude that there is a trade-off between the accuracy and time response objectives, regarding the optimal value of δ. More tags provide more information for the location determination process but on the other hand more time is required for detecting them.

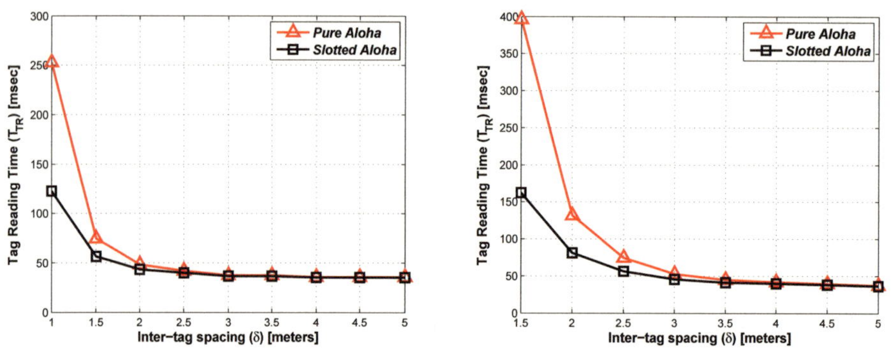

(a) Tag reading time vs δ when $R_{max} = 3m$. (b) Tag reading time vs δ when $R_{max} = 5m$.

Fig. 5. Impact of system design parameters on *Time Response*.

Figure 6 depicts the processing time T_{pr} (specified in flops[1]) of each positioning algorithm as the inter-tag spacing increases, for $R_{max} = 3m$ and $R_{max} = 5m$ in figures 6(a) and

[1] The execution time of a program depends on the number of floating-point operations (FLOPs) involved. Every computer has a processor speed which can be defined in flops/sec. Knowing the processor speed and how many flops are needed to run a program gives us the computational time required: Time required (sec) = Number of FLOPs/Processor Speed (FLOP/sec) (Canale, n.d.).

6(b), respectively. The main observation is the high processing time of the Multi-Lateration approach for dense tag deployments. The most interesting remarks, however, can be made if Figure is taken into account. The W-AVG approach has the best performance if both objectives are considered. Moreover, for $R_{max} = 5m$ and $\delta = 5m$, the accuracy of the ML technique is high without considerable processing cost. Therefore, more sophisticated techniques can alleviate the need for carefully designed systems.

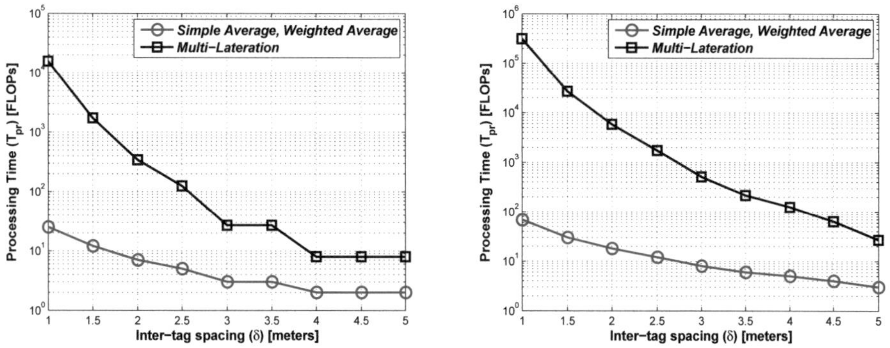

(a) Processing time vs δ when $R_{max} = 3m$. (b) Processing time vs δ when $R_{max} = 5m$.

Fig. 6. Impact of positioning algorithm on *Time Response*.

Finally in Table 2 we summarize the main advantages and disadvantages of the system design parameters regarding their accuracy, time response, complexity and behavior under different environmental situations.

6. Conclusion

The growing popularity of the RFID technology and the increasing demand for intelligent location-aware services in indoor spaces motivated exploring its potential for providing accurate and time efficient localization with low deployment cost. However, despite the great benefits RFID can offer, the interference among its components and some materials are its main limiting factors. Therefore the impact of the RFID interference problem on the positioning performance should be extensively studied before the deployment of RFID-assisted location systems.

In this chapter, we explore the applicability of the RFID technology in location sensing and the main design and environmental factors that should be considered before developing an RFID-based localization scheme. We focused on a scenario when the location of multiple reader-enabled terminals needs to be estimated based on the information retrieved from low cost passive tags, which are deployed in an area. We proposed a mathematical model for taking into account all implicating factors which affect the accuracy performance of the system, that is all types of collisions among its components, interference from materials, and temporal environmental changes. Extensive simulations were conducted to evaluate the impact of these parameters. More precisely, when reader collisions is not an issue, a low dense ($\delta \leq 4$ meters) deployment of passive tags can provide an accurate location information with error less than 1 meter. However, in a highly colliding environment, passive tags should be deployed with spacing of 1 meter in order to have similar location error resilience. Interesting remarks can be drawn regarding the communication range of readers.

Design Parameter		Pros	Cons
Reference Tag Deployment	$\delta : [5 \rightarrow 1]m$	- MLE ↓ - Robustness .	- MLT ↑
Maximum Read Range	$R_{max} : [5 \rightarrow \delta]m$	- MLE ↓ for multi-user case - MLT ↓	- MLE ↑ for single-user case
Positioning algorithm	S-AVG	- Lowest complexity - Good MLE resilience as shadowing increases	- Highest MLE - Suffers the most from all interference types
	W-AVG	- Moderate complexity - Best performance when shadowing is high	- When interference is high, its increased complexity over SA doesn't provide accuracy advantage
	ML	- Best accuracy - Best MLE resilience against all interference types	- Highest complexity - Bad performance when shadowing is high
Tag Reading activity	$\beta : [1 \rightarrow 0]$	- MLE ↓	- Less users are simultaneously localized

Table 2. System Design Guide.

In the absence of collisions, short read range (2 meters) is beneficial. In contrast when readers attempt simultaneously access to the medium, a higher range (3-4 meters) results in better accuracy.
To summarize, RFID technology is suitable for positioning, but its performance degrades in highly populated environments and thus a denser tag deployment or/and a mechanism for controlling reader transmissions are required.

7. References

Bahl, P. & Padmanabhan, V. (2000). RADAR: An In-Building RF-Based Location and Tracking System, *IEEE INFOCOM 2000*.

Baudin, M. & Rao, A. (2005). RFID applications in manufacturing.

Bekkali, A., Sanson, H. & Matsumoto, M. (2007). RFID Indoor Positioning Based on Probabilistic RFID Map and Kalman Filtering., *WiMob*, IEEE Computer Society, p. 21.

Burdet, L. A. (2004). Rfid multiple access methods, *Technical report*, ETH Zürich.

Caffery, J.J., J. (n.d.). A new approach to the geometry of toa location, *52nd IEEE Vehicular Technology Conference (IEEE VTS-Fall VTC'00)*, Vol. 4.

Canale, C. (n.d.). *Numerical Methods for Engineers.*
EPCglobal (n.d.). http://www.epcglobalinc.org.
GP, J. & SW, K. (2008). Survey, Nomenclature and Comparison of Reader Anti-Collision Protocols in RFID, *IETE Tech Rev* 25(5): 285–292.
Han, S., Lim, H. & Lee, J. (2007). An efficient localization scheme for a differential-driving mobile robot based on rfid system, *Industrial Electronics, IEEE Transactions on* 54(6): 3362 -3369.
Hightower, J. & Borriello, G. (2001). Location Systems for Ubiquitous Computing., *IEEE Computer* 34(8): 57–66.
Hightower, J., Want, R. & Borriello, G. (2000). SpotON: An Indoor 3D Location Sensing Technology Based on RF Signal Strength, *Technical report*, University of Washington, Department of Computer Science and Engineering, Seattle, WA.
Hinckley, K. & Sinclair, M. (1999). Touch-Sensing Input Devices, *CHI*, pp. 223–230.
Ingram, S., Harmer, D. & Quinlan, M. (2004). Ultra-wideband Indoor Positioning Systems and their use in emergencies, *PLANS 2004: Position Location and Navigation Symposium.*
King, T., Kopf, S., Haenselmann, T., Lubberger, C. & Effelsberg, W. (2006). COMPASS: A probabilistic indoor positioning system based on 802.11 and digital compasses, *WiNTECH '06: Proceedings of the 1st international workshop on Wireless network testbeds, experimental evaluation & characterization*, ACM, New York, NY, USA, pp. 34–40.
Klair, D., Chin, K.-W. & Raad, R. (2009). On the energy consumption of pure and slotted aloha based rfid anti-collision protocols, *Comput. Commun.* 32(5): 961–973.
Lee, H.-J. & Lee, M. C. (2006). Localization of mobile robot based on radio frequency identification devices, *SICE-ICASE, 2006. International Joint Conference*, pp. 5934–5939.
Matlab (n.d.). http://www.mathworks.com.
Ni, L. M., Liu, Y., Lau, Y. C. & Patil, A. P. (2004). LANDMARC: Indoor Location Sensing Using Active RFID, *Wireless Networks* 10(6): 701–710.
Pahlavan, K. & Levesque, A. H. (2005). *Wireless Information Networks*, second edn, John Wiley & Sons, New York.
Papapostolou, A. & Chaouchi, H. (2009a). Exploiting multi-modality and diversity for localization enhancement: WiFi & RFID usecase, *IEEE 20th International Symposium on Personal Indoor and Mobile Radio Communications (PIMRC)*, pp. 1903 – 1907.
Papapostolou, A. & Chaouchi, H. (2009b). WIFE: Wireless Indoor Positioning Based on Fingerprint Evaluation, *Networking 2009*, pp. 234–247.
Priyantha, N. B., Chakraborty, A. & Balakrishnan, H. (2000). The Cricket Location-Support system, *Proceedings of the 6th ACM International Conference on Mobile Computing and Networking*, ACM, Boston, MA, USA, pp. 32–43.
Schwartz, M. (1986). *Telecommunication networks: protocols, modeling and analysis*, Addison-Wesley Longman Publishing Co., Inc., Boston, MA, USA.
Stelzer, A., Pourvoyeur, K. & Fischer, A. (2004). Concept and application of lpm - a novel 3-d local position measurement system, *Microwave Theory and Techniques, IEEE Transactions on* 52(12): 2664 – 2669.
Ubisense (n.d.). http://www.ubisense.net/default.asp.
Wang, C., Wu, H. & Tzeng, N.-F. (2007). RFID-Based 3-D Positioning Schemes., *INFOCOM*, IEEE, pp. 1235–1243.
Want, R. (2006). An Introduction to RFID Technology, *IEEE Pervasive Computing* 5(1): 25–33.
Want, R., Hopper, A., Falcao, V. & Gibbons, J. (1992). The Active Badge Location System, *ACM Transactions on Information Systems* 10(1): 91–102.

Wellenhoff, B. H., Lichtenegger, H. & Collins, J. (1997). *Global Positions System: Theory and Practice*, forth edn, Springer Verlag.

Xu, B. & Gang, W. (2006). Random sampling algorithm in rfid indoor location system, *Electronic Design, Test and Applications, 2006. DELTA 2006. Third IEEE International Workshop on*.

Yamano, K., Tanaka, K., Hirayama, M., Kondo, E., Kimuro, Y. & Matsumoto, M. (2004). Self-localization of mobile robots with rfid system by using support vector machine, *Intelligent Robots and Systems, 2004. (IROS 2004). Proceedings. 2004 IEEE/RSJ International Conference on*, Vol. 4, pp. 3756 – 3761.

Youssef, M. & Agrawala, A. (2005). The Horus WLAN location determination system, *MobiSys '05: Proceedings of the 3rd international conference on Mobile systems, applications, and services*, ACM Press, New York, NY, USA, pp. 205–218.

12

Use of Active RFID and Environment-Embedded Sensors for Indoor Object Location Estimation

Hiroaki Fukada, Taketoshi Mori, Hiroshi Noguchi and Tomomasa Sato
The University of Tokyo
Japan

1. Introduction

Indoor object localization system has become more and more important in various fields these days. For example, people not only feel stress but also waste precious time when they cannot find what they want in the expected place. If we can provide people with information about the object location, people will save lots of time and lead a comfortable daily life. Furthermore, if we can detect object movement and estimate object location online, we will be able to know life patterns of people by analyzing the behavior of objects in everyday life. Efficient online object localization system should be able to identify the object a user wants and to determine its location. In our work, we focus on object's "location" in the environment (e.g. Table, Bed, Sofa, etc.) instead of object's 3-dimensional "position", because we think the only object location is sufficient to achieve our application. Various technologies have been used to construct such systems up-to-date, but most of them have difficulty in recognition of the objects.

Against this problem, many studies have focused on radio frequency identification (RFID) technology due to its strong identification ability(Hightower et al., 2000; Mori et al., 2007; 2005; Ni et al., 2004; Shih et al., 2006). In general, RFID system is composed of two devices: 1) RF readers and 2) RFID tags. Data including RFID tags' identification information are communicated between RF readers and RFID tags via RF signals, which can be transmitted even if obstacles stand between RF readers and RFID tags. At this point, RFID technology is superior to other technologies such as camera vision in identifying objects. Another important characteristic of RFID technology is that signal strength indicator received by RF readers, which we call RSSI, has a certain dependency on the distance between RFID tags and RF readers. This relationship can suggest us an effective clue to estimate the distance from each RF reader to the target RFID tag(Hightower et al., 2000).

RFID technology can be divided into two types depending on the mechanism of data transmission: 1) Passive RFID and 2) Active RFID. The main difference of these two RFID systems is the way of data transmission. Because passive RFID tags do not contain any batteries inside them, they utilize the power of passive readers to activate themselves. As a result, the data transmission range is short, 1 meter at best. In contrast, because active RFID tags contain batteries inside themselves, they utilize their own power for data transmission. Consequently, the data transmission range of active RFID is much longer than that of passive one, some active RFID systems can achieve data transmission range up to 100 meters.

We adopt active RFID instead of passive one as our key technology for the following reasons. One reason is its long transmission range. Since we aim to develop an indoor object localization method, long transmission range is more convenient than short one. Another reason is the number of RF readers required for object localization. As the transmission range of active RFID is much longer than that of passive RFID, the required number of RF readers is much less than that of passive ones. This advantage of active RFID plays a great role in reducing the total introduction cost of the system. The other reason is for the potential of active RFID tag. One remarkable characteristic of active RFID tag compared with passive one is that active RFID tag can attach sensors inside. In fact, every active RFID tag, which we used in our work, contains a vibration sensor to detect object motion. It is certain that users have to exchange battery of active RFID tags regularly in about one year or so. However, the battery itself is inexpensive and the benefits provided by the system are much greater than the exertion spared for the exchange. Also, rapid technology progress will definitely expand the battery life in the near future.

Several researchers have focused on developing indoor localization methods based on active RFID up-to-date(Hightower et al., 2000; Ni et al., 2004; Shih et al., 2006; yao Jin et al., 2006; Zhao et al., 2007). For example, Hightower et al. (2000) applied triangulation algorithm to the SSIs received by several RF readers to estimate the 3-dimensional position of tag indoors. This estimation method works well under the condition that few obstacles exist in the environment, however it fails to localize objects once too often in the environment where various obstacles exist like actual human living space. The main reason for the failures is that received SSI, which we call RSSI, is quite sensitive to environmental factors such as the presence and the location of people and furniture because the radio waves are weak against those factors. To reduce the environmental influences on RSSI, some researches introduced the concept of reference tags as an indicator of object position(Ni et al., 2004; Shih et al., 2006). It is certain that reference tags are useful for reducing the influences on RSSI to a certain extent, still it cannot be evaluated as the perfect solution to indoor object localization. In those researches, the authors also conducted some experiments in the environment where obstacles exist to show the robustness of their methods. However, the complexity of their experimental environment is far from that of our target environment. Human living space is full of various obstacles not only static ones such as furniture, but also dynamic ones such as human beings. To estimate object location robustly in such an environment, we have to confront with more difficult problems than those researches.

To improve the robustness of object localization, our previous work(Mori et al., 2007) focused on the idea that any objects' movements were connected with human behavior. In other words, human position in the environment would be an important clue in estimating object location. Therefore, we introduced a kind of position sensors underneath the floor in the previous work, which we call floor sensors, so as to detect human position in the environment. As a result, floor sensors played an effective role in detecting human position, however, some challenges still remained unsolved, such as the number of sensors required for human localization. To achieve high-resolution human localization, the position sensors need to cover the whole area of the environment. As a matter of fact, 356 position sensors were embedded in the environment. Because each position sensor is not cheap, to cover the whole area costs a great deal. In addition, it is troublesome to repair those position sensors in case of breakdown. To reduce the cost and maintenance burden caused by floor sensors, we have combined active RFID technology with various types of switch sensors. The main advantage of these sensors against floor sensors is that they are inexpensive and easy to install into any kinds of environment. In addition, because these sensors are generally used for human monitoring and crime prevention nowadays, it is quite natural to have these sensors embedded in human living space. Substitution of simple sensors for floor sensors makes it difficult to detect human position accurately in the environment, which will cause a decline in estimation accuracy of object location. To solve this problem, we use an integrated algorithm in compensation for the lack of human position information. By taking this approach, we have proposed a method for indoor object location estimation based on active RFID and simple environment-embedded sensors, which achieves sufficient accuracy even without using any costly sensors designed for detecting human position.

2. Hardware composition

In this section, we introduce our active RFID system and various sensors embedded in our experiment environment.

2.1 Active RFIS system

In our research, we adopted *Spider V Active RFID System* (Fig. 1) produced by RF Code as the key technology for the following two reasons.

- Capability of measuring received signal strength indicator (RSSI) between a tag and reader
- Vibration sensor attachment on each Active RFID tag

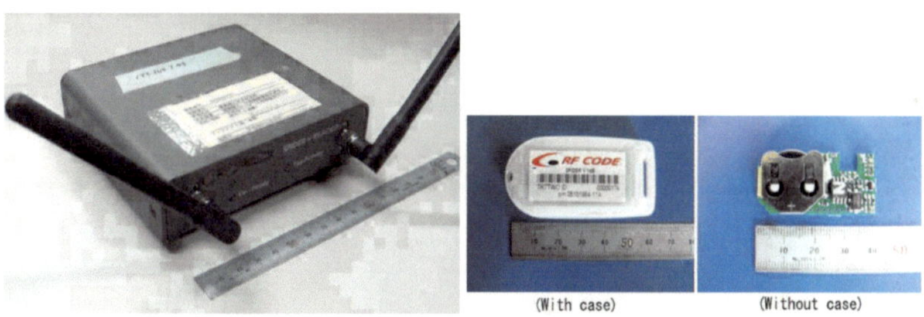

(a) RF Reader (b) RFID Tag

Fig. 1. Spider V Active RFID System

The specifications of the RF reader and the RFID tag are summarized in Table 1 and Table 2.

Item	Specification
Operating Temperature	−20 °C to +70 °C
Read Range	Over 10m
Dimensions	127mm × 130mm × 40mm
Operating frequency	303.8MHz

Table 1. Specifications of Spider V Active RFID Reader

2.2 Environment-embedded sensors

Sensing Room(Mori et al., 2006) is a typical residential environment embedded with various types of sensors in different spots such as high resolution pressure sensors under the floor,

Item	Specification
Battery type	Replaceable coin cell (CR2032)
Battery life	Up to 3 years
Operating Temperature	$-20\,^{\circ}C$ to $+70\,^{\circ}C$
Weight	20g
Dimensions	60mm × 30mm × 10mm
Operating frequency	303.8MHz
Group code & tag ID code	Over 1 trillion IDs
Signal transmission period	1 sec

Table 2. Specifications of Spider V Active RFID Tag

and micro switch sensors in a cabinet and a refrigerator. Retrieved from these sensors, the data are accumulated into the database system and then used to learn users' behavior pattern in a daily life to provide them with appropriate supports. Figure 2 shows the overview picture of *Sensing Room* and its sensor modules. The following sections will describe each sensor module's structure in details.

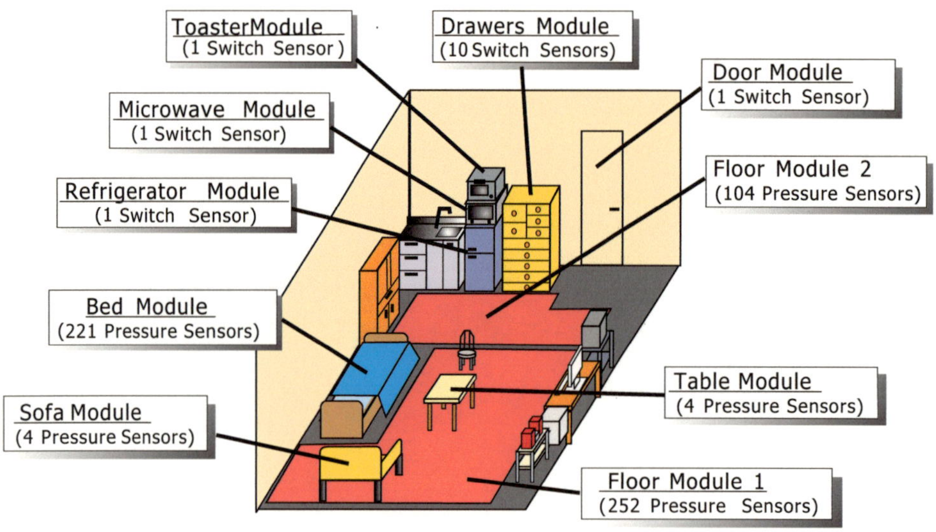

Fig. 2. Experiment Environment (Sensing Room)

- **Floor Sensor Module**

  Sensing Room's floor is embedded with a number of high resolution pressure sensors underneath. By analyzing the pressure values obtained from these sensors, the locations of persons in the room can be estimated. The arrangement of pressure sensing units and screenshots taken during the execution time of floor sensor monitor program are illustrated in Fig. 3.

  In floor sensor module, a Force Sensing Register (FSR) manufactured by INTERLINK co. is used as a basic pressure sensing element. FSR is a pressure-sensitive resistor that has a characteristic of changing its resistance value when a load is charged on its surface. This characteristic is utilized to detect person's weight charging on the floor. Floor sensor

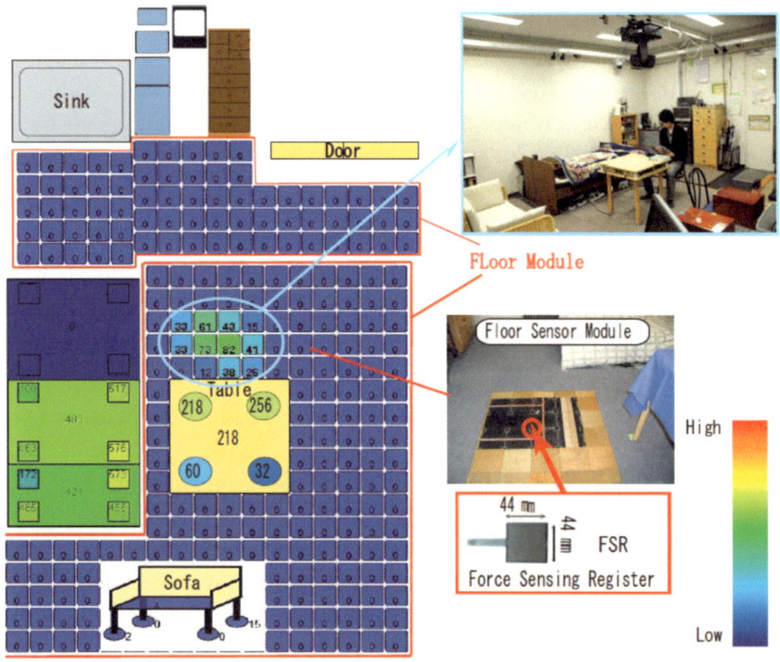

Fig. 3. Floor Sensor Module

module has 356 pieces of FSR in total embedded under a large number of 200mm x 200mm veneer tiles spread thorough the floor.

Firstly all pressure values are retrieved from each FSR; these data are analyzed to derive a pressure load distribution on the floor. Next, a labeling process will be taken on the retrieved pressure value data that exceed a determined threshold value to compute approximate weight and the CG (Center of Gravity) point of a load distribution corresponding to each labeling area. This process will produce the output as pairs of (x, y) coordinates corresponding to estimation result of a user's location in the room. Because the interval between FSRs is 200mm, the resolution of location estimation is also 200mm at best. Sensor data sampling rate of floor module is 10 Hz, a sufficient speed to track a person's location smoothly in a daily situation.

- **Switch Sensor Module**

  Our switch sensor module utilizes a number of micro typed switch sensors equipped to a cabinet, a toaster, a microwave, and a refrigerator to detect a cabinet drawers', a toaster door's, a microwave door's, and a refrigerator doors' open-closed state. Ten micro switches are embedded to each drawer of a cabinet, while two are equipped to the refrigerator, one to the toaster, and one to the microwave oven. Same as floor sensor module, the switch sensor data's sampling rate is 10 Hz. Although we did not target those locations in the evaluation experiment, switch sensor module can be used for locating objects in the locations where RFID signal is not accessible such as "in the cabinet drawer" or "in the refrigerator".

- **Table Sensor Module**

Table sensor module measures weight on it with pressure sensors embedded at its four corners. As well as floor sensor module and switch sensor module, the sampling rate of table sensor module is also 10 Hz. By analyzing the reaction of the acquired sensor data, the system can detect the moment smoothly that an object is placed on the table or the moment that an object is taken from the table.

- **Sofa Sensor Module**

  Sofa sensor module has the same structure as table sensor, which has four pressure sensors embedded at its four corners. The sampling rate of sofa sensor module is also the same as other sensors, which is 10 Hz. Analyzing the reaction of the acquired sensor data provides the system with the information such as human is sitting on the sofa or human moves from the sofa.

3. Location estimation methods

To provide a robust object localization method in residential environment, we integrate the following two approaches in our research: 1) RSSI-based Localization and 2) Sensor-based Localization. RSSI-based Localization regards plural received signal strength indicators (RSSIs) as a unique RSSI pattern to classify into particular object location. Whereas, sensor-based localization estimates possible location candidates based on the information about human behavior and location detected with various sensors. We describe each estimation approach in the following part, and then we demonstrate how to integrate these two approaches into one localization method.

3.1 Location estimation based on RSSI

As mentioned above, RSSI has a certain dependency on the distance between RF readers and RFID tags, which is shown in Fig. 4. According to Fig. 4 however, the dependency does not demonstrate linear relationship due to the nature of RSSI and environmental noises such as furniture. Still, RSSI does not change dramatic unless the layout of the environment changes dramatically. In other words, RSSI shows rather constant value under static environmental condition. Therefore, we attempt to use plural RSSIs instead of single RSSI in order to provide a reliable location indicator.

Based on the idea described above, several RF readers are installed at different spots in the environment. Because active RFID shows long data transmission range, only five RF readers are sufficient to cover the whole area of the environment. By attaching an RFID tag to each object, these RF readers can receive RF signals from RFID tags, which means there are the same number of clues to estimate the distance between the object and the surrounded RF readers. Pattern recognition approach regards these location clues as one pattern and determines the object's location by comparing the acquired pattern with typical patterns of each supposed object location.

In our work, we adopted three kinds of pattern recognition methods such as k-nearest neighbor (KNN)(Cover & Hart, 1967), distance-weighted k-nearest neighbor (DKNN)(Pao et al., 2008), and three-layered neural network (NN)(Shimodaira, 1994) algorithms. The main reason for these choices is the high flexibility in dealing with multidimensional data. In addition, KNN and DKNN algorithms in particular are strong at pattern discrimination of high dimensional data within short processing time.

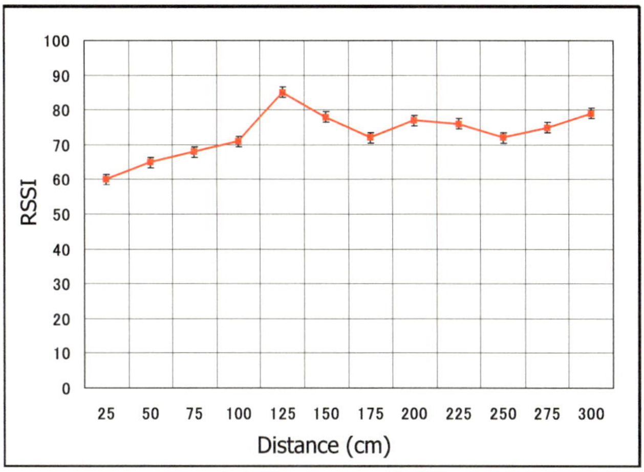

Fig. 4. Dependency of RSSI on Distance

In general, pattern recognition process can be divided into three phases: 1) Preparation Phase, 2) Learning Phase, and 3) Classification Phase. In preparation phase, the system needs a learning database including representative training datasets of each class. Then, in learning phase, the system determines some parameters required to discriminate each class from others. Lastly, in classification phase, the system compares input data with training datasets in the learning database and determines the most similar class according to the defined degree of similarity.

In KNN algorithm, for example, the pattern recognition process is demonstrated as Fig. 5. The first step when the target object is placed at unknown place, is to collect SSIs transmitted from the tag attached with target object and received by several RF readers. In the Fig. 5, we supposed five RF readers in the environment. We can regard this five SSIs, which we call data set, as one pattern of SSI. Next step is to compare the pattern with training data set stored in learning database. In learning database, we have sufficient data sets, which store both SSI pattern and object location, which is called class, as one set. What we explained so far is common process about pattern recognition. The unique process to KNN is called voting process which we will mention below. In KNN algorithm, we used euclid distance as an indicator represents the similarity between one data set pattern and another. In other words, the smaller the euclid distance is, the more similar the data sets are. We calculated the euclid distance between the new data set and every data set stored in learning database and sorted the training data set in increasing order. Then we choose 'k' data sets from the top of the sorted learning database. What we call voting process is to determine object location by counting the number of locations contained in the selected data sets and choosing the most one as estimated result.

Our learning database is constructed as follows. RFID tags are placed at each supposed object location to acquire RSSIs between surrounding RF readers with themselves. The typical scene of RSSI calibration is shown in Fig. 6. Thus, RSSI calibration is conducted at 13 locations with five RF readers installed in the environment, which is shown in Fig. 7. As 1420 datasets are acquired at each object location, the total number of datasets saves into the learning database is;

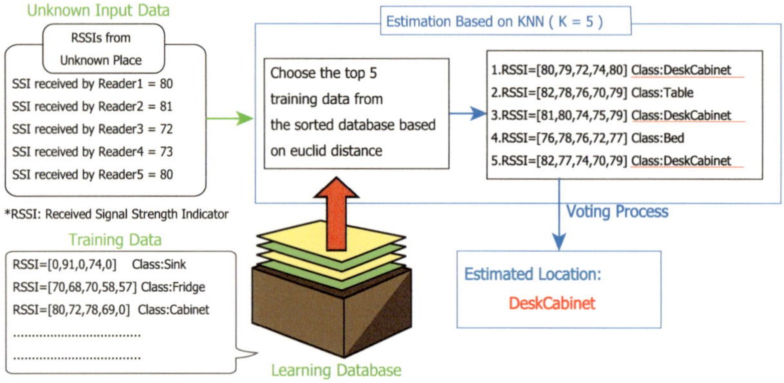

Fig. 5. Location Estimation Based on KNN

$$13(locations) \times 1420(datasets/location) = 18460(datasets) \tag{1}$$

The time required for conducting the whole RSSI calibration is about 13 minutes since our system can collect RSSIs from 24 RFID tags at the same time in one second.

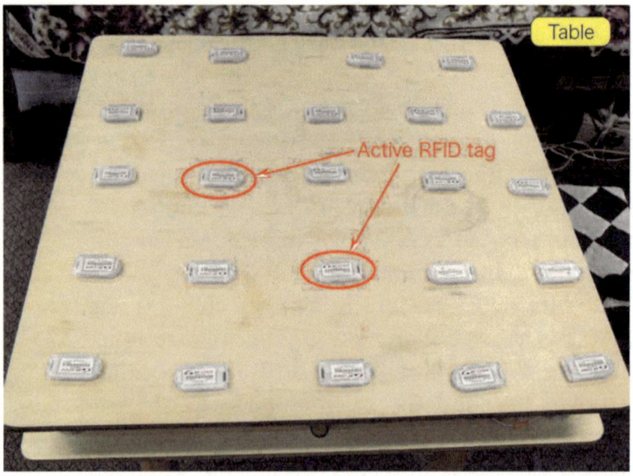

Fig. 6. RSSI Calibration at Table

Based on this learning database, we conducted a cross validation to evaluate the effectiveness of our pattern recognition approach. In the cross validation, the whole learning database is divided into two parts, one is called testing dataset, and the other is called training dataset. A testing dataset consists of one RSSI dataset from every object location in the environment, whereas, a training dataset consists of the rest part of the learning database. After the first evaluation, another testing dataset composed of the next 13 RSSI datasets from every object location will be chose for evaluation. Thus, 1420 times of evaluation are conducted in total. The estimation performance based on KNN algorithm is demonstrated in Fig. 8.

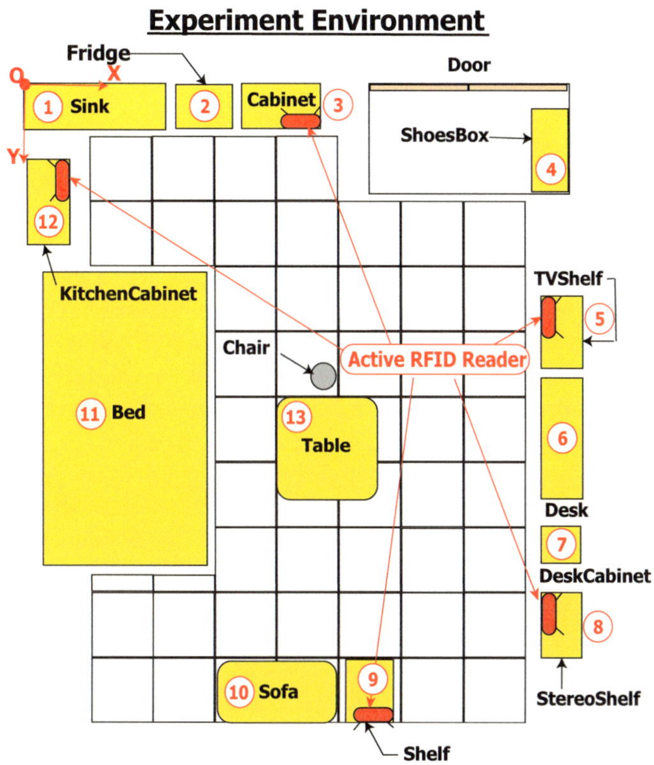

Fig. 7. Supposed Object Locations and RF Readers

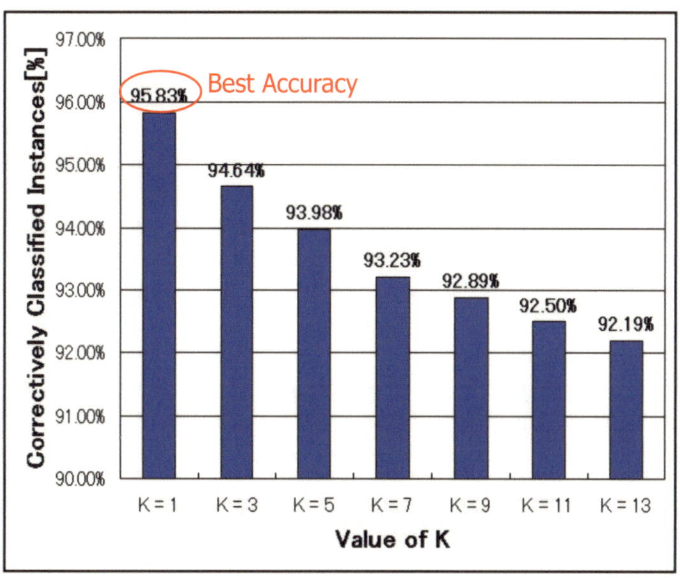

Fig. 8. Location Estimation Performance by KNN Algorithm

According to Fig. 8, the pattern recognition approach works effectively in discriminating each class from others, although there is slight dispersion in estimation performance between different k values.

3.2 Location estimation based on object motion and human behavior

Another approach to improve object localization performance is to make the best use of sensing information. As mentioned before, several kinds of sensors are used in our work. Vibration sensors attached inside RFID tags are supposed to provide the system with the information about object motion state, whereas, sensors embedded in the environment are supposed to provide the information about human behavior and location.

It is important to perceive the moment that an object is placed for estimating its location with sensors in the environment effectively. The vibration sensor on each RFID tag offers a great solution to meet this requirement by detecting object motion state. However, to integrate the vibration sensors into our system needs another problem to be solved.

Generally, active RFID tags are produced under the following policies, 1) saving the battery, 2) miniaturizing the size, and 3) cutting down the cost. To follow these policies, the frequency of data transmission and the performance of vibration sensor inside are set up to be low. These restrictions cause some significant problems. For example, the system cannot detect the moment that object motion state changes in real-time because vibration sensor data requires a moment, which is the sampling rate, to convey its reaction to the system. In addition, vibration sensor often fails to detect object motion in the case that the movement is faint. However, object motion detected with vibration sensor is considered as the most important information in our system because the system uses vibration information to determine the timing to estimate object location. To deal with the time delay between actual object movement and vibration detection, we stagger a few seconds in our algorithm to estimate the exact moment that an object starts to move.

The concrete location estimation algorithm based on environment-embedded sensors and vibration sensor is constructed as follows. Our system can estimate the following three cases individually online by combining detected reaction of each sensor. a) Object is put on and taken away from a table. b) Object is put on and taken away from a sofa. c) Object is put into and taken out of a drawer. That is to say, as long as the movement of object is concerned about the area where we installed embedded sensors, we can estimate its behavior. To be concrete, our system can detect not only the final location where object is placed, but also the state of object in starting and quitting movement. The system estimates the two kinds of object state as follows.

3.2.1 Estimation of movement start

In this section, we describe an algorithm to detect the start of object movement and to estimate the original location from which object begins to move. On the occasion of estimation, we assume that target object is in a still state before the system receives any change of sensor state.

1. **Check the state of environment-embedded sensors**

   According to the embedded sensors. if an object starts to move from a place where sensors are installed, the system can detect the exact moment with the related sensors. Even if the object moves from a place where no sensors are installed, the system can also recognize the moment by referring to the reaction of the vibration sensor and other embedded sensors.

2. **Check the state of vibration sensor**

   If a vibration sensor also reacts soon after the embedded sensor reaction, the system estimates that object movement should have something to do with the sensor-embedded place. In other words, the object is very likely to be moved from that place.

3. **Recheck the state of environment-embedded sensors**

   After the vibration sensor reaction, if the system receives the reaction of the same embedded sensor, it indicates that the object must be moved from the place.

To make the general rules mentioned above clearer, we pick up a typical scene to demonstrate the estimation rules in Fig. 9. Figure 9 shows the scene that an object is moved from the table. Firstly, the system can detect the state that something is on the table by checking the reaction existence of the table sensor. Secondary, when the object moves, the vibration sensor reaction will inform us of the timing of motion start. If the object does move from the table, the change of table sensor data will indicate the strong relativity of the object and the table. Thus, the system can estimate the object has been moved from the table in good possibility.

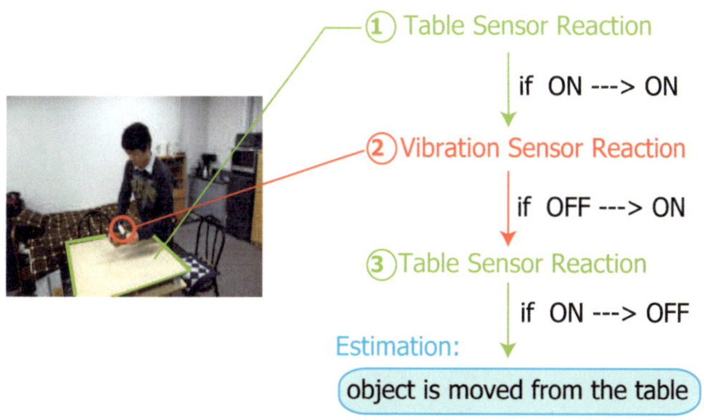

Fig. 9. Sample of Movement Start Estimation

Whereas, the process of object location estimation based on sensors is described as follows.

3.2.2 Estimation of movement end

In this process, we describe an algorithm to detect the end of object movement and to estimate the final location where the object is placed. The system estimates the object location on the assumption that target object has been moving until the vibration sensor reaction disappears.

1. **Receive the change of state of environment-embedded sensors**

   If the system receives the reaction of environment-embedded sensor on the condition that the object is in the moving state, it will suggest that the object is close to the place where the sensor is embedded because of the presupposition that only one user is in the environment.

2. **Check the change of state in vibration sensor**

   The phenomenon that vibration sensor's reaction vanishes under the condition of the embedded sensor being active indicates the high relativity between the object and the place where the sensor is embedded.

3. **Recheck the state of environment-embedded sensors**
   The second time reaction after the vibration sensor becomes inactive allows us to determine that the object is placed on the place.

To make the general rules mentioned above clearer, we pick up a typical scene to demonstrate the estimation rules in Fig. 10. Figure 10 shows the scene that an object is placed in a drawer of a cabinet. Firstly, the system will receive a reaction from the related switch sensor in addition to the continuous reaction from the vibration sensor on the RFID tag, which means the user opens the drawer with the object gripped in his or her hand. Soon after that, if the reaction of the vibration sensor disappears, the possibility of the object being put into the drawer suddenly increases. However, this does not give the confirmation because the location where the object is placed might have no relationship with the drawer at all. Still, if the system receives another reaction from the same switch sensor before long the vibration sensor's reaction vanishes, the connection between the object's location and the drawer becomes even deeper than ever.

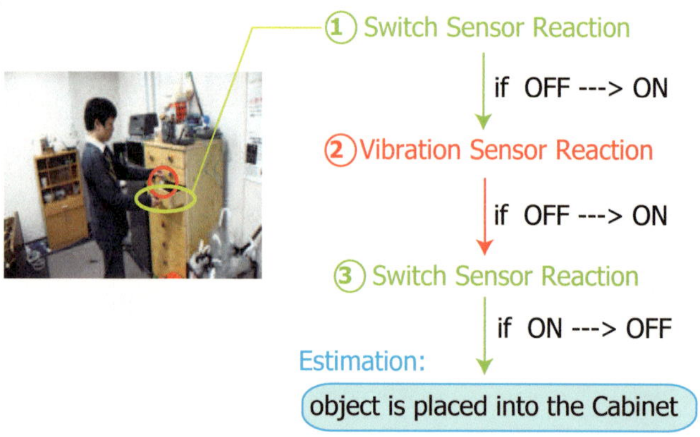

Fig. 10. Sample of Movement End Estimation

In this way, the system estimates the motion and the location of the object by combining the information from vibration sensor and environment-embedded sensors. The concept of the algorithm is easy to follow, but we have to overcome some difficulties to make the estimation algorithm work well. One of the difficulties is to deal with the time delay caused by limited sampling rate, which we used to collect sensor data. For example, an object must have moved before the reaction of the vibration sensor and must have been placed before the reaction disappeared from the system. We estimate the length of time delay from actual experiments and conquer the difficulty by taking the time lag into consideration in estimating object motion.

Another difficulty about the vibration sensor is that sometimes it does not work well. For example, if an object is moved roughly, vibration sensor will keep reacting throughout the movement, however, if an object is moved silently, the vibration reaction will sometimes disappear. This means that the system should not expect continuous vibration reaction during the object movement. Therefore, we defined a time interval to estimate the state of object movement more accurately. If the period from the last reaction of vibration sensor is within that interval, the system still regards the object as moving. Because the length of that time interval depends on the way a user moves object, we decide the parameter from actual experiments.

Although the solution mentioned above works well in estimating object motion, it also has a problem in other aspect. That solution makes it difficult to decide the timing when an object is moved or when an object is placed in real-time because the system has to wait for the time interval to make the decision. It matters when we combine the reaction of a vibration sensor with those of environment-embedded sensors to estimate where the object is placed. According to the estimation algorithm mentioned above, the real-time detection of the object being placed is essential in determining the final location of the object. However, the information that object is placed will be clarified for the first time a few seconds later after the actual point in time. Toward this problem, the system saves a series of sensor reactions into a temporary buffer and applies the proposed estimation rules to those data after the state of object motion fixes. The weakness of this solution is that the system cannot estimate object location in real-time. However, we can know the correct time about the object being placed from the object movement history into which the system stores the object estimation results every sampling rate. In case that the system cannot estimate object location in real-time, it saves the estimated result until the state of object is settled.

3.3 Integration method

So far we explained two estimation algorithms, one is based on pattern recognition, the other is based on sensing technology. Each approach has its own strength and weakness. In our work, as we have mentioned, we integrated these two approaches into one estimation method as shown in Fig. 11. First, the algorithm processes the data from the vibration sensor and embedded sensors to decide whether the target object is in the sensor-embedded area or not(Case 1 in Fig. 11). If the object is in the area, the system uses the data from the vibration sensor and embedded sensors (except for the floor sensor) for the estimation. If the object is not in the area, the algorithm estimates the candidates for object location by using the human position and object motion detected with floor and vibration sensors. In this case, the system determines the most probable object location by integrating the locations estimated on the basis of the RSSI data with those estimated on the basis of the human position data.

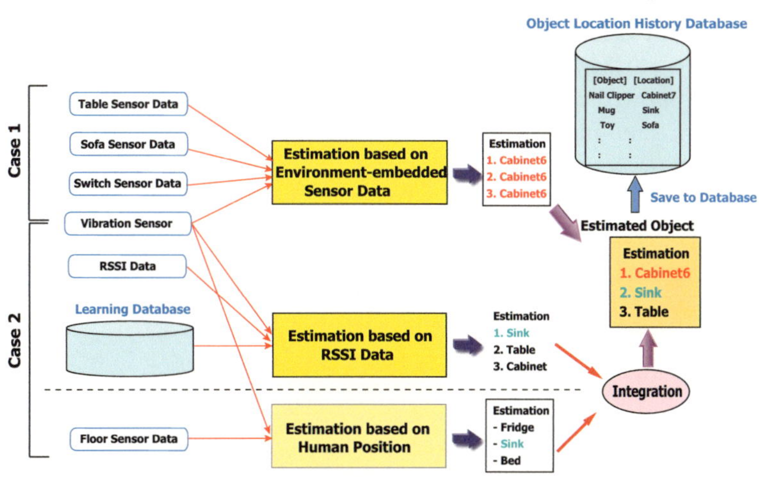

Fig. 11. Object Localization Algorithm

4. Experiments

In this section, we describe the design of our experiments to evaluate the proposed system effectively and the conditions which we used throughout the experiments.

4.1 Experimental design

To evaluate our estimation algorithm from different aspects, various experiments were conducted based on different conditions. First, we conducted exactly the same experiment as many times as the number of pattern recognition methods used in our research, which are k-nearest neighbor (KNN), distance-weighted k-nearest neighbor (DKNN), and three-layered neural network (NN). The purpose is to examine the effect of each method on the estimation performance. In general, classification performance highly depends on the parameters used in each pattern recognition algorithm. For example, the performance of KNN or DKNN is dependent on parameters such as the value of k, whereas the performance of neural network depends on parameters such as the number of nodes in hidden layer. In our experiments, various combination of parameters were examined to find out the best one that presents the highest estimation performance.

Besides, we divided experiment conditions into three types, 1) Estimation only based on RSSI data, 2) Estimation based on RSSI data and sensor data that contains floor sensor data, and 3) Estimation based on RSSI data and sensor data except for floor sensor data. This division enables us to evaluate not only the efficiency of estimation based on RSSI, but the effectiveness of our proposed integration of estimation algorithm.

4.2 Experimental conditions

Our experimental conditions are listed in Fig. 12. As introduced before, *Sensing Room*, shown at the left part of Fig. 12, was our test environment. Throughout the experiments, four objects, shown at the top left part of Fig. 12, were selected as typical daily objects, which were a nail clipper, a mug, a coffee mill, and a stuffed animal. On each object, an active RFID tag including a vibration sensor was attached. Also we assumed 13 locations where objects would be placed and five readers installed at different places. For pattern recognition, we constructed a learning database with about 18,000 data sets stored in it. In more detail, the same amount of RSSI datasets of each location of the labeled 13 locations were stored as training datasets.

In the experiment, a participant leaded a typical daily life using four objects with active RFID tags attached shown in Fig. 13. The system was supposed to estimate the location of each object every sampling frame. The total number of targeted frames was 2520. To provide the localization performance through a sequence of daily activity, we defined the ratio of the number of correctly estimated frames to the total number of targeted frames as the performance metric (Eq.2). In this case, "correct frame" means the frame that both identification and localization succeeded. Furthermore, throughout the experiment, we only adopted first location candidate and ignored the second and third location candidates in order to provide a more reliable indicator of object localization.

$$Accuracy[\%] = \frac{CorrectNumberofFrames}{TotalNumberofFrames} \quad (2)$$

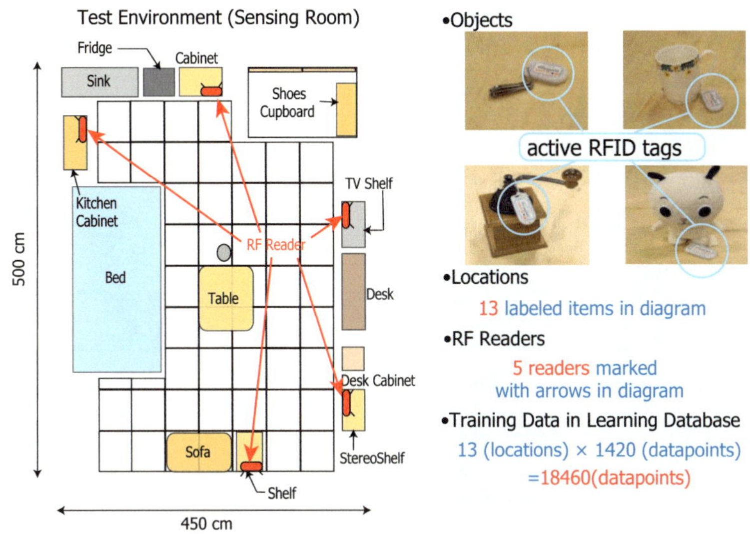

Fig. 12. Experiment Conditions

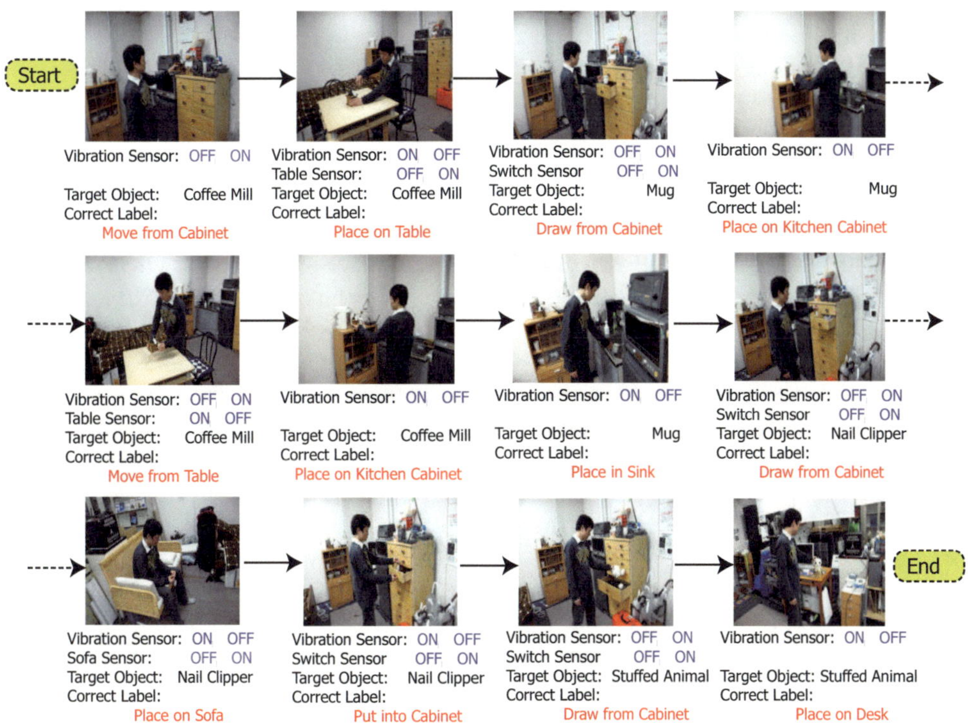

Fig. 13. Experiment Scenes

4.3 Results and discussion

We classified the estimation results by the pattern recognition method used for the localization and by the types of information used for the estimation, as shown in Table 3. There was

(Data from floor, table, sofa, switch, and vibration sensors)

	RSSI Data Only	RSSI and Sensor Data	RSSI and Sensor Data (w/o Floor Sensor)
KNN	50.2%	97.0%	95.3%
DKNN	49.6%	97.0%	95.3%
3-layered NN	22.0%	98.6%	90.6%

Table 3. Location Estimation Results (Only first location candidate is allowed)

(Data from floor, table, sofa, switch, and vibration sensors)

	RSSI Data Only	RSSI and Sensor Data	RSSI and Sensor Data (w/o Floor Sensor)
KNN	60.3%	100.0%	95.3%
DKNN	61.2%	100.0%	95.3%
3-layered NN	36.5%	100.0%	92.6%

Table 4. Location Estimation Results (Up to third location candidate is allowed)

little difference in the results among the pattern recognition method used: KNN, DKNN, and three-layered NN algorithm. There was a substantial difference in the results among the pattern recognition methods used for the estimation. Localization accuracy with only RSSI of the active RFID was 50% at best, whereas when we combined these two approaches followed our proposed estimation algorithm, the accuracy reached 97.0% regardless of the pattern recognition method. With the three-layered NN algorithm, it reached 98.6% at best. Although we used floor sensors for the estimation in the best case, the system still recorded 95.3% even without floor sensors as shown in the table.

The results shown in Table.3 suggest two things in particular. One is that the pattern recognition method used has little effect on the location estimation accuracy. Although we used three kinds of methods such as k-nearest neighbor (KNN), distance-weighted k-nearest neighbor (DKNN), and three-layered neural network (NN), none of them achieved sufficient accuracy in object localization. The main cause of estimation mistakes we suppose is that the object location is far from all the RF readers. As the radio wave is sensitive to environmental noises, the further the distance between tag and reader is, the more unreliable RSSI becomes. The other thing which we noticed from the results is that the lack of estimation accuracy caused by not using floor sensor data can be approximately compensated for by using the proposed algorithms and other simple sensors instead of floor sensors. Although floor sensors can detect human position accurately, they are costly and require complicated maintenance. To reduce the cost and maintenance burden, we estimated object location by using only the RSSI data and data from other simple sensors (table, sofa, and switch sensors). The results indicate that data from a combination of these sensors can achieve accuracy almost equal to that of using floor sensors.

To make a comparison, we conducted another experiment using exactly the same data as the previous experiment. In this case, not only the first location candidate but also the second and the third location candidates were counted. The result is shown in Table 4.

The result shown in Table 4 indicates that the estimation performance does not make a big difference between single location candidate and plural location candidates. Of course, when we allow the second and the third location candidates, the estimation performance improves to some extent. However the improvement is too slight to make a significant impact on the estimation performance of our system.

Although we conducted all the experiments in *Sensing Room*, our object location estimation method does not rely on either the experimental environment or the kinds of sensors. That is to say, our method can work well in any houses as long as the sensors embedded in the house can detect the same kinds of human behavior.

5. Conclusion

In conclusion, we have developed an indoor object localizing method by using active RFID tags and simple switch sensors embedded in the environment. Our system uses 1) a pattern recognition approach to classify the RSSIs collected from several RF readers into a particular location, and 2) the information detected by vibration sensors and environment-embedded sensors to improve the robustness of the method. Although position sensors used in our previous work can detect accurate human position in the environment, we attempted to eliminate them because of their disadvantages by combining simple switch sensors. The results show that our method can be used to estimate the location of daily objects with sufficient accuracy without the use of the position sensors.

One of future work is to reduce the number of RF readers. In our work, we use five active RFID readers placed at different locations so as to cover the whole environment. However, because the unit cost of RF readers is quite expensive, we have to reduce the number of RF readers to ease the economical burden on introducing our system without lowering the performance of object location estimation.

6. References

Cover, T. & Hart, P. (1967). Nearest neighbor pattern classification, *Information Theory, IEEE Transactions on* 13(1): 21 – 27.

Hightower, J., Borriello, G. & Want, R. (2000). SpotON: An indoor 3d location sensing technology based on rf signal strength, *Technical Report UW CSE 2000-02-02*, University of Washington.

Mori, T., Noguchi, H., Takada, A. & Sato, T. (2006). Sensing room environment: Distributed sensor space for measurement of human dialy behavior, *Transaction of the Society of Instrument and Control Engineers* E-S-1(1): 97–103.

Mori, T., Siridanupath, C., Noguchi, H. & Sato, T. (2007). Active rfid-based object management system in sensor-embedded environment, *FGCN2007Workshop: International Symposium on Smart Home (SH'07)*, Jeju Island, Korea, pp. 25–30.

Mori, T., Takada, A., Noguchi, H., Harada, T. & Sato, T. (2005). Behavior prediction based on daily-life record database in distributed sensing space, *International Conference on Intelligent Robots and Systems*, pp. 1833–1839.

Ni, L. M., Liu, Y., Lau, Y. C. & Patil, A. P. (2004). Landmarc: Indoor location sensing using active rfid, *Wireless Networks* 10(6): 701–710.

Pao, T.-L., Cheng, Y.-M., Yeh, J.-H., Chen, Y.-T., Pai, C.-Y. & Tsai, Y.-W. (2008). Comparison between weighted d-knn and other classifiers for music emotion recognition, *International Conference on Innovative Computing Information and Control(ICICIC'08)*, pp. 530–530.

Shih, S.-T., Hsieh, K. & Chen, P.-Y. (2006). An improvement approach of indoor location sensing using active rfid, *the First International Conference on Innovative Computing, Information and Control(ICICIC'06)*, pp. 453–456.

Shimodaira, H. (1994). A weight value initialization method for improving learning performance of the backpropagation algorithm in neural networks, *International Conference on Tools for Artificial Intelligence (ICTAI)*, pp. 672–675.

yao Jin, G., yi Lu, X. & Park, M.-S. (2006). An indoor localization mechanism using active rfid tag, *IEEE International Conference on Sensor Networks, Ubiquitous, and Trustworthy Computing(SUTC'06)*, pp. 40–43.

Zhao, Y., Liu, Y. & Ni, L. M. (2007). Vire: Active rfid-based localization using virtual reference elimination, *International Conference on Parallel Processing (ICPP'07)*, pp. 56–63.

13

RFID Sensor Modeling by Using an Autonomous Mobile Robot

Grazia Cicirelli, Annalisa Milella and Donato Di Paola
Institute of Autonomous Systems for Automation (National Research Council)
Italy

1. Introduction

Radio Frequency Identification (RFID) technology has been available for more than fifty years. Nevertheless, only in the last decade, the ability of manufacturing the RFID devices and standardization in industries have given rise to a wide application of RFID technology in many areas, such as inventory management, security and access control, product labelling and tracking, supply chain management, ski lift access, and so on.

An RFID device consists of a number of RFID tags or transponders deployed in the environment, one or more antennas, a receiver or reader unit, and suitable software for data processing. The reader communicates with the tags through the scanning antenna that sends out radio-frequency waves. Tags contain a microchip and a small antenna. The reader decodes the signal provided by the tag, whereas the software interprets the information stored in the tagŠs memory, usually related to its unique ID, along with some additional information. Compared to conventional identification systems, such as barcodes, RFID tags offer several advantages, since they allow for contactless identification, cheapness, reading effectiveness (no need of line of sight between tags and reader). Furthermore, passive tags work without internal power supply and have, potentially, a long life run. Owing to these advantageous properties, RFID technology has recently attracted the interest of the mobile robotics community that has started to investigate its potential application in critical navigation tasks, such as localization and mapping. For instance, in (Kubitz et al., 1997) RFID tags are employed as artificial landmarks for mobile robot navigation, based on topological maps. In (Tsukiyama, 2005), the robot follows paths using ultrasonic rangefinders until an RFID tag is found and then executes the next movement according to a topological map. In (Gueaieb & Miah, 2008), a novel navigation technique is described, but it is experimentally illustrated only through computer simulations. Tags are placed on the ceiling in unknown positions and are used to define the trajectory of the robot that navigates along the virtual line on the ground, linking the orthogonal projection points of the tags on the ground. In (Choi et al., 2011) a mobile robot localization technique is described, which bases on a sensor fusion that uses an RFID system and ultrasonic sensors. Passive RFID tags are arranged in a fixed pattern on the floor and absolute coordinate values are stored in each tag. The global position of the mobile robot is obtained by considering the tags located within the reader recognition area. Ultrasonic sensors are used to compensate for limitations and uncertainties in RFID system.

Although effective in supporting mobile robot navigation, most of the above approaches either assume the location of tags to be known a priori or require tags to be installed in order to form specific patterns in well-structured environments. Nevertheless, in real environments this is not always possible. In addition, due to the peculiarities of these approaches, no sensor model is presented, because they use only the identification event of RFID tags, without considering at what extent. On the other hand, modelling RFID sensors and localizing passive tags is not straightforward. RFID systems are usually sensitive to interference and reflections from other objects. The position of the tag relative to the receivers also influences the result of the detection process, since the absorbed energy varies accordingly. These undesirable effects produce a number of false negative and false positive readings that may lead to an incorrect belief about the tag location and, eventually, could compromise the performance of the overall system (Brusey et al., 2003; Hähnel et al., 2004).

Algorithms to model RFID system have been developed by a few authors. They use different approaches that varies depending on the type of sensor information used and the method applied to model this information. Earlier works model the sensor information considering only tag detection event. More recent ones, instead, consider also the received signal strength (RSSI) value. This difference is principally due to the evolution of new RFID devices. Nevertheless, in some cases the RSSI is simulated by means of the different power levels of the antenna (Alippi et al., 2006; Ni et al., 2003). (Alippi et al., 2006), for example, suggest a polar localization algorithm based on the scanning of the space with rotating antennas and several readers. At each angular value the antenna is provided with an increasing power by the reader. At the end of each interrogation campaign from each reader, the processing server obtains, for each tag, a packet containing the reader ID, the angular position, the tag ID and the minimum detection power.

One of the first works dealing with RFID sensor modeling is the one proposed in (Hähnel et al., 2004). The sensor model is based on a probabilistic approach and is learnt by generating a statistics by counting the frequency of detection given different relative position between antenna and tag. In (Liu et al., 2006) the authors present a simplified antenna model that defines a high probability region, instead of describing the probability at each location, in order to achieve computational efficiency. In (Vorst & Zell, 2008) the authors present a novel method of learning a probabilistic RFID sensor model in a semi-autonomous fashion.

A novel probabilistic sensor model is also proposed in (Joho et al., 2009). RSSI information and tag detection event are both considered to achieve a higher modelling accuracy. A method for bootstrapping the sensor model in a fully unsupervised manner is presented. Also, in (Milella et al., 2008) a sensor model is illustrated. The presented approach differs from the above in that they use fuzzy set theory instead of probabilistic approach.

In this chapter we present our recent advances in fuzzy logic-based RFID modelling using an autonomous robot. Our work follows in principle the work by (Joho et al., 2009), since we use both signal strength information and tag detection event for sensor modelling. However, our approach is different in that is based on a fuzzy reasoning framework to automatically learn the model of the RFID device. Furthermore, we consider not only the relative distance between tag and antenna, but also their relative orientation.

The main contribution of our work concerns supervised learning of the model of the RFID reader to characterize the relationship between tags and antenna. Specifically, we introduce the learning of the membership function parameters that are usually empirically established by an expert. This process can be inaccurate or subject to the expert's interpretation. To overcome this limitation, we propose to extract automatically the parameters from a set of

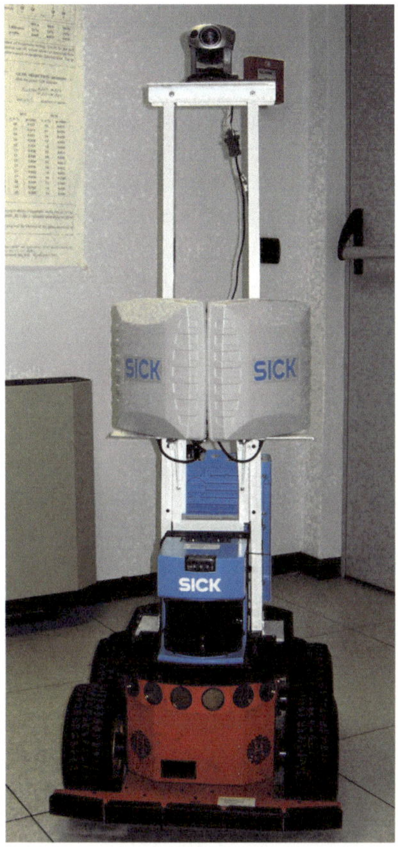

Fig. 1. The mobile robot Pioneer P3AT equipped with two RFID antennas and a laser range scanner.

training data. In particular, Fuzzy C-Means (FCM) algorithm is applied to automatically cluster sample data into classes and also to obtain initial data memberships. Next, this information is used to initialize an ANFIS neural network, which is trained to learn the RFID sensor model. The RFID sensor model is defined as combination of an RSSI model and a Tag Detection Model. Experimental results from tests performed in our Mobile Robotics Lab are presented. The robot used in the experimental session is a Pioneer P3AT equipped with two RFID antennas and a laser range scanner (see Fig. 1). The RFID system is composed by a SICK RFI 641 UHF reader and two antennas, whereas tags are passive UHF tag ŞDogBone" by UPM Raflatac.
The rest of this chapter is organized as follows. Section 2 describes the sensor modelling approach. Experimental results are shown in Section 3. Finally, conclusions are drawn in Section 4.

2. Learning the Sensor Model

In our work, modeling an RFID device means to model the possibility of detecting a tag given its relative position and angle with respect to the antenna. Building this sensor model involves two phases: data acquisition and model learning. The former refers to the strategy we apply in order to collect data. The latter, instead, refers to the construction of the model actually learnt by using recorded data. To model the RFID device we use a Fuzzy Inference System and then to learn it the Adaptive Neuro-Fuzzy Inference System (ANFIS) is applied: the membership function parameters and the rule base are automatically learnt by training an ANFIS neural network on sample instances removing, in this way, the subjectivity of an observer. First sample data are automatically clustered into classes by using the Fuzzy C-Means (FCM) algorithm that at the same time gives an initial fuzzy inference system. Next this information is used to initialize the ANFIS neural network. In the subsequent, both algorithms FCM and ANFIS will be briefly reviewed before the sensor model description.

2.1 Data recording

Past approaches to data recording, presented in related works (Hähnel et al., 2004; Milella et al., 2008), fix a discrete grid of different positions and count frequencies of tag detections for each grid cell. These detections are collected by moving a robot, equipped with one or more antennas, on this grid in front of a tag attached to a box or a wall. This way of proceeding is advantageous in that measurements are taken at known positions and detection rates are computed as tag detection frequencies on a grid. However, this procedure could be tedious and slow if a huge quantity of measurements has to be taken. We follow a slightly different approach to collect the data useful for the sensor model construction. After having deployed a number of tags at different positions in our corridor-like environment, the robot, equipped with the antennas, is manually moved up and down the corridor, continuously recording tag measurements. With tag measurements we refer to the relative distance and relative orientation of the antenna with respect to the tag and RSSI value for each tag detection. Notice that, for each detected tag, the reader reports the tag ID, the RSSI value and which antenna detected the tag. True tag locations are computed by using a theodolite station, whereas the robot positions, in a map of the environment, are estimated applying an accurate self-localization algorithm called Mixture-Monte Carlo Localization (Thrun et al., 2000) by using laser data. Then the relative position between tags and robot are known. Notice that more tags can be simultaneously read by the antenna, therefore the recording phase is, at the same time, rich in data and faster with respect to the above reported ones. In addition, the proposed approach skips the tedious effort of choosing grid points, since a variety of positions for the robot (or antennas) is guaranteed during the guided tour of the environment.

2.2 Fuzzy C-Means (FCM)

FCM is one of the most popular family of clustering algorithms that is C-Means (or K-Means), where C refers to the number of clusters. These algorithms base on the minimum assignment principle, which assigns data points to the clusters by minimizing an objective function that measures the distance between points and the cluster centers. The advantages of these algorithms are their simplicity, efficiency and self-organization. FCM is a variation of C-Means. It was introduced in (Bezdek, 1981). The peculiarity of fuzzy clustering is that data points do not belong to exactly one cluster, but to more than one cluster since each point has associated a membership grade which indicates the degree to which it belongs to the different clusters.

Given a finite set of data point vectors $Z = \{Z_1, Z_2, ..., Z_N\}$, FCM algorithm partitions it into a collection of $C \leq N$ clusters such that the following objective function is minimized:

$$J_q = \sum_{i=1}^{C} \sum_{k=1}^{N} w_{ik}^{q} \|Z_k - V_i\|^2$$

where V_i are the cluster centers for $i = 1, ...C$; w_{ik} is the membership value whit which point Z_k belongs to the cluster defined by V_i center and $q > 1$ is the fuzzification parameter. This parameter in general specifies the fuzziness of the partition, i.e. larger the value of q greater is the overlap among the clusters.
Starting by an initial guess for the cluster centers, FCM algorithm alternates between optimization of J_q over the membership values w_{ik} fixed the cluster centers V_i and viceversa. Iteratively updating w_{ik} and V_i, FCM moves the cluster centers to the optimal solution within the data set. Membership values and cluster centers are computed as follows:

$$w_{ik} = \left[\frac{D_{ik}^2}{\sum_{j=1}^{C} D_{jk}^2} \right]^{\frac{-1}{q-1}} \quad \text{under the constraint } \sum_{i=1}^{C} w_{ik} = 1 \ \forall k$$

$$V_i = \frac{\sum_{k=1}^{N} w_{ik}^q Z_k}{\sum_{k=1}^{N} w_{ik}} \qquad \text{for } i = 1, ..., C$$

where D_{ik} is the distance between i-th cluster center and k-th sample point. The iterative process ends when the membership values and the cluster centers for successive iterations differ only by a predefined tolerance ϵ.

2.3 Adaptive Neuro Fuzzy Inference System (ANFIS)

ANFIS (Jang, 1993) implements a Sugeno neuro-fuzzy system making use of a hybrid supervised learning algorithm consisting of backpropagation and least mean square estimation for learning the parameters associated with the input membership functions.
A typical $i - th$ if-then rule in a Takagi and Sugeno fuzzy model is of the type:

$$\text{if } x_1 \text{ is } A_i \text{ and } x_2 \text{ is } B_i \text{ then } f_i = p_i x_1 + q_i x_2 + r_i$$

where A_i and B_i are the linguistic terms associated with the input variables x_1 and x_2. The parameters before the word Şthen" are the premise parameters, those after Şthen" are the consequent parameters. Thereafter the case of two input variables x_1 and x_2 and two if-then rules is considered for simplicity. The main peculiarity of a Sugeno fuzzy model is that the output membership functions are either linear or constant.
The architecture of the ANFIS network is composed by five layers as shown in figure 2.

Layer 1 The first layer is the input layer and every node has a node function defined by the membership functions of the linguistic labels A_i and B_i. Usually the generalized bell membership function:

$$\mu_{A_i}(x) = \frac{1}{1 + (\frac{x - c_i}{a_i})^{2b_i}}$$

or the Gaussian function is chosen as node function:

$$\mu_{A_i}(x) = e^{-(\frac{x-c_i}{a_i})^2}$$

where a_i, b_i and c_i are the *premise parameters*. The same holds for $\mu_{B_i}(x)$.

Layer 2 In the second layer each node computes the firing strength or weight of the corresponding fuzzy rule as product of the incoming signals.

$$w_i = \mu_{A_i}(x_1)\mu_{B_i}(x_2) \quad i = 1, 2$$

Each node of this layer represents the rule antecedent part.

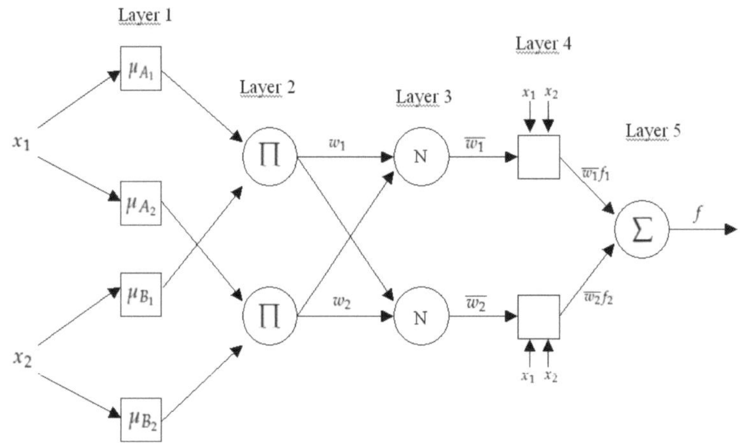

Fig. 2. The ANFIS architecture.

Layer 3 The third layer normalizes the rule weights considering the ratio between the *i*-th weight and the sum of all rule weights:

$$\overline{w}_i = \frac{w_i}{\sum_i w_i} \quad i = 1,2$$

Layer 4 In the fourth layer the parameters of the rule consequent parts are determined. Each node produces the following output:

$$\overline{w}_i f_i = \overline{w}_i (p_i x_1 + q_i x_2 + r_i)$$

where $\{p_i, q_i, r_i\}$ are the *consequent parameters*.

Layer 5 Finally the fifth layer computes the overall output as following:

$$f = \sum_i \overline{w}_i f_i = \frac{\sum_i w_i f_i}{\sum_i w_i}$$

In this work we use Gaussian membership functions and their parameters, the premise parameters, are initialized by using the FCM algorithm described in the previous section. Training the network consists of determining the optimal premise and consequent parameters. During the forward pass the consequent parameters of layer 4 are identified by least square estimate. In the backward pass, instead, the premise parameters are updated applying gradient descent. For more details see (Jang, 1993).

2.4 Sensor Model

Our RFID system, at each tag detection event returns two pieces of information: the tag unique ID and its signal strength. Note that receiving a signal strength measurement implicitly involves that a tag has been detected, but we consider both information in order to make a distinction among the different tags deployed in the environment. However in the rest of the paper, for simplicity, all the variables that will be defined will refer to a generic unique tag, assuming that only relative pose between tag and antenna is relevant. This last assumption is

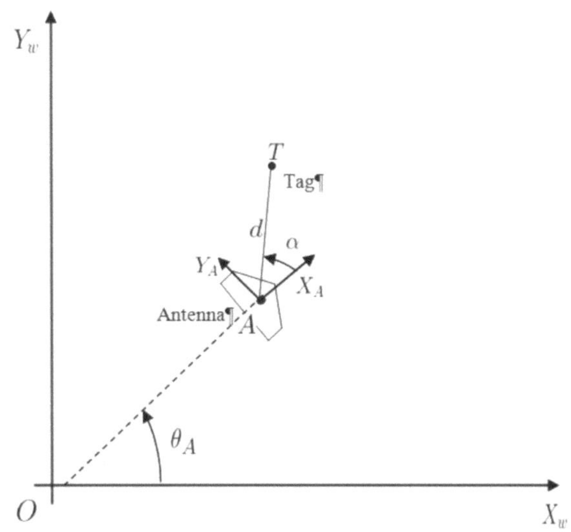

Fig. 3. Relative pose between tag and antenna.

a strong one since, as discussed in the introduction section, the propagation of an RFID signal is influenced by a set of factors dependent on the particular location of each tag: for example the materials the tag is attached on or the surface materials around the tag that can reflect or absorb the electromagnetic waves or the orientation of the tag. While location-dependent models certainly provide more precision they involve a high computational cost. In this work we tried to find a good trade-off between computational overhead and precision, developing a model that bases on both the relative location and the relative orientation of the antenna with respect to the tag.

First of all some variable definitions are needed: we define α the relative orientation between antenna and tag (see Figure 3). As shown in figure 3 points A and T are antenna and tag position in the world reference system X_wOY_w, whereas d is the distance between T and A. Angle θ_A is the absolute orientation of the antenna in the world reference system. Each antenna is mounted on the robot and its pose with respect to the robot is known, therefore θ_A as well as each antenna position is simply obtained by using the absolute pose of the robot in the X_wOY_w system.

As introduced before the sensor model specifies the possibility of detecting a tag given the relative position between antenna and tag. This is modelled by multiplying the expected signal strength $f_s(d,\alpha)$ and the frequency $f_T(d,\alpha)$ of detecting a tag T given a certain distance d and a certain relative orientation α between tag and antenna. In formula:

$$\rho = f_s(d,\alpha)f_T(d,\alpha) \tag{1}$$

In other words the sensor model is obtained combining an RSSI Model (SSM) and a Tag Detection Model (TDM). These two models are learnt by using Fuzzy Inference System, applying ANFIS networks. Both models are detailed in the next two subsections.

2.4.1 RSSI Model (SSM)

RSSI Model is learnt applying the ANFIS network with two inputs, d and α, and one output f_s. Data samples used as input to FCM and ANFIS are the ones stored during the data acquisition

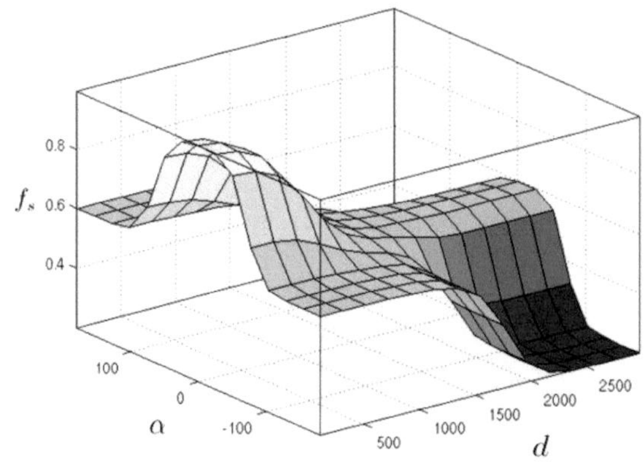

Fig. 4. Input-Output surface for RSSI Model.

phase, as described in section 2.1. First FCM algorithm is applied to initialize the membership function parameters of the input variables considering $C = 3$ clusters (see section 2.2), then ANFIS is trained by using an additional training data set with 12395 samples. Each training data sample is composed by the couple of input variables (d, α) and by the relative signal strength s, stored during data acquisition, opportunely normalized in [0,1]. For simplicity data with distance $d < 3meters$ has been considered. Figure 4 shows the surface that models the if-then rules of the obtained fuzzy inference system. Lighter areas denote higher received signal strength.

2.4.2 Tag Detection Model (TDM)

Tag Detection Model has been built similarly to RSSI model. The input variables are the same (d, α), whereas the output variable is the frequency f_T of detecting a tag given d and α. In order to build the training set, the proper f_T value must be associated to each couple (d, α). This has been done by first discretizing the space into a grid of cells and then counting the number of tag detection events (n_T^+) and the number of no-tag detection events (n_T^-). For each cell the frequency value f_T is evaluated by using its definition formula:

$$f_T = \frac{n_T^+}{n_T^+ + n_T^-}$$

FCM, with $C = 3$ (see section 2.2), is then applied on a first training set of data to obtain an initial fuzzy inference system used as input for ANFIS network. A second training set with 12395 sample data is used to train the network. In this case each sample is composed by the input couple (d, α) and the output value f_T. The obtained input-output surface is displayed in figure 5.

3. Experiments

Some tests have been carried out in our laboratory by using the Pioneer P3AT robot shown in figure 1. The robot has been moved randomly in front of a tag. During navigation a number of points P_i for $i = 1, ..., M$ have been generated uniformly distributed within a circular area around each robot pose. Knowing the absolute position and orientation of the robot and the

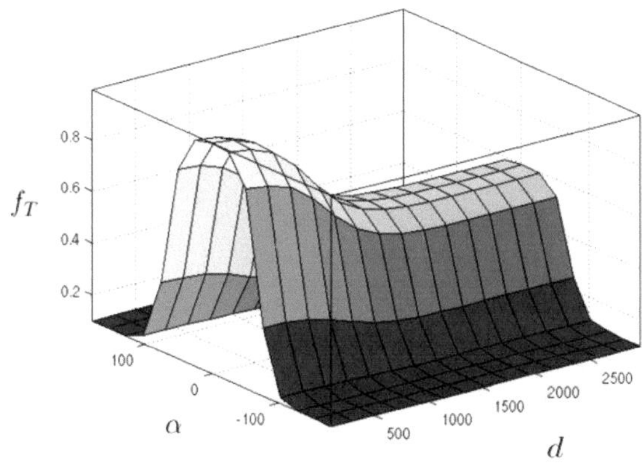

Fig. 5. Input-Output surface for Tag Detection Model.

absolute position of each generated point, the distance and relative orientation between each point P_i and each antenna can be estimated. These data are used as input to the RFID sensor model and the output ρ_i is obtained for each P_i. Figure 6 shows some plots of the described procedure in different poses of the robot. For clarity of display, data relative to only one antenna are plotted. In particular in each plot the green points refer to the set of randomly generated points, the red oriented triangle is the antenna, the blue star point denotes the position of one tag. The green area of each point changes depending on the confidence value ρ_i defined by the sensor model. Higher ρ_i larger the green area around point P_i. As can be seen in figure 6 larger areas are for those points close to the antenna current position and in front of it. Points located behind the antenna have very low ρ_i values and then are represented by smaller green areas.
At the same time, during navigation, the signal strengths s_j received by the RFID reader have been stored and compared with the f_s values returned by the RSSI model. More specifically a path of 200 robot poses Q_j, for $j = 1, ..., 200$, has been considered and for each pose the average $\bar{f}_s^j$ has been estimated considering only those points localized close to the tag:

$$\bar{f}_s^j = \frac{\sum\limits_{k \in \mathbb{P}} f_s^k}{|\mathbb{P}|}$$

where $\mathbb{P} = \{P_i \,:\, \|P_i - T\| < 10cm\}$. Figure 7 shows the error $Error = |\bar{f}_s^j - s_j|$ estimated in each robot pose. As can be noticed the error is below 20% which is a good result considering the high fluctuations of RSSI signals. Furthermore this proves the high reliability of RSSI model and then of RFID sensor model which combines both SSM and TDM.

4. Conclusion

In this chapter an approach for developing an RFID sensor model has been presented. The model is a combination of an RSSI model and a tag Detection model. The main contribution of our work concerns the supervised learning of the model to characterize the relationship between tags and antenna. FCM and ANFIS networks have been used to learn the Fuzzy Inference Systems describing both SSM and TDM. Experimental tests prove the reliability of

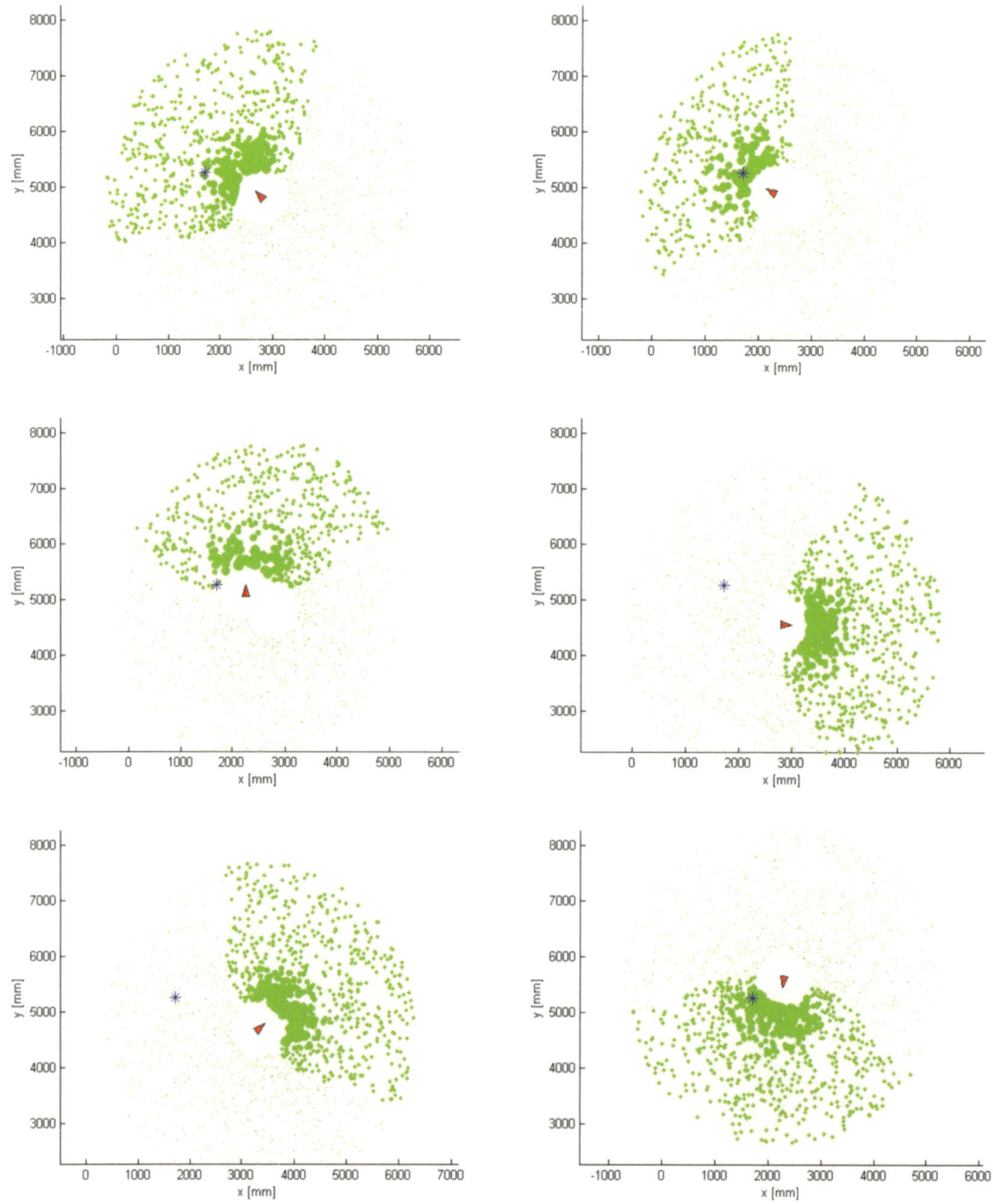

Fig. 6. Sample pictures of points randomly deployed around different robot poses with plotted importance weights (green blobs). The red oriented triangle is one antenna placed on the robot, the blue star point is the tag.

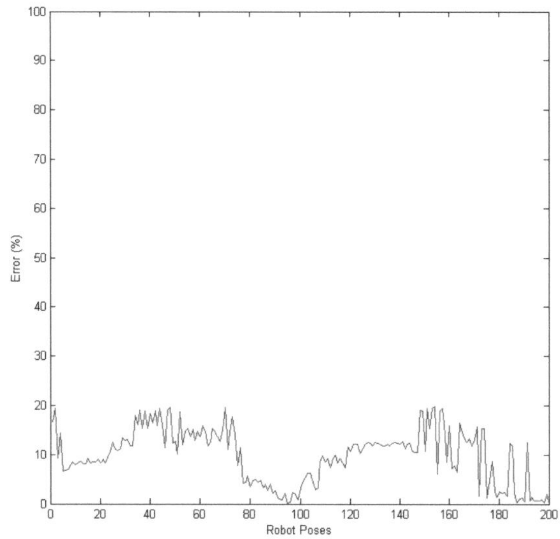

Fig. 7. Percentage average error on f_s values vs. robot poses.

the obtained model. Constructing a reliable sensor model is very important for successive applications such as tag localization, robot localization, just to mention a few. Our future work, in fact, will address these two problems: automatic localization of tags displaced in unknown positions of the environment and, successively, absolute robot localization.

5. References

Alippi, C., Cogliati, D. & Vanini, G. (2006). A statistical approach to localize passive rfids, *IEEE International Symposium on Circuits and Systems*, Island of Kos, Greece.

Bezdek, J. C. (1981). *Pattern Recognition with Fuzzy Objective Function Algorithms*, Plenum, New York.

Brusey, J., Floerkemeier, C., Harrison, M. & Fletsher, M. (2003). Reasoning about uncertainty in location identification with rfid, *IJCAI-03 Workshop on Reasoning with Uncertainty in Robotics*.

Choi, B. S., Lee, J. W., Lee, J. J. & Park, K. T. (2011). A hierarchical algorithm for indoor mobile robots localization using rfid sensor fusion, *IEEE Transactions on Industrial Electronics* to appear.

Gueaieb, W. & Miah, M. S. (2008). An intelligent mobile robot navigation technique using rfid technology, *IEEE Transactions on Instrumentation and Measurement* **Vol. 57**(No. 9).

Hähnel, D., Burgard, W., Fox, D., Fishkin, K. & Philipose, M. (2004). Mapping and localization with rfid technology, *IEEE International Conference on Robotics and Automation (ICRA2004)*, New Orleans, LA, USA.

Jang, S. R. (1993). Anfis: adaptive-network-based fuzzy inference system, *IEEE Trnas. on Systems, Man and Cybernetics* Vol. 23(No. 3): 665–685.

Joho, D., Plagemann, C. & Burgard, W. (2009). Modeling rfid signal strength and tag detection for localization and mapping, *IEEE International Conference on Robotics and Automation (ICRA2009)*, Kobe, Japan.

Kubitz, O., Berger, M., Perlick, M. & Dumoulin, R. (1997). Application of radio frequency identification devices to support navigation of autonomous mobile robots, *IEEE 47th Vehicular Technology Conference*, Phoenix, Arizona, USA, pp. 126–130.

Liu, X., Corner, M. & Shenoy, P. (2006). Ferret: Rfid localization for pervasive multimedia, *8th UbiComp Conference*, Orange County, California, USA.

Milella, A., Cicirelli, G. & Distante, A. (2008). Rfid-assisted mobile robot system for mapping and surveillance of indoor environments, *Industrial Robot: An International Journal* **Vol. 35**(No. 2): 143–152.

Ni, M. L., Liu, Y., Lau, Y. C. & Patil, A. P. (2003). Landmarc: Indoor location sensing using active rfid, *IEEE International Conference on Pervasive Computing and Communications*, Fort Worth, Texas, USA.

Thrun, S., Fox, D., Burgard, W. & Dellaert, F. (2000). Robust monte carlo localization for mobile robots, *Artificial Intelligence* Vol. 128(No. 1-2): 99–141.

Tsukiyama, T. (2005). World map based on rfid tags for indoor mobile robots, *Proceedings of the SPIE*, Vol. Vol. 6006, pp. 412–419.

Vorst, P. & Zell, A. (2008). Semi-autonomous learning of an rfid sensor model for mobile robot self-localization, *European Robotics Symposium*, Vol. Vol. 44/2008 of *Springer Tracts in Advanced Robotics*, Springer, Berlin/Heidelberg, pp. 273–282.

14

Location of Intelligent Carts Using RFID

Yasushi Kambayashi and Munehiro Takimoto
Nippon Institute of Technology & Tokyo University of Science
Japan

1. Introduction

This chapter addresses optimizing distributed robotic control of systems using an example of an intelligent cart system designed to be used in common airports. This framework provides novel control methods using mobile software agents. In airport terminals, luggage carts used by traveler are taken from a depot but are left after use at arbitrary points. It would be desirable that carts be able to draw themselves together automatically after being used so that manual collection becomes less laborious. In order to avoid excessive energy consumption by the carts, we employ mobile software agents and RFID (Radio Frequency Identification) tags to identify the location of carts scattered in a field and then cause them to autonomously determine their moving behavior using a clustering method based on the ant colony optimization (ACO) algorithm.

When we pass through terminals of an airport, we often see carts scattered in the walkway and employees manually collecting them one by one. It is a laborious task and not a fascinating job. It would be much easier if carts were roughly gathered in any way before the laborers begin to collect them. Multiple robot systems have made rapid progress in various fields, and the core technologies of multiple robot systems are now easily available (Kambayashi & Takimoto, 2005). Employing such technologies, it is possible to give each cart minimum intelligence, making each cart an autonomous robot. We realize that for such a system cost is a significant issue and we address one of those costs, the power source. A big, powerful battery is heavy and expensive; therefore such intelligent cart systems with small batteries are desirable to save energy (Takimoto, Mizuno, Kurio & Kambayashi, 2007; Nagata, Takimoto & Kambayashi, 2009; Oikawa, Mizutani, Takimoto & Kambayashi, 2010; Abe, Takimoto & Kambayashi, 2011).

Travelers pick up carts at designated points and leave them in arbitrary places. It is desirable that intelligent carts (intelligent robots) draw themselves together automatically. A simple implementation would be to give each cart a designated assembly point to which it automatically returns when free. That solution is easy to implement, but some carts would have to travel a long way back to their own assembly point, even though they are located close to other assembly points. That strategy consumes unnecessary energy.

To improve efficiency, we employ mobile software agents to locate carts scattered in a field, e.g. an airport, and enable them to determine their moving behavior autonomously using a clustering algorithm based on ant colony optimization (ACO). ACO is a swarm intelligence-based method and a multi-agent system that exploits artificial stigmergy for the solution of combinatorial optimization problems. Preliminary experiments yield a favorable result. Ant colony clustering (ACC) is an ACO specialized for clustering objects. The idea is inspired by the collective behaviors of ants, used by Deneubourg to formulate an algorithm that simulates the ant corps gathering and brood sorting behaviors (Deneuburg, Goss, Franks, Sendova-Franks, Detrain & Chretien, 1991).

We have studied the base idea for controlling mobile multiple robots connected by communication networks (Kambayashi, Tsujimura, Yamachi, Takimoto, & Yamamoto, 2010; Kambayashi & Takimoto, 2005). Our framework provides novel methods to control coordinated systems using mobile agents. Instead of physical movement of multiple robots, mobile software agents can migrate from one robot to another so that they can minimize energy consumption for aggregation. In this chapter, we describe the details of implementation of the multi-robot system using multiple mobile agents and static agents that implement ACO as well as the location system using RFID. The combination of the mobile agents augmented by ACO and mobile multiple robots with RFID opens a new horizon of efficient use of mobile robot resources. We report here our experimental observations of our robot cart system.

Quasi-optimal cart collection is achieved in three phases. The first phase collects the positions of the carts. One mobile agent issued from the host computer visits scattered carts one by one and collects the positions of them. The precise coordinates and orientation of each robot are determined by interrogating RFID tags under the floor carpet. Upon the return of the position collecting agent, the second phase begins wherein another agent, the simulation agent, performs the ACC algorithm and produces the quasi-optimal gathering positions for the carts. The simulation agent produces not only the assembly positions of the carts but also the moving routes and waiting timings for avoiding collisions; i.e. the entire behaviors of all the intelligent carts. The simulation agent is a static agent that resides in the host computer. In the third phase, a number of mobile agents, the driving agents are issued from the host computer. Each driving agent migrates to a designated cart, and drives the cart to the assigned quasi-optimal position that was calculated in the second phase.

The behaviors of each cart are determined by the simulation agent. It is influenced, but not determined, by the initial positions and the orientations of scattered carts, and is dynamically re-calculated as the configuration of the field (positions of the carts in the field) changes. Instead of implementing ACC with actual carts, one static simulation agent performs the ACC computation, and then mobile agents distribute the sets of produced driving instructions. Therefore our method eliminates unnecessary physical movement and provides energy savings.

The structure of the balance of this paper is as follows. In the second section, we review the history of research in this area. In the third section, we describe the controlling agent system that performs the arrangement of the intelligent carts. The agent system consists of several static and mobile agents. The static agents interact with the users and compute the ACC algorithm and the simulation of the intelligent carts' behaviors. The other mobile agents gather the initial positions of the robots and drive the carts to the assembly positions. The fourth section describes how each robot determines its coordinates and orientation by sensing RFID tags under the floor carpet. The fifth section describes the ACC algorithm we have employed to calculate the quasi-optimal assembly positions and moving instructions for each cart. Finally, in the sixth section, we summarize the work and discuss future research directions.

2. Background

Kambayashi and Takimoto have proposed a framework for controlling intelligent multiple robots using higher-order mobile agents (Kambayashi & Takimoto, 2005). The framework helps users to construct intelligent robot control software using migration of mobile agents. Since the migrating agents are of higher order, the control software can be hierarchically assembled while they are running. Dynamic extension of control software by the migration of mobile agents enables the controlling agent to begin with relatively simple base control software, and to add functionalities one by one as it learns the working environment. Thus we do not have to make the intelligent robot smart from the beginning or make the robot learn by itself. The controlling agent can send intelligence later through new agents. Even though the dynamic extension of the robot control software using the higher order mobile agents is extremely useful, such a higher order property is not necessary in our setting. We have employed a simple, non-higher-order mobile agent system for our intelligent cart control system. We previously implemented a team of cooperative search robots to show the effectiveness of such a framework, and demonstrated that that framework contributes to energy savings for a task achieved by multiple robots (Takimoto, Mizuno, Kurio & Kambayashi, 2007; Nagata, Takimoto & Kambayashi, 2009; Oikawa, Mizutani, Takimoto & Kambayashi, 2010; Abe, Takimoto & Kambayashi, 2011). Our simple agent system should achieve similar performance.

Deneuburg formulated the biologically inspired behavioral algorithm that simulates the ant corps gathering and brood sorting behaviors (Deneuburg, Goss, Franks, Sendova-Franks, Detrain, & Chretien, 1991). His algorithm captured many features of the ant sorting behaviors. His design consists of ants picking up and putting down objects in a random manner. He further conjectured that robot team design could be inspired from the ant corps gathering and brood sorting behaviors (Deneuburg, Goss, Franks, Sendova-Franks, Detrain, & Chretien, 1991). Wang and Zhang proposed an ant inspired approach along this line of research that sorts objects with multiple robots (Wang & Zhang, 2004).

Lumer improved Deneuburg's model and proposed a new simulation model that was called Ant Colony Clustering (Lumer, & Faieta, 1994). His method could cluster similar object into a few groups. He presented a formula that measures the similarity between two data objects and designed an algorithm for data clustering. Chen et al have further improved Lumer's model and proposed Ants Sleeping Model (Chen, Xu & Chen, 2004). The artificial ants in Deneuburg's model and Lumer's model have considerable amount of random idle moves before they pick up or put down objects, and considerable amount of repetitions occur during the random idle moves. In Chen's ASM model, an ant has two states: active state and sleeping state. When the artificial ant locates a comfortable and secure position, it has a higher probability of being in the sleeping state. Based on ASM, Chen has proposed an Adaptive Artificial Ants Clustering Algorithm that achieves better clustering quality with less computational cost.

Algorithms inspired by behaviors of social insects such as ants that communicate with each other by the stigmergy are becoming popular (Dorigo & Gambardella, 1996). Upon observing real ants' behaviors, Dorigo et al found that ants exchanged information by laying down a trail of a chemical substance (pheromone) that is followed by other ants. They adopted this ant strategy, known as ant colony optimization (ACO), to solve various optimization problems such as the traveling salesman problem (TSP) (Dorigo & Gambardella, 1996). Our ACC algorithm employs pheromone, instead of using Euclidian distance to evaluate its performance.

Retrieving accurate location information of an object is a key to enabling a robot to perform any task. Recently, to solve the localization problem in indoor environment, various RF-based techniques have been developed (Hightower & Borriello, 2001; Finkenzeller, 2003). One of the often used methods is to use reference stations with well-known locations, where the transponder position is reported by nearby stations (Werb & Lanzl, 1998). A criticism of this approach is that the number of stations and the interval between the stations affects the accuracy (Kim & Chong, 2007). In order to deal with the problem of object localization, Kim and Chong have developed a location sensing system incorporating 315 MHz active RFID devices. Their system enables a robot to gather information and understand the context of the environment using the RFID transponder's location as well as stored information. They use a directional antenna to determine the arrival direction of the RF signal. Even though their system successfully provides guidance to mobile robots to a stationary target, the RFID transponder fails to provide precise coordinate of a robot. Also the system is costly because of the use of active RFID devices.

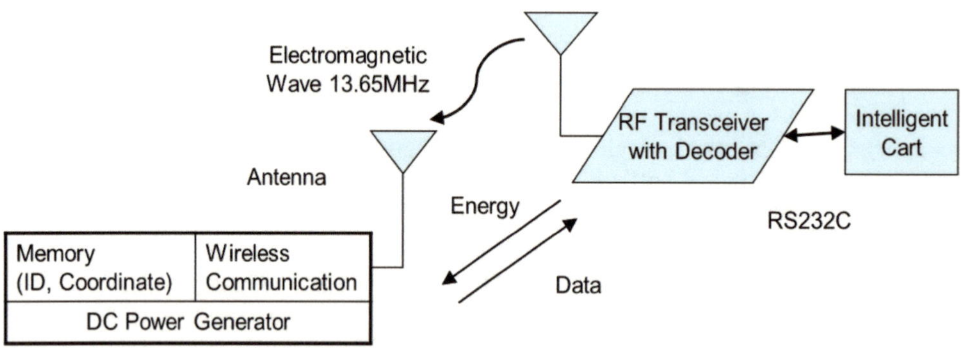

Fig. 1. Architecture of a passive RFID tag

In order to get precise coordinates for the robot, we utilized another well-known and well-analyzed approach. We embedded a very large number of passive RFID devices that emit very weak signals under the floor carpet of our test environment so that only robots extremely near the RFID tag can receive the positional signal. Since 13.56 MHz passive RFID devices are very inexpensive, we believe our rather primitive method is practical enough to retrieve the location of a robotic intelligent cart. Fig. 1 shows the architecture of a passive RFID tag.

3. Controlling agents

Our system model consists of carts and a few kinds of static and mobile software agents. All the controls for the intelligent carts as well as ACC computation performed in the host computer are achieved through the static and mobile agents. They are: 1) user interface agent (UIA), 2) operation agents (OA), 3) position collecting agent (PCA), 4) clustering simulation agent (CSA), and 5) driving agents (DA). All the software agents except UIA and CSA are mobile agents. A mobile agent traverses carts scattered in the field one by one to collect their coordinates. After receiving the moving instructions computed by a static agent, many mobile agents migrate to the carts and drive them to the assembly positions. Fig. 2 shows the interactions of the cooperative agents to control an intelligent cart.

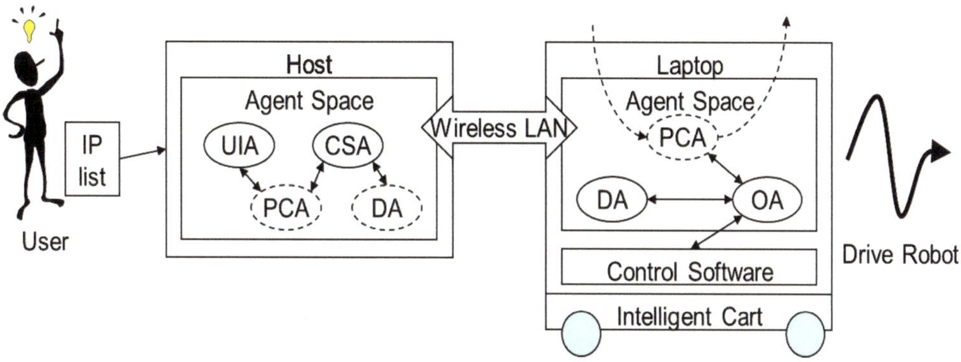

Fig. 2. Cooperative agents to control an intelligent cart

The followings are details of each agent:

1. User Interface Agent (UIA): The user interface agent (UIA) is a static agent that resides on the host computer and interacts with the user. It is expected to coordinate the entire agent system. When the user creates this agent with a list of IP addresses of the mobile robots, UIA creates PCA and passes the list to it.
2. Operation Agent (OA): Each intelligent cart has at least one operation agent (OA). It has the task that the cart on which it resides is supposed to perform. Each intelligent cart has its own OA. Currently all operation agents (OA) have a function for collision avoidance and a function to sense RFID tags embedded in the floor carpet to detect its precise coordinates in the field.
3. Position Collecting Agent (PCA): A distinct agent called position collecting agent (PCA) traverses intelligent carts scattered in the field one by one and collects their coordinates. PCA is created and dispatched by UIA. Upon returning to the host computer, it hands the collected coordinates to the clustering simulation agent (CSA) for ACC.
4. Clustering Simulation Agent (CSA): The host computer houses the static clustering simulation agent (CSA). This agent actually performs the ACC algorithm by using the coordinates collected by PCA as the initial positions, and produces the quasi-optimal assembly positions of the carts, and then performs yet another simulation to produce instructions each cart follows to reach its assigned goal position. Upon terminating the simulation and producing the procedures for all the intelligent carts, CSA creates a number of driving agents (DA).
5. Driving Agent (DA): The quasi-optimal arrangement coordinates, as well as procedures to reach them, produced by the CSA are delivered by driving agents (DA). One driving agent is created for each intelligent cart, and it contains the set of procedures for the cart. The DA drives its intelligent cart to the designated assembly position. DA is the intelligent part of the intelligent cart.

OA detects the current coordinate of the intelligent cart on which it resides. The coordinate information is retrieved from the RFID tag under that cart. Each cart has its own IP address and UIA hands in the list of the IP addresses to PCA. First, PCA migrates to an arbitrary cart and starts hopping between them one by one. It communicates locally with OA, and writes the coordinates of the cart into its own local data area. When PCA gets all the coordinates of the robots, it returns to host computer. Upon returning to the host computer, PCA creates CSA and hands in the coordinate data to CSA which computes the ACC algorithm.

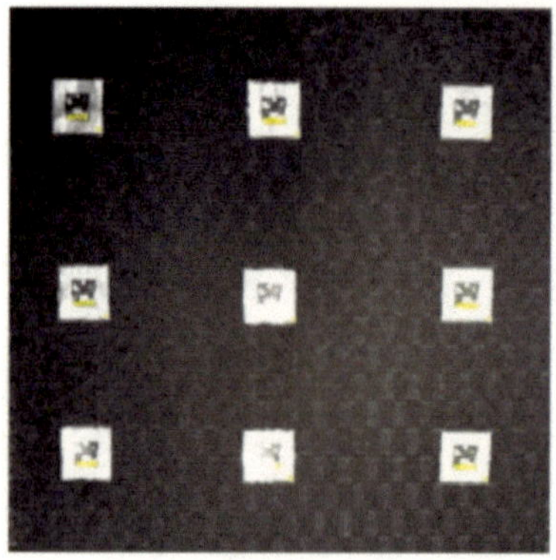

Fig. 3. RFID tags under a carpet tile

Each intelligent cart employs RFID (Radio Frequency Identification) to get precise coordinates. We set RFID tags in a regular grid shape under the floor carpet tiles (Fig. 3). The tags we chose have a small range so that the position-collecting agent can obtain fairly precise coordinates from the tag. Also, the cart itself has a basic collision avoidance mechanism using infrared sensors.

CSA is the other static agent and its sole role is ACC computation. When CSA receives the coordinate data of all the carts, it translates the coordinates into coordinates for simulation, and then performs the clustering. When CSA finishes the computation and produces a set of assembly positions, it then creates the set of procedures for intelligent cart movements by simulating the behaviors of all the carts. The simulator produces one set of moving procedures for each cart including not only the course it takes but also timing to avoid collisions so that each cart can know exactly what it should do just for a few steps. The reason why the procedures are for a few steps is that it is impractical to forecast the entire behaviors of all the carts from the initial positions to the final destinations. Therefore the system repeats the cycle of collecting positions, performing ACC, and driving robots for short distance for several times to accomplish the clustering task.

Then CSA creates enough DA so that there is one for each cart. The DA convey the sets of procedures to the intelligent carts. Each DA receives its destination IP address from PCA, and the set of procedures for the destination cart, and then migrates to the destination cart. Each DA has a set of driving procedures that drives its assigned cart to the tentative destination, while it avoids collision. OA has the basic collision detection and avoidance procedures, and DA has task-specific collision avoidance procedures.

We have implemented the entire multiple mobile software agents for mobile robot control using Agent Space (Satoh, 1999). Agent Space is a library for constructing mobile agents developed by Satoh. By using its library, the user can implement a mobile agent environment with the Java language.

4. Determining the positions and directions

In order to get the coordinate of each robot, each robot employs an RFID reader/writer module ARW13T-RS01 by Niigata Seimitsu Co. Ltd. (Niigata Seimitsu, 2011). The module is based on ISO 15693 and uses carrier frequency 13.56MHz. Fig. 4 shows the effective scope of sensing. As the diagram shows the effective scope of the RFID module is quite narrow so that the robot can sense the RFID tag exactly under its body.

For driving carts on a quasi-optimal route, one needs not only the precise coordinates of each robot but also the direction each cart faces. In order to determine the direction that it is facing, each cart moves straight ahead in the direction it is currently facing and obtains two positions (coordinates) from RFID tags under the carpet tiles. Determining current orientation is important because there is a high cost for making a cart rotates through a large angle. It is desirable that each cart be assigned rather simple forward movements rather than complex movement with several direction-changes when there are obstacles to avoid. Therefore whenever OA is awake, it performs the two actions, i.e. obtaining the current position and calculating the current direction.

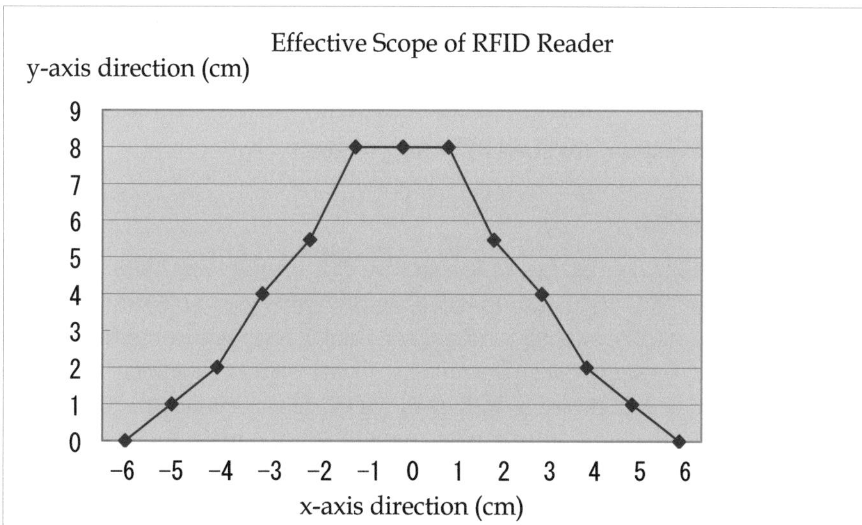

Fig. 4. Effective scope of the RFID reader/writer module

Since each carpet tile has nine RFID tags, as shown in Fig. 3, the cart is supposed to obtain the current position as soon as OA gets the sensor data from the RFID module. If OA can not obtain the position data, it means the RFID module can not sense a RFID tag, OA moves the cart forward until the RFID module senses a RFID tag. Once OA obtains the current position, it drives the cart a short distance until the RFID module detects the second RFID tag. Upon obtaining two positions, it is a simple computation to determine the direction to which the cart is moving, as shown in Fig. 5.

In a real situation, we may find the cart close to the wall and facing it. Then we can not use the simple method just described above since the cart would move forward and collide with the wall. In order to accommodate such situations, we gave RFID tags near the wall a special signal to the cart that tells it that it is at the end of the field (near the wall) so that the cart that senses the signal can rotate to the opposite direction. This is only required in our experimental implementation because the current collision detection mechanism does not otherwise work well enough to avoid collision with the wall. When the cart finds wall while moving to obtain the direction, it arbitrarily changes the direction, and starts obtaining two positions again.

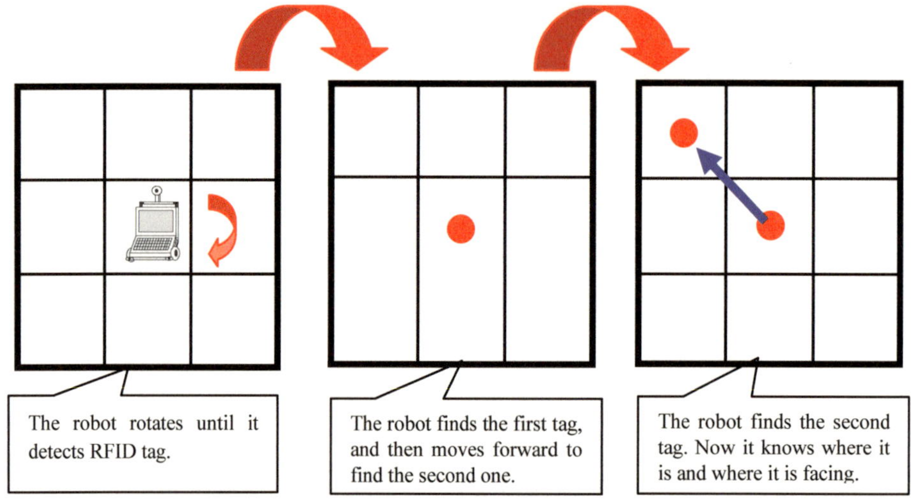

Fig. 5. Determining orientation from two RFID tag positions

5. Ant colony clustering

PCA collects the positions of the carts obtained by OA on the carts, and passes them to CSA that performs the ACC algorithm to calculate assembly points to which the carts are directed. In traditional ACO systems, artificial ants leave pheromone signals so that other artificial ants can trace the same path (Deneuburg, Goss, Franks, Sendova-Franks, Detrain, & Chretien, 1991). In our ACC system, however, we made objects have a pheromone such that more objects are clustered into a place where strong pheromone is sensed. Our randomly walking artificial ants have high probability to pick up an object with weak pheromone, and to put the object where they sense strong pheromone. They are not supposed to walk long distances so the artificial ants tend to pick up a solitary object and produce many small clusters of objects. When a few clusters are generated, they tend to grow.

Since the purpose of traditional ACC is clustering or grouping objects into several different classes based on selected properties, it is desirable that the generated chunks of clusters grow into one big cluster such that each group has distinct characteristics. In our system, however, we want to produce several roughly clustered groups of the same type, and move each cart a minimal distance. (We assume we have one kind of cart, and that we do not want carts to move long distances.) Therefore our artificial ants have the following behavioral rules:

1. An artificial ant's basic behavior is random walk, and when it finds an isolated object, it picks up the object.
2. When the artificial ant finds a cluster with certain number of objects, it tends to avoid picking up an object from the cluster. This number can be updated later.
3. When the artificial ant with an object finds a cluster, it put down the object so that the object is adjunct to one of the objects in the cluster.
4. If the artificial ant cannot find any cluster with certain strength of pheromone, it just continues a random walk.

By the restrictions defined in the above rules, the artificial ants tend not to convey objects long distances, and produce many small heaps of objects at the first stage. In order to implement the first feature, the system locks objects with certain number of adjoining objects, and no artificial ant can pick up such a locked object. The number for locking will be updated later so that artificial ants can bring previously locked objects in order to create larger clusters. When the initially scattered objects are clustered into a small number of heaps, then the number of objects that makes objects locked is updated, and the further activities of the artificial ants re-start to produce smaller number of clusters. We have implemented a simulator to evaluate our ACO algorithm, and succeeded in producing not only the quasi-optimal gathering positions but also the precise behaviors of all the carts to reach the calculated gathering positions. Fig. 6 shows the behavior of an artificial ant. The details of the ACO algorithm we have designed and implemented are reported in (Kambayashi, Tsujimura, Yamachi, Takimoto & Yamamoto, 2010).

Even though the ACC algorithm achieves some quasi-optimal clustering, it is hard to have confidence that we can make all the carts autonomously move to the gathering positions so that they form the quasi-optimal clusters. As the carts move toward the assigned positions, the configuration of the entire cart system changes and also each cart may perform unexpected behaviors, such as slipping tires over-stirring as well as under-stirring. Therefore we need to dynamically re-perform ACC to re-calculate the new goal position for each cart based on the current position after all the carts move independently. On the other hand, we found from the preliminary experiments that excessive re-computation of ACC might produce one large cluster, and that was not what we desired. In this section, we discuss how frequently we perform ACC to guide carts so that they form quasi-optimal clusters. In order to give each cart not only the goal position but also the procedure to reach it, we have implemented the simulator to execute a simulation of all the carts so that we can assign one precise behavior for each cart as well as perform ACC algorithm, and confirmed that it is feasible to produce the behavioral instructions for all the carts so that ultimately they can reach the quasi-optimal positions.

Upon receiving the positions of all the carts, the CSA immediately starts ACC simulation and produces the quasi-optimal clusters. Fig. 7b show the calculated clusters the CSA proposed from the initial cart positions in Fig. 7a. At this moment, none of the carts know how to behave, i.e. which direction and how far each should go, because each cart has not get been assigned its goal position. Upon obtaining the goal clusters, CSA performs yet another simulation. At this time, the simulation imitates the behaviors of all the carts from the initial positions to the tentative goal positions. This simulation produces the moving routes and wait timing for avoiding collisions. Fig. 7c shows one of the best samples of the simulated clusters that can be actually formed by the moving carts. Surprisingly the simulated clusters are quite similar to the calculated clusters by the ACC.

This second simulation produces the precise coordinate and wait timing for each cart, thus generating a set of moving procedures for each cart. One procedure consists of not only a route for the cart but also the timing when the cart stops and how long it waits to avoid collision against other colleague carts.

Upon constructing all the instructions for all the carts, a number of DAs are created to convey the procedures. Each DA drives corresponding cart to the simulated gathering positions. At a certain time, all the carts move toward the assigned positions through the instructions given by DA. After that period, the configuration of the field changes, then we need to re-perform the ACC again so that it reflects the current configuration (positions of all the carts).

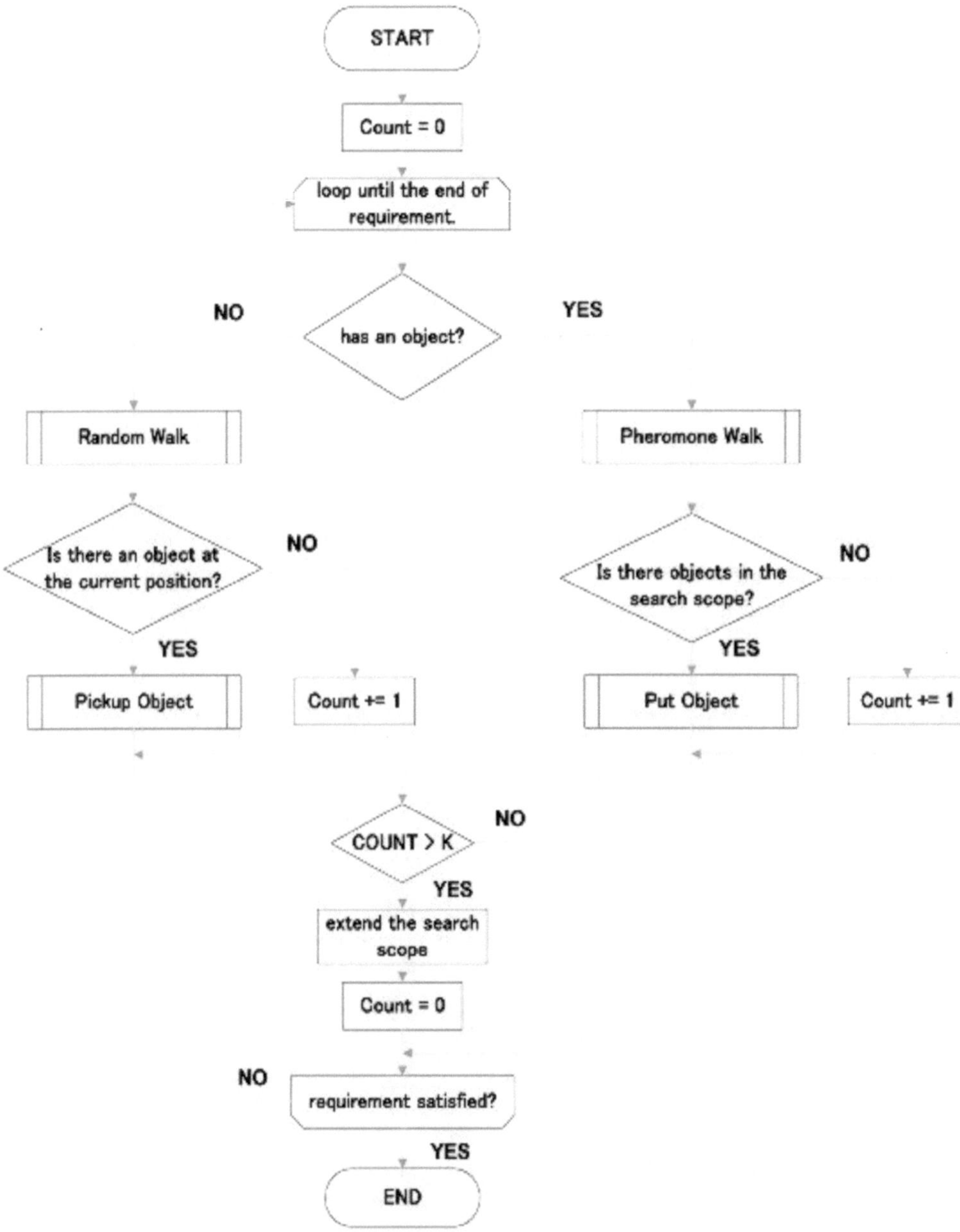

Fig. 6. The behavior of the artificial ant

Table 1 shows the summary of the numerical experiments. We have set the field size to be 100 times 100, and performed three trials of the number of ACC we perform to achieve final clustering, i.e. 1, 3, and 5. We can observe that performing ACC for five times produced the best result, i.e. the least moving distance of aggregate of all the carts. One means the simulator performs ACC only once, then all the carts try to form the given clusters. They form clusters anyway, but the number of clusters is relatively larger than in the case of larger numbers of repetitions of ACC. For 300 carts, about four clusters seem to be optimal number of clusters (see Fig. 6b). The figures Fig. 6a through 6c are typical simulation results obtained in our 300-cart example.

Performing the ACC three times produces a near optimal number of clusters, but the moving total distance is not optimal. It may be that small number (one and three in our experiments) of ACC performances can only give each cart rough idea of what to do and the carts execute futile movements. Performing the ACC five times drastically improves efficiency. We confirmed our conjecture that repetition of the ACC produces better results. The average moving distance becomes optimal. Fig. 6c shows such an optimal clustering case. The lines denote the trace; we can see each carts moves almost optimal route to form the clusters.

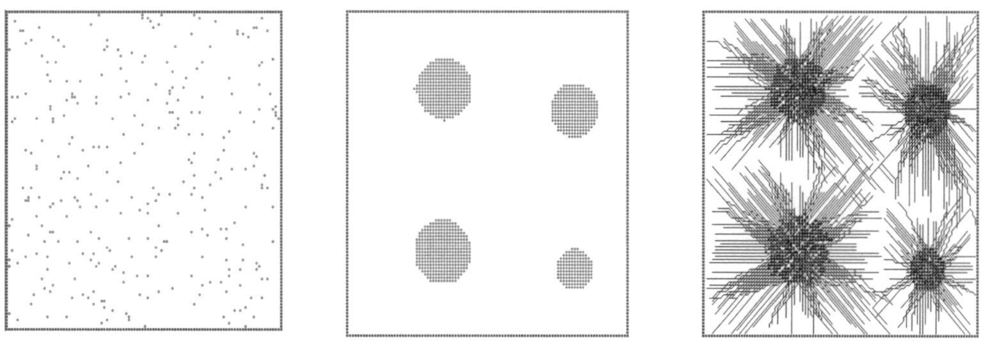

a) Initial positions of the cart

b) Clusters produced by ACC

c) Formed clusters and correspondence to the initial positions

Fig. 7. Simulated results of the intelligent cart behaviors

6. Conclusion and future directions

We have presented a framework for autonomous intelligent carts connected by communication networks. Mobile and static software agents collect the coordinates of scattered carts and implement the ant colony clustering (ACC) algorithm in order to find quasi-optimal positions to assemble the carts. Making mobile multiple robots perform the ant colony optimization is enormously inefficient. Therefore a static agent performs the ACC algorithm in its simulator and computes the quasi-optimal positions for the intelligent carts. Then other mobile software agents carrying the requisite set of procedures migrate to the carts, and drive the carts using the sequence of control commands that is constructed from the computed set of procedures.

No. of ACC	Average Cluster Size	Average Moving Distances
1	6.3	11.63
3	3.3	11.83
5	3.7	2.92

Table 1. Averages of Calculated Moving Distances and Simulated Moving Distances

Since our control system is composed of several small static and mobile agents, it shows an excellent scalability. Our control framework can be applied not only intelligent cart system but also any general purpose multiple mobile robot systems. Then the number of mobile robots increases, we can simply add the increases number of mobile software agents to direct the mobile robots. The user can enhance the control software by introducing new features as mobile agents so that the multiple mobile robot system can be extended dynamically while the robots are working. Also mobile agents decrease the amount of the necessary communication. They make mobile multiple robot applications possible in remote sites with unreliable communication. In unreliable communication environments, the multiple mobile robot system may not be able to maintain consistency among the states of the robots in a centrally controlled manner. Since a mobile agent can bring the necessary functionalities with it and perform its tasks autonomously, it can reduce the necessity for interaction with other sites. In the minimal case, a mobile agent requires that the connection be established only when it performs migration (Binder, Hulaas & Villazon, 2001). The concept of a mobile agent also creates the possibility that new functions and knowledge can be introduced to the entire multi-agent system from a host or controller outside the system via a single accessible member of the intelligent multiple mobile robot system (Kambayashi & Takimoto, 2005). While our current application is simple cart collection, the system should have a wide variety of applications.

We have implemented a team of mobile robots that simulate intelligent carts to show the feasibility of our model (see Fig. 8.) In the current implementation, an agent on the robot can obtain fairly precise coordinates of the robots from RFID tags.

The ACC algorithm we have proposed is designed to minimize the total distance intelligent carts move. We have analyzed and demonstrated the effectiveness of our ACC algorithm through simulation, performing several numerical experiments with various settings. Although we have so far observed favorable results from the experiments in the simulator, we must apply the results of the simulation to a real multiple mobile robot system, and we are aware of its difficulty.

Although the intelligent carts are roughly gathered, if they are not serially aligned, the human laborers would have to align them one by one. The work is still hard and must be facilitated.

We are now re-implementing the ACC algorithm to use not only the sum of moving distances but also the orientation of each robot so that the mobile robots that are facing similar direction tend to get together. This can be achieved through employing vector values for pheromone values to compute ACC simulation.

Even though ACC computation with robots' orientations make the calculation more complex, compared with the time for robot movements, the computation time for the ACC algorithm is negligible. Even if the number of artificial ants increases, the computation time will increase linearly, and the number of objects should not influence the computation's complexity. Because any one step of each ant's behavior is simple, we can assume it takes constant execution time. Even though apparently obvious, we need to confirm this with quantitative experiments. As we mentioned in the previous section, we need to design the artificial ants to have certain complex features that change their ability to adapt to circumstances. We defer this investigation to our future work.

Fig. 8. A team of mobile robots work under control of mobile agents

For another investigation, we are designing a completely different intelligent cart assembly system where entire multiple mobile robot system performs the ACC by using mobile software agents (Oikawa, Mizutani, Takimoto & Kambayashi, 2010; Abe, Takimoto & Kambayashi, 2011). We call the system distributed ant colony clustering where two new mobile software agents are introduced to control the driving agents. They are ant agents and pheromone agents. The ant agents represent the artificial ants. They see the mobile robots and influence the driving agents to the quasi-optimal positions. The pheromone agents represent pheromone and diffuse the effects by migrations. In general, making mobile multiple robots perform the ant colony optimization has been impossible due to enormous inefficiency. Our approach, however, does not need the ant-like robots and other special devices, because those agents are just software agents and do not require any physical movements. So far we are not aware of any multiple robot system that integrates pheromone as a control means as Deneuboug envisaged in his seminal paper (Deneuburg, Goss, Franks, Sendova-Franks, Detrain & Chretien, 1991).

By using pheromone agents, we can implement the serialization of clustered carts (Abe, Takimoto & Kambayashi, 2011). In many ways, we have room to improve our automatic cart collection system before integrating everything into one working multiple robot system.

7. Acknowledgment

We acknowledge our colleagues Yasuhiro Tsujimura and Hidemi Yamachi. We enjoyed fruitful discussions with them. We appreciate Kimiko Gosney who gave us useful comments. This work is partially supported by Japan Society for Promotion of Science (JSPS), with the basic research program (C) (No. 20510141), Grant-in-Aid for Scientific Research.

8. References

Abe, T., Takimoto M. & Kambayashi, Y. (2011). Searching Targets Using Mobile Agents in a Large Scale Multi-robot Environments. *Proceedings of the Fifth KES International Conference on Agent and Multi-Agent Systems: Technologies and Applications, Lecture Notes in Artificial Intelligence 6682,* Berlin, Heidelberg, New York Springer-Verlag, pp. 211-220, ISBN 978-3-642-21999-3, Manchester, UK, Jun. 2011

Binder, W.J., Hulaas, G. & Villazon, A. (2001). Portable Resource Control in the J-SEAL2 Mobile Agent System. *Proceedings of International Conference on Autonomous Agents,* pp. 222-223, ISBN 1-58113-326-X, Montreal, Canada, May 2001

Chen, L., Xu, X. & Chen, Y. (2004). An adaptive ant colony clustering algorithm, *Proceedings of the Third IEEE International Conference on Machine Learning and Cybernetics,* pp. 1387-1392, ISBN 0-7803-8403-2, Shanghai, China, Aug. 2001

Deneuburg, J., Goss, S., Franks, N., Sendova-Franks, A., Detrain, C. & Chretien, L. (1991). The Dynamics of Collective Sorting: Robot-Like Ant and Ant-Like Robot, *Proceedings of First Conference on Simulation of Adaptive Behavior: From Animals to Animats,* pp. 356-363, MIT Press, Cambridge, MA, U.S.A., ISBN 0-262-63138-5,

Dorigo, M. & Gambardella, L.M. (1996). Ant Colony System: a Cooperative Learning Approach to the Traveling Salesman, *IEEE Transaction on Evolutionary Computation,* (Apr. 1997), pp. 53-66, Vol.1, No.1, ISSN 1089-778X

Evolution Robotics Inc. Homepage (2008). http://www.evolution.com/.

Finkenzeller, K. (2003). *RFID Handbook: Fundamentals and Applications in Contactless Smart Cards and Identification,* (2nd ed.), John Wiley & Sons Ltd., ISBN 0-470-84402-7, Chichester, U.K.

Hightower, J. & Borriello, G. (2001). Location Systems for Ubiquitous Computing, *IEEE Computer Magazine,* Vol. 34, No. 8, Aug. 2001, pp. 57-66, ISSN 0018-9162

Kambayashi, Y. & Takimoto, M. (2005). Higher-Order Mobile Agents for Controlling Intelligent Robots, *International Journal of Intelligent Information Technologies*, Vol.1, No.2, pp. 28-42, ISSN 1548-3657

Kambayashi, Y., Tsujimura, Y., Yamachi, H., Takimoto M. & Yamamoto, H. (2009). Design of a Multi-Robot System Using Mobile Agents with Ant Colony Clustering, *Proceedings of the 42nd Hawaii International Conference on System Sciences*, IEEE Computer Society, ISBN 0-7695-3450-3, CD-ROM, Hawaii, U.S.A., Jan. 2008

Kambayashi, Y., Tsujimura, Y., Yamachi, H., Takimoto M. & Yamamoto H. (2010). Integrating Ant Colony Clustering Method to Multi-Robots Using Mobile Agents, In: *Multi-Agent Applications with Evolutionary Computation and Biologically Inspired Technologies: Intelligent Techniques for Ubiquity and Optimization,* S. H. Chen, Y. Kambayashi & H. Sato, (Ed.), pp. 174-191, IGI Global, ISBN 953-7619-81-7, Hershey, PA, U.S.A.

Kim, M. & Chong, N. Y. (2007). RFID-based mobile robot guidance to a stationary target. *Mechatronics*, Vol.17, No.4-5, May-Jun. 2007, pp. 217-229, ISSN 0957-4158

Lumer, E. D. & Faieta, B. (1994). Diversity and Adaptation in Populations of Clustering Ants, From Animals to Animats 3: *Proceedings of the 3rd International Conference on the Simulation of Adaptive Behavior*, pp. 501-508, Cambridge, MA, U.S.A., MIT Press, ISBN 0-262-53122-4

Nagata, T., Takimoto, M. & Kambayashi, Y. (2009). Suppressing the Total Costs of Executing Tasks Using Mobile Agents. *Proceedings of the 42nd Hawaii International Conference on System Sciences*, IEEE Computer Society, ISBN 0-7695-3450-3, CD-ROM, Hawaii, U.S.A., Jan. 2008

Niigata Seimitsu Co. Ltd. Homepage (2011). http://www.niigata-s.co.jp/products/module/reader_writer/arw13t.htm

Oikawa, R., Mizutani, M., Takimoto M. & Kambayashi, Y. (2010). Distributed Ant Colony Clustering Using Mobile Agents and Its Effects. *Proceedings of the 14th KES International Conference on Knowledge-based and Intelligent Information and Engineering Systems: Part I, Lecture Notes in Artificial Intelligence 6276*, Berlin, Heidelberg, New York Springer-Verlag, pp. 198-208, ISBN 978-3-642-15386-0, Cardiff, UK, Sep. 2010

Satoh, I. (1999). A Mobile Agent-Based Framework for Active Networks. *Proceedings of IEEE Systems, Man, and Cybernetics Conference*, pp. 161-168, ISBN 0-7803-5731-0, Tokyo, Japan, Dec. 1999

Takimoto, M., Mizuno, M., Kurio, M. & Kambayashi, Y. (2007). Saving Energy Consumption of Multi-Robots Using Higher-Order Mobile Agents, *Proceedings of the First KES International Symposium on Agent and Multi-Agent Systems: Technologies and Applications, Lecture Notes in Artificial Intelligence 4496*, Berlin, Heidelberg, New York Springer-Verlag, pp. 549-558, ISBN 3-540-72830-6, Wroclaw, Poland, May, 2007

Wang T. & Zhang, H. (2004). Collective Sorting with Multi-Robot. *Proceedings of the First IEEE International Conference on Robotics and Biomimetics*, pp. 716-720, ISBN 0-7803-8614-8, Shenyang, China, Aug. 2004

Werb, J. & Lanzl, C. (1998). Designing a Positioning System for Finding Things and People Indoors. *IEEE Spectrum*, Vol.35, No. 9, pp. 71-78, ISSN 0018-9235

15

Services, Use Cases and Future Challenges for Near Field Communication: the StoLPaN Project

Carlo Maria Medaglia[1], Alice Moroni[1], Valentina Volpi[1], Ugo Biader Ceipidor[1], András Vilmos[2] and Balázs Benyó[3]

[1]CATTID- "Sapienza" University of Rome,
[2]SafePay
[3]Budapest University of Technology and Economics
Italy

1. Introduction

Over the last couple of decades, the mobile phones have become more and more integrated in everyday people's lives. According to the International Telecommunication Union (ITU), at the end of 2009 the penetration of mobile phones in the developed economies was 97% (ITU, 2009 as cited in European Payments Council [EPC], 2010). Not only the penetration has grown, but also functions and services accessible from mobile phones have improved, thanks to the growing availability of communication technologies and to the miniaturization of electronic components inside consumer's devices.

As an example, thanks to location technologies such as GPS, the mobile phone can nowadays be used to locate a person's position and, thanks to wireless communication technologies, such as Wi-Fi, GPRS and UMTS, personalized content can be delivered on the person's device. Automatic identification technologies such as RFID are not excluded from this process of integration and convergence of communication interfaces in the worldwide most popular electronic device. In fact, one of the latest short-range auto-ID technologies, named Near Field Communication (NFC), can be described as the integration of an RFID HF reader into a mobile phone, moreover allowing the device to act as a contactless smart card. NFC originates from RFID technology, but differently from the latter it supports bidirectional communication, making possible to overcome the distinction among tag and reader device. From the technical point of view, NFC operates within the unlicensed Radio Frequency band of 13,56 MHz and it is used to provide easy short-range connectivity to different electronic devices. As described in the standards (ISO/IEC 18092, ECMA-340 and ETSI 102.190), the communication distance is up to 20 cm but the real operating distance is strictly related to the antenna dimension and design: if integrated in a mobile phone, the antenna has to be very small and so the communication distance is typically 2-4 cm. The standard for contactless smart cards (ISO/IEC 14443) is also related to NFC operational mode: data stored on the NFC secure chip can be read in the same way proximity cards OF proximity cards.

As mobile phones are the most popular personal devices worldwide, extending them with an RFID reader and a "card emulation mode" makes it possible to create a wide set of applications and services, from mobile payments and ticketing, to mobile social networking and pervasive advertising services. The main goal of companies and merchants is to give people services they really need, moreover improving their experience as consumers or users.

2. NFC services and use-cases

As it enables several ways of use, NFC is a really adaptable technology. It can operate in three communication modes, based on three different types of interaction between the mobile phone and other NFC-enabled devices (Figure 1).

Fig. 1. NFC communication modes

The first one is the above mentioned "card emulation mode", that is compatible with existing contactless infrastructure (based on ISO/IEC 14443 standard). In a card emulation mode scenario, the mobile phone communicates the sensitive information stored inside an internal secure, tamper-resistant chip (Secure Element - SE) linked to the NFC module by moving itself close to a reader, for example a validation machine on a bus or a POS terminal in a shop, etc. In this way the mobile device acts as an authentication token for enabling services that require high level of security, such as mobile payment, mobile ticketing, mobile identity, access control and so on. Compared to a traditional card support normally used for enabling the above mentioned services, the mobile device offers additional capabilities, first of all a display and a keyboard, as well as the possibility to connect to the Internet by a mobile network, via GPRS/UMTS or via Wi-Fi.

The second type of interaction is peer-to-peer communication between two NFC-enabled devices (for example two NFC mobile phones, or an NFC phone and a printer, or a camera). As they touch together, they can exchange data and information such as the business card or the identification key necessary to quickly initiate a configuration (e.g. pairing) with Bluetooth or Wi-Fi connections.

The third and last type is the read/write mode that enables the mobile phone to initiate a service by reading the information stored in a RFID tag, maybe added to a smart poster situated in a strategic place, for example the bus stop, the shopping centre or the pub. The information stored in the tag consists of a few kilobyte: it can be a URL address, a phone

number or a short text message. When the mobile phone touches the tag and reads the data inside, the related application on the device can connect the mobile browser to a web page that can also be a social network profile.
The interoperability and the easy integration with different wireless and wired technologies favor the use of NFC in a multi-application scenario. Moreover, if used within a smart poster or combined on a kiosk or a totem, NFC can be a very useful technology to clear the information overload giving the right information in the right place at the right moment.
Over the last half-decade, several pilots involving services based on NFC technology have been conducted all over the world. One of the first pilot was hosted in Caen, France, in 2005: it enabled two-hundred mobile phone users to interact with NFC smart posters, as well as with car parking machines and ticket terminals. Once the NFC was tested from a technical point of view, the consumers acceptance was checked and the results have showed that end users like the quickness and convenience of NFC technology (Kannainen, 2009).
Currently, the most tested services are those involving NFC in card emulation mode, such as proximity payments and ticketing. They usually follow a client-side payment and validation model based on offline micro-payment transactions using the existing contactless infrastructure.

3. The role of the StoLPaN consortium in the development of NFC technology

3.1 Research challenges and objectives

Although NFC is one of the most promising technology in the near future, one of the main problems in creating an NFC mass market is the lack of application level standardization and interoperability: while the low-level standardization process has been already completed by standardization bodies such as ISO/IEC, ETSI and also by NFC Forum, as detailed in the following paragraph, there are still significant differences between NFC implementations (devices, operating systems, etc.) that have to be considered.
The StoLPaN (Store Logistics and Payment with NFC) consortium, which includes companies and research centers all over Europe, has worked on overcoming standardization and interoperability issues, mainly dealing with application level standardization, creating in this way a transparent technical environment for the Service Providers and a homogeneous user experience for the customers.
The two major research challenges the consortium faced during a three-year project (2006-2009) co-funded by the European Commission within the 6th Framework Programme were related to the multi-application operation in the mobile handset and the elaboration of a smart retail procedure and payment process based on auto-ID technologies such as RFID and NFC. The whole project aimed to reach a consistent user experience contributing to the industry progress.
The StoLPaN Project was based on three main research questions:

- What is the technical environment that can ensure the integration of NFC based services and applications provided by different Service Providers into a single device, irrespective of its features and operating system?
- How can a smart retail scenario including payment process be implemented making use of auto-ID technologies such as RFID and NFC?
- What business model can support a mass adoption of NFC based services?

Besides investigating the research challenges and related questions, the following objectives were part of the defined goals of the StoLPaN project:

- To elaborate transparent logistical and technical processes that can be relied on in the various business interactions that provides a tool for dynamically managing individual service portfolios even with international scope.
- To develop a handset-independent JME-based mobile host application in order to provide seamlessly multiple services.
- To demonstrate the effectiveness of the proposed proof-of-concept solution in a smart retail environment.

3.2 Overview of the NFC standardization process

In July 2006, when the StoLPaN Project started, the NFC low-level standards already completed were the Near Field Communication Interface and Protocol-1 (NFCIP-1), about "modulation schemes, codings, transfer speeds, and frame format of the RF interface, as well as initialization schemes and conditions required for data collision control during initialization" [ISO/IEC 18092 (ECMA-340), 2004] and the Near Field Communication Interface and Protocol-2 (NFCIP-2), about "the mechanism to detect and select one communication mode" between Card Emulation, Peer-to-Peer and Reader/Writer modes [ISO/IEC 21481 (ECMA-352)].

The ECMA International started to work on Near Field Communication standard in 2002. An apposite Task Group was charged to define signal interfaces and protocols. In December 2002 Near Field Communication Internet Protocol-1 (NFCIP-1) was adopted as Standard ECMA-340, which came to a second edition on December 2004. ISO/IEC adopted the NFCIP-1 as a standard in December 2003.

On the other side, the first, historical edition of ECMA-352 that specifies the mechanism to select one communication mode between Card Emulation, Peer-to-Peer and Reader/Writer modes was published in December 2003 and approved as an ISO/IEC standard (ISO/IEC 21481) in 2005. ECMA published the second edition of the ECMA-352 (Near Field Communication Interface and Protocol-2) standard on June 2010.

Also the European Telecommunications Standards Institute (ETSI) is involved in the standardization of NFC technology. More in detail, the ETSI's Smart Card Platform group (ETSI/SCP), which deals with the SIM card specifications, has worked on specifying the interface between the SIM card (but in this context it is better to refer to the UICC - Universal Integrated Circuit Card, which is the physical support on which the logical module known as Subscriber Identity Module, or SIM, is present) acting as a Secure Element and the NFC chipset stored in the phone (ETSI TS 102 622, ETSI TS 102 613). The first standard adopted by ETSI SCP, approved in 2007, was related to the physical connection between the UICC and the NFC chip: as there was only one free contact in the UICC, the connection with the NFC chipset was required to use one single wire and, due to this reason, was named "Single Wire Protocol" (SWP). In 2008 ETSI also approved a protocol standard that specifies how chips embedded in NFC mobile phones communicate between each other. This standard is called "Host Controller Interface" (HCI).

Nevertheless, the interoperability remains a crucial issue in the NFC ecosystem. Some of the most used contactless technology compatible with NFC, like MIFARE system developed by NXP or FeliCa by Sony, are currently proprietary standards, and use their own security solutions. Anyway, since Nokia, NXP and Sony were the first to developed the NFC lower layer communication, when the open standard NFCIP was developed, the backward compatibility with the proprietary solutions was assured (Mayes & Markantonakis, 2008).

Although the physical layer of the NFC technology could refer to well-known established international standards available to enhance interoperability, the application standardization scenario was more uncertain. A number of researchers and associations has worked on defining a standard in implementing NFC applications and services. The NFC Forum, a non-profit industry association that promotes the use of NFC short-range wireless interaction in consumer electronics, mobile devices and PCs (NFC Forum, http://www.nfc-forum.org/home/) has defined a common format for message encapsulation, called NFC Data Exchange Format (NDEF), for exchanging data between an NFC Forum Device and another NFC Forum Device or an NFC Forum Tag (NFC Forum, 2006). In 2007 the GSM Association (GSMA), a global trade association representing more than 700 mobile network operators across 218 countries of the world launched two initiatives for the development of NFC applications into a common ecosystem: the Mobile NFC initiative, supported by nineteen MNOs, which have worked together to develop a common vision on Mobile NFC services, promoting the development of a stable and efficient ecosystem and preventing market fragmentation (GSMA, 2007a, 2007c) and the Pay-Buy-Mobile project, supported by thirty-four of the world's largest MNOs (GSMA, 2007b), focused on contactless and mobile payment scenario, trying to standardize the operational approach with NFC technology.

Another relevant initiative for promoting the development of mobile payments based on contactless and NFC technology in Europe was conducted by the AEPM (Association Européenne Payez Mobile), an association established in October 2008 in France. The AEPM has published a set of specifications that define a common approach for enabling mobile contactless proximity payments. The technical solution proposed by both the GSMA and the AEPM is based on the UICC as the Secure Element for a mobile payment transaction.

The Mobey Forum (Mobey Forum, 2010) and the Global Platform (Global Platform, 2006) have respectively published guidelines and technical documentation focused on possible alternatives and multi-application architecture for the Secure Element.

Focusing on the evolution of the market scenario during the years covered by the StoLPaN project, the first commercial NFC-enabled mobile phone was launched on the market by Nokia (Nokia 6212, which supports UMTS connectivity) in 2008. One year earlier, in 2007, Nokia launched the Nokia 6131 NFC, a fully integrated NFC mobile phone with GPRS connectivity, which was still a prototype. Even Motorola, Samsung, LG and Sagem, other stakeholders of the sector, developed their own NFC prototype models. At that time there was still uncertainty about the Secure Element's position. Nokia first built it into the handset (embedded Secure Element). Now, encouraged by GSMA, it seems that most of the handset manufacturers accepted to put the SE on UICC. The NFC ecosystem moves all around this issue and the related business and operating models driven forward from competing forces (manufacturers, MNO's, banks, etc.).

4. How the StoLPaN consortium contributed to the industry progress

Under the scenario described above, the StoLPaN consortium contributed to the ecosystem and industry progress by working on the management and distribution of services in a dynamic and open scenario, presenting a proposal for the post-issuance procedures for multi-application SEs (StoLPaN consortium, 2008a). Moreover, the consortium has detailed the technical environment necessary for the dynamic management of NFC services, building a proof-of-concept prototype of the NFC wallet application (StoLPaN consortium, 2008b) and demonstrated the effectiveness and efficiency of the solution in a smart retail environment (StoLPaN consortium, 2009a).

In the following sections, we will give an overview on the main findings of the StoLPaN project in reference to the three abovementioned issues.

4.1 Dynamic application management on SEs

The StoLPaN consortium identified the post issuance and application management as the key issues to be faced to offer users a variety of NFC applications on the same device and building so a real ecosystem. In the current section we are describing the technical model for the dynamic card content management of Secure Elements placed in a mobile handset.

The StoLPaN model provides a solution for dynamic application management that can be uniformly used in local, as well as in global operations, both between parties with consolidated contractual relationship, but also between ad hoc business partners.

One of the main challenges of the new mobile NFC service environment is that the present card issuance models are not designed to support the dynamic post issuance personalization process because the Service Providers:

- have absolutely no control over the cards - we also refer to them as Secure Elements (SEs) in the following - on which their application should be stored, except making a decision of using them or not;
- have no control over the other applications stored in the same Secure Element;
- may not know personally their clients, and may not have the chance for a physical contact with either the Secure Element Issuer or with the user.

The existing technical diversity calls for early standardization of the post issuance and personalization process, otherwise local island solutions will prevail and the technology will not be capable of adequately serving several hundred million users and thousands of Service Providers expected when the NFC services will reach critical mass.

The new logistical and technical model that ensures the necessary openness and interoperability fulfils the following criteria:

- open relationship between the Service Providers, the Secure Element issuers and the Users;
- technical transparency for the Service Providers;
- service homogeneity for the user.

It is possible to establish one single logistical process for loading, personalization and life cycle management of applications that is technologically agnostic and supports all types Secure Elements, even multiple ones, in the communication devices. In this environment the user can freely decide which Service Providers and what services to use, and can even enjoy the services of multiple Service Providers. The result is free access to the customer base of the multiple SE issuers, and improved economics of developing NFC services.

4.1.1 Issues to consider

The first issue that has to be taken into account for providing dynamic card content management of Secure Elements is related to the complexity of the mobile NFC value ecosystem. In fact, main characteristics of the service environment are as follows:

- There are potentially many Service Providers who would place their applications on the Secure Element in the mobile handsets and there are potentially multiple Secure Element issuers in any countries.
- The Secure Element is an external condition for all the Service Providers, without any possibility of influencing its technical parameters, with only a "take it or leave it" choice.

- Users are mobile and may wish to use NFC services even if they are abroad. They may also wish to dynamically change the service portfolio they use even after the issuance of the Secure Element, adding services here and there and deleting others when they are not needed any more.
- A number of Service Providers are global and prefer to have uniform solutions for the applications deployment and operation, irrespective of the specific market where the application is delivered.
- Even if the various NFC applications have their own specifics and requirements, they need to share the same Secure Element and must coexist side-by-side, and eventually interoperate.

There are many constrains in the mobile NFC world which are unknown for either the Service Providers or the Secure Element issuers in their current operations. This is a new way of doing business, without anyone being able to substantially influence the service environment and with the necessity of cooperating with even unknown partners. There is a need for a transparent logistical model and a technical solution that can ensure uniform procedures for the parties involved, where they do not necessarily have to negotiate and elaborate the details of each and every interaction and where even previously unknown business partners can seamlessly realize the procedures of application deployment and management. Without such an approach, the NFC ecosystem will not prevail, and will not be satisfactory business model and an user friendly, valuable service for the customers.

Another relevant issue is related to the already discussed need of application-level interoperability: the industries working with NFC technology such as ETSI and GSMA are now busy addressing the many different technical issues. However, application interoperability has not been set as a target by any standardization body. Being able to hide handset and NFC platform specifics, so that any application can be loaded on any handset, will allow NFC services to be easily deployed worldwide, addressing millions of consumers. Just this aspect makes a good enough business case for the majority of the Service Providers to launch their services on NFC-enabled devices and will lead to the success of NFC.

The service distribution needs to be defined, too. There is a number of actors involved in the NFC value chain but their roles and form of cooperation is not adequately defined. It means that the distribution of any NFC service application requires special, individual agreements between the partners involved.

The target of the StoLPaN research and development activities is to support the market to develop the application environment to a level where all interoperability issues are solved. We have reviewed the majority, if not all, NFC related standards, use cases and business models. We have then condensed the wide range of requirements into a few preconditions, processes and interfaces and presented our findings in white papers (StoLPaN Consortium, 2008a, 2008b). The research carried out by the StoLPaN consortium led us to conclude that, to support quick proliferation of NFC services, the industry has to achieve a homogeneous, dynamic service environment which would mean that even after the issuance of the cards any services can be loaded onto virtually any Secure Element and managed through the whole life cycle of the application. In the subsequent sections we will introduce a logistical and technical process that provides a solution for these requirements.

4.1.2 Dynamic card content management and roles within the ecosystem

Before describing the logistical process that will contribute to the establishment of a truly global, interoperable NFC service environment based on a standardized dynamic card

content management process, we need to clarify what we mean for card content management and to describe the roles necessary to build up the NFC ecosystem.
First of all, let's set what we mean for card content management. In general, there are two types of content management:

- **card content management** which includes the establishment/deletion of the new Security Domain, as well as the application loading and personalization of the smart card application;
- **application content management** which covers the product/portfolio management of the Service Provider.

This section is focusing on card content management, while the complementary application content management process will be discussed in par. 4.2. The solution elaborated below is explained in reference to a Secure Element, which can indifferently be a SIM card, an embedded chip or an SD card as well. The concept discussed in this section also provides the algorithm for a selection process in case of multiple Secure Elements deployed in the same mobile handset. It is the industry's - industries' - task to elaborate such standards in the NFC domain that, if followed, would provide a transparent environment for both the service providers and for the users. Right now, when there are only few commercial handsets on the market, when only trial and pre-commercial NFC services are operating it is the right time to work on these standards without hurting the commercial or financial interest of the parties involved. It would be a great mistake to miss the present opportunity for standardization, for the elaboration of uniform solutions. If not done properly the result will be a more complex and more expensive NFC service environment.
The proposed card content management process is quite complex, but also very flexible with various roles/functions included. We have identified the functions necessary to complete the process, but the actual actors in these roles will always be situation driven.
First of all, we have defined a set of *primary roles*. The complete service scenario cannot be performed without these roles (actors) however one single actor may assume more than one role in the process. The roles (actors) and their relationships are shown in Figure 2.
The primary roles are defined as follows:

- *User*: The User is the person who initiates the request for the post issuance and personalization by selecting the application/service for use.
- *Secure Element Issuer*: The Issuer of the Secure Element is the entity who controls the SE, it has the right to decide over the utilization of the storage capacity of the SE. To exercise these rights the SE Issuer needs to be in possession of the secret key(s) that allow general control over the management of SE. It has the right to define the rules about who, when and under what conditions may utilize storage space on the SE, or may deploy card content onto the SE.
- *Service Provider*: A Service Provider may be anyone wishing to deploy/manage a service application on the Secure Element. There should not be made any distinctions between the Service Providers if they comply with the industry standard security and SE Issuer specific business conditions. Service Providers can be large service operators, like banks, or transport operators for ticketing applications, but they can also be retailers for their loyalty and other programs as well as authorities for various ID cards, etc.

Besides the primary roles described, in order to provide the full functioning, economic and convenient service, the following *support roles* needs to be considered:

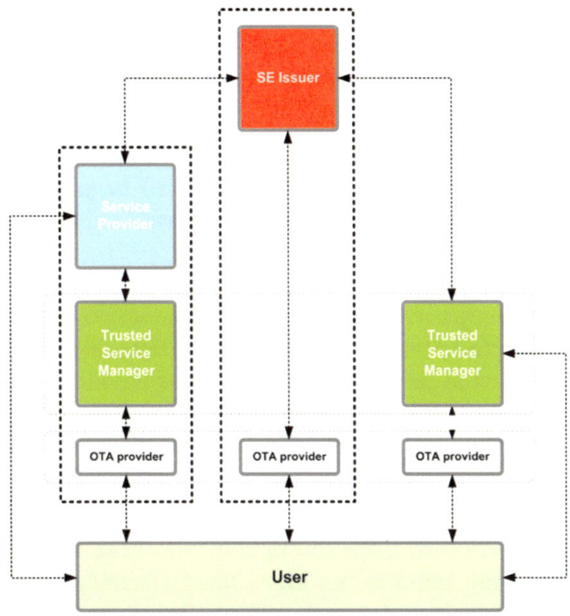

Fig. 2. Roles within the NFC ecosystem

- *OTA Provider*: The Over-The-Air (OTA) Provider is an entity who provides remote access to the Secure Element, enabling the key value added feature of the post issuance and personalization procedure. OTA identifies a service, but at the same time it is also used as a common name for various communication technologies all enabling secure data communication between a Secure Element and a back office architecture. From our perspective, the technical implementations of OTA services are transparent and do not affect the proposed solution.
- *Trusted Service Manager/Trusted 3rd Party (TSM):* As we have already discussed, the NFC technology will be able to support such added value services that are not possible, not even considered in case of the traditional card based contact or contactless (RFID) applications. Service Providers having performed their activities for years may not be able to change the way they act, the functions they provide, but still may want to participate in the new form of service operation or to enhance the services they offer without the need to change existing core processes. The way to solve this conflict is to involve a Trusted Service Manager, who can provide the technology and service support that is necessary for accomplishing these objectives. The Users are also facing a challenge presented by the availability of a large number of applications on a single or in a more complex situation on multiple Secure Elements. The services need to be managed, and to be protected: this is a time consuming and sometimes potentially difficult activity that many Users do not want to bother with. Again, a 3rd party such as the Trusted Service Manager can help the User supporting this activity. The two roles, TSM for the Service Provider and TSM for the User, have different requirements and specifics but they are not exclusive, even the same TSM may act in either position for different parties. It is important to keep in mind that we strictly treat the TSM as a service support function and not as an entity whose tasks would be to solve technical imperfections in the provisioning of the service.

- *Application Issuer*: The application issuer supplies the application that implements and fulfils the business requirements of Service Providers. It is able to guarantee secure interoperability between the card and the card acceptance device. Sometimes the Service Provider itself is the application issuer too.

In reality, there will be many Service Providers offering contactless services and requiring online application management support. There will also be many SE issuers, but most probably a number of other actors too. Some roles will need to be filled, in case of each and every post issuance personalization interaction, for example there is always a User for the service and a Secure Element Issuer providing access to the SE. Although the involvement of the additional actors in the support roles is optional, there is nothing that would prevent the involvement of even multiple actors in one single transaction, all acting in various capacities for either the User, and/or the Service Provider and/or the SE Issuer. While the support functions are required, they do not necessarily need the involvement of additional actors as simple technical infrastructures can perform the OTA and the TSM activities.

4.1.3 The proposed card content management process

While it is not realistic to expect that one concept will satisfy all the service needs, or all the preferences of the Users and Service Providers, most probably there will be several co-existing business models; it is important that all the actors involved can be served andsupported all can be served and supported with the below described technical process, resulting in a uniform service environment.

The process initialization can have different forms, as in the visionary NFC world users can find information about services they like in many ways on multiple channels:

1. The user opens up a newspaper and finds an interactive advertisement promoting a service on which there is an RFID tag. A simple touch of this smart poster hands over all the necessary info to initiate registration and deployment of the service.
2. The user may also browse the Internet with his phone and when he finds something he is interested in, a link helps him in initiating a service relationship.
3. The same service can also be located using a PC. While browsing, the user opens up the advertisement, enters his phone number, which triggers an SMS containing the service specific information.

The ways are endless. However, one important remark is that the originator of these requests is always the User, in a pull-based interaction model. This is important to avoid unsolicited services pushed on the User's mobile phone.

Once received the service request, before the application installation can take place, the Service Provider has to collect information on the targeted device in order to be able to perform the remote card content management procedure. In this phase the proposed service environment needs to be evaluated, the SE Issuer needs to be identified, and the potentially available remote support services need to be defined as well.

More in detail, the information required contains details about the:

1. *NFC device*: The Service Provider and the Secure Element issuer need to identify the end-user device for providing remote management.
2. *Secure Element*: The Service Provider needs information on Secure Element's Card Product Life Cycle (CPLC) to find out the Issuer of the targeted Secure Element and also to evaluate the security environment of the SE itself.
3. *Secure Element Issuer*: The automated contact information of the SE Issuer or a pointer to it are also required.

In the described procedure, it is supposed to store the reference data respectively on the Secure Element (in case of multiple SEs, each SE stores its own specific information) and in the handset's operating system. This information is sent to the Service Provider for evaluation through a message generated by an application on the User's device that is addressed to a specific address of the SP (for example an URL), or to its associated TSM partner, where it can be processed automatically. This relationship is transparent for the users, they do not need to know how the Service Provider delivers the service.

Considering the message received from the User, the Service Provider can decide whether the User's technical environment satisfies its requirements and, in case of multiple SEs, which one it prefers as a storage space/runtime environment for its application on the base of technical, security and financial considerations. The message received may also contain the User's preference in terms of SE selection, which the Service Provider should take into account. Following the evaluation of the technical information, the SP either starts the card content management procedure for the selected SE or informs the User that for some - identified - reasons the NFC service application cannot be loaded onto its mobile handset.

At this point, the Service Provider or its TSM partner can identify the Issuer of the Secure Element selected for use on the basis of the information contained in the service initiating message sent from the User's mobile device. This piece of information is actually the only data element necessary for starting the automated card content management process that is not available at this moment either on the Secure Element or in the mobile phone.

On the base of the information received, the SE Issuer can perform the requested post issuance processes. These processes include the generation of Security Domains (SD), the application loading installation and deletion. The SE Issuer also generates specific keys for the Service Provider to ensure exclusive access to the new SD and to the application. To deliver these tasks to the User, the Issuer may use third party service providers - OTA providers, Certification authorities, TSM - but may also perform these tasks itself using its own in-house infrastructure. Once the requested operations are performed and the required data is loaded onto the card, the Service Provider or its TSM receives from the SE Issuer a confirmation response, together with the specific keys to access the Security Domain.

Alternatively, depending from the SE Issuer policy, the Service Provider may get an exclusive access to its application and assigned Security Domain in order to manage its own application without any interaction of the Card Issuer. This requires special management rights that are described in the Global Platform specifications (Global Platform, 2006).

We are describing a process where, by providing the necessary technical information about the User's environment to the Service Provider, this is enabled to launch an automated process with the designated Card Issuer for the seamless establishment of a new Security Domain and for the loading of a new application.

Figure 3 gives us an overview about the entire card content management process.

Technical cornerstone of the dynamic card content management process is that there is a set of technical parameters and information in possession of the User that could facilitate an automated procedure to establish a new Secure Domain for any selected Service Provider. If the necessary information is provided to the SPs, as well as to the Issuer of the SE, they will be able to manage between themselves a seamless deployment process.

According to our proposal, the SE shall contain a reference (for example an URL) to the current Issuer of the specific Secure Element. This could be a pointer to a database which maintains the list of SE Issuers or even a direct access information to the Issuer itself.

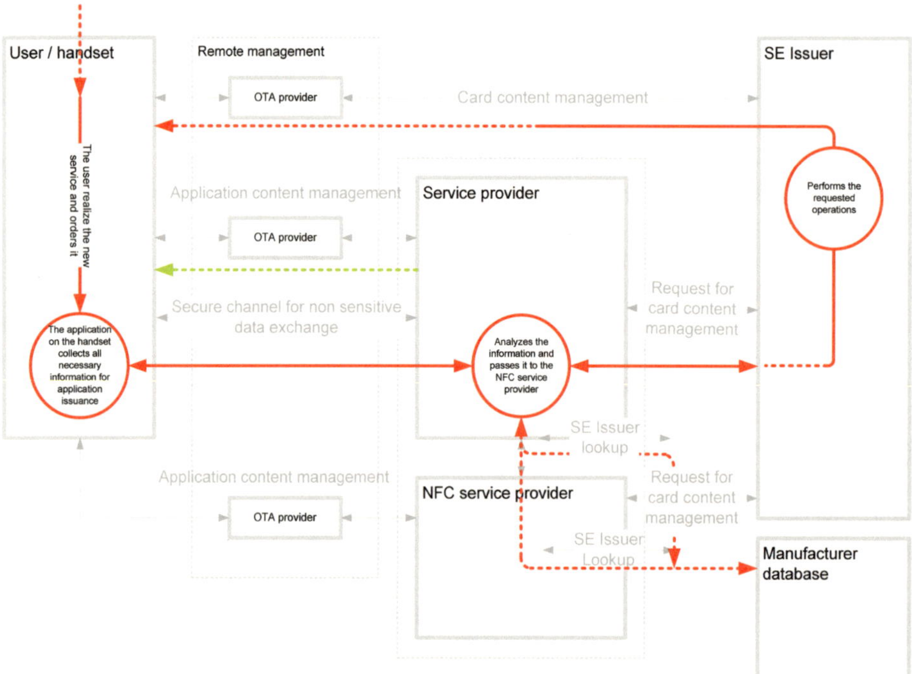

Fig. 3. Overview of the card content management process

Another issue to consider is the SE selection in multi SE environments. Although the currently available NFC handsets support only one Secure Element, we also clearly see the potential that in one NFC handset there may be multiple SEs hosted. We think that including more than one Secure Element provides more flexibility, allows differentiation of security levels and also increases the business potential of the technology. At this point, given a free choice between a SIM storage and another SE, we cannot judge which alternative will be preferred by the Service Providers. There may be a number of technical, security and business issues which will influence the decision.

The actual choice between the SEs will partly be influenced by technical factors, but currently unknown business conditions will have a major impact on these decisions. If a simple algorithm should drive the selection, the following issues need to be considered in the sequence listed here:

- SE capacity availability;
- Security level of Secure Element;
- Controllability of the Secure Element;
- Cost;
- Business considerations/existing business relationship between the parties;
- User preferences.

Last but not least, the customer support is a crucial issue on which attention is needed. As described among the roles listed, the present model introduces a new TSM to support the customer in situations where the management of multiple applications stored in the SE(s) may be just too complex or time consuming. The best example for such a situation is when the handset is lost or when migration is necessary from one SE to another one. Instead of

letting the User do this task alone, which is practically blocking and reordering each and every application again, a simple request to the TSM may solve the problem. However, to get to this point, two aspects have to be clearly seen.
First, the User needs to decide that he needs such support for himself, because the Service Providers' various TSMs will not be able to provide him this function, because each of them will only have information about the application(s) it manages. Second, the application(s) need to contain some sort of summary information that, if provided to the TSM, will describe the application in satisfactory details that allows to identify the Service Provider, the User and his technical environment, and also the application itself, but it still does not provide details that could be misused.

4.2 NFC wallet application or the HOST application

In the current pilot operations, the service portfolio contains only a limited number of services (use cases) hard coded into the mobile handset. These implementations do not allow the removal or the insertion of any new or unused NFC service. Without this service portfolio dynamism, these operations effectively limit the penetration of NFC services.
According to the StoLPaN consortium, in order to quickly spread the adoption of NFC services among end-users, they need a simple way of downloading and removing NFC services to and from their mobile device. People want a dynamic NFC platform on their handset that hides the complexity of changing NFC services. They also want a generic, simple and easy way to manage their NFC service portfolio. Service Providers also need NFC platforms in handsets which can dynamically accept their applications, to minimize the barriers to their services. Secure Element issuers need a platform that help them sell space on their SE in a dynamic manner. Technically, this can be managed by a dynamic NFC wallet (also referred to as HOST) application stored on the mobile phone regardless of the model used and based on a modular architecture that provides a transparent and seamless environment to the Users, the various Service Providers and the Secure Element issuers.
A proof-of-concept prototype of this wallet application, along with a related smart retail scenario demonstration, has been designed and implemented by the StoLPaN consortium (StoLPaN consortium, 2008b, 2009a).
In this section we are describing the technical implementation of the StoLPaN HOST application.

4.2.1 Seamless NFC environment enablers

The StoLPaN project has defined the functional and non-functional description of a dynamic NFC environment and reviewed its potential lifecycle. This complex analysis resulted in the development of a prototype where dynamic application management can be demonstrated and further analysed. The analysis determined three important elements which need to be defined in an open, dynamic NFC environment.
Common issuance processes - The Global Platform defines the smartcard content management procedures implemented in most Secure Elements, explaining how a card issuer can manage card content. However, it does not explain how an application provider can contact the card issuer. This is a clear issue in a dynamic environment. In addition, none of the current standards define how the interoperability of the User Interface elements of the service and the actual hosting of the wallet platform can be assured. Section 4.1 describes a procedure for resolving these issues. It shows how the relationship between a Secure Element Issuer

and a Service Provider can be determined using existing protocols and standards and how this offers a communication channel for exchanging wallet compatibility information. An official standard addressing these issues would help the industry to make dynamic NFC wallets commercially available.

Application selection - Once applications are installed on a mobile handset, they are registered in the Card Manager. Each application in this registry is active by default. This means that they are able to respond to a call from the card acceptance device. Today, the decision on which application to use in a defined context, for example which banking card to use for a transaction, is made by the acceptance device, based on the matched priority list of the Secure Element and the acceptance device. Application selection by the user can be fulfilled by creating a single element list of the available applications. The procedure for application selection is not fully defined in any standard or recommendation. What is even more disquieting is the expected growth in complexity caused by introducing multiple service types on one card, the plug and play use of multiple Secure Elements and the presence of multiple wallets in the same system. As a result, application selection is a very complex subject. The current section does not intend to cover the topic in detail, but without a detailed standard on application selection it will be impossible to build a smoothly functioning NFC wallet service.

Application development - Most NFC services have a software element that contains the user interface and/or its structure. This code is hosted on a certain platform (a midlet based wallet core, a Smart Card Web Server implementation or similar) which represents the user interface. The link between the user interface and the application resides in the Secure Element. The developer creates the user interface element and the non-sensitive application logic of the service so that it works smoothly with the hosting platform. This is only possible if the developer knows all the hosting platform interfaces in detail. There are many ways to ease and speed up the developers' work. Out of the many good practices we would like to emphasize two things. One is that the hosting platform should contain many pre-coded modules that the developer can use as building blocks via open APIs. This has the additional benefits that it makes the certification process much faster and controllable. The other is the creation of Software Developer Toolkits to make the use of these building blocks even easier and at a higher quality level.

The development of a well-defined platform can significantly decrease the cost of NFC service development and hence bringing faster penetration of NFC services.

4.2.2 General wallet requirements

Here below are described the general requirements to build up the NFC wallet application:

- **Remote management**: The platform/wallet must provide the necessary functions to enable remote application management. This covers all the functions which are necessary for remote or proximity service delivery and deletion and for the continuous operation of the services as well.
- **NFC events handling:** Several use cases require the handling of RFID/NFC hardware events from the application runtime environment. Therefore, the hosting platform must contain an application programming interface that allows NFC applications to access information on external contactless targets such as RFID tags or other NFC devices.
- **Security features**: The targeted application environment for NFC applications should provide a reusable set of functionality that encompasses the security needs of the

various use cases implementations. It should contain authentication, security policy settings and cryptographic services. In the dynamic service portfolio, each use case implementation must comply with the specification and security rules provided by the wallet manager to ensure a homogeneous wallet environment. This goal can only be achieved through well specified usage of the pre-certified security services embedded in the hosting platform.

- **User Interface** (UI): The wallet/platform should provide a homogeneous NFC user experience with similar design principles and opportunities for accessing all known added value functions. It must be possible to personalize the interface on multiple levels, to reflect the preferences and the specifics of service providers and of the wallet manager.

4.2.3 Solutions for seamless NFC issues – the StoLPaN HOST

In order to define the requirements of different NFC services, the StoLPaN consortium has analysed various use cases from different industry segments. As a result of this analysis, we were able to create the system architecture and to define its boundaries. After detailing the functional description, we made an implementation to check its viability.

When we started our implementation we had to choice the most suitable platform. The selected platform had to be able to support a seamless user experience for downloading, using and deleting NFC services. The platform also needed to support the dynamic wallet concept and, finally, we wanted to create a friendly, open environment for developers.

The conclusion was that embedded chip handsets with MIDP 2.0 support already had all the mandatory features we needed to meet our main objectives. As they are commercially available, we can hopefully state that the features they carry represent the minimum that we will see in all future models. This ensures that our work can be easily reused in the future. Our middleware should work on any future MIDP 2.0 handset based on UICC or embedded chips without modification. For non-MIDP 2.0 based handsets (e.g. MIDP3, Android, SCWS, JC3, etc.), it may be possible to implement the concepts with a short or none software code. This is because it is not necessary to implement a wallet application if the underlying platform provides all the features necessary for a homogeneous NFC application environment.

The StoLPaN HOST wallet has a modular architecture and is based on a component model, which is the industry trend: both the next generation of mobile java, and MIDP 3.0's architecture are going to be component based. The component structure enables Service Providers to integrate NFC applications as components in the HOST dynamically and efficiently. In this way, Service Providers only have to deal with their business logic and can use pre-coded platform services for standard functions instead of implementing the whole application with all the related technical concerns of compatibility and portability. The code will be handset agnostic when run on the StoLPaN HOST, as the HOST hides all handset specifics. The StoLPaN Platform provides loose coupling between the individual third party components and the HOST core. As the APIs (that represent this coupling) to the HOST core services are open and available for the programmers, development of new NFC services can be carried out without changing the HOST core platform.

There are two types of components in the StoLPaN HOST:

- **Host Core Components**: they are part of the HOST. These components are required for the HOST to function correctly as they provide low-level functions to the second type of components which are implemented by third parties.

- **Third Party Service Components (TPSCs)**: they are not part of the HOST. They are installed, replaced or uninstalled without disturbing the HOST or other components that are not dependent on the replaced or uninstalled components.

The relations of the HOST, Third Party Service Components (TPSCs), and the Third Party Cardlet Applications (TPCs) are shown in Figure 4.

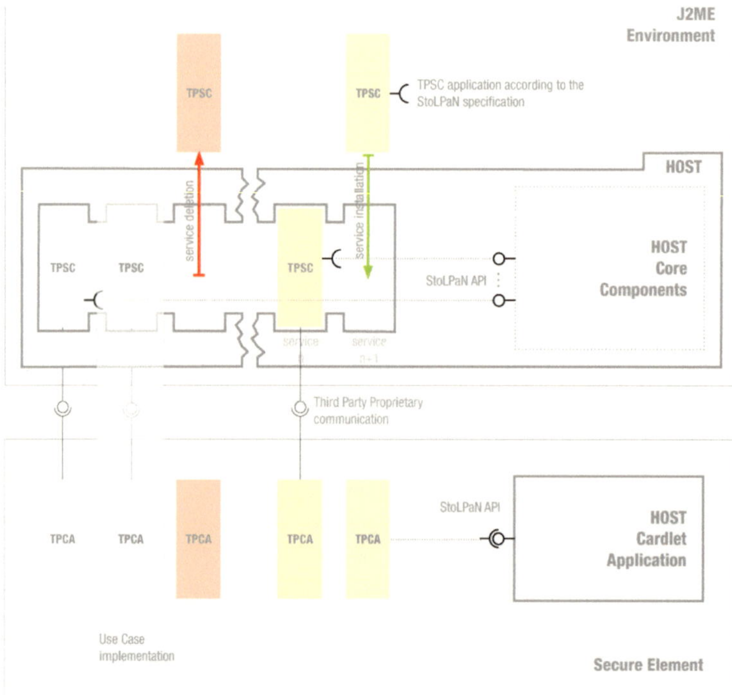

Fig. 4. The HOST structure and the use of Third Party Software Components (TPCs)

Among HOST Core Components, the StoLPaN Cardlet (HOST Cardlet Application) is a smartcard application residing on the default Secure Element in the handset. It mainly addresses shortcomings of the security and smartcard application management in the StoLPaN framework. It provides cryptographic support and can store keys, certificates and authentication schemes such as PIN or password for granting access to applications.

On the other side, the Third Party Service Components realize user interfaces and business logic on the client side for managing third party service-specific workflow and all third party related functions. The Third Party Cardlet Applications contain the sensitive application logic and any user related data in a secure environment provided by the applied Secure Element. The legacy cardlets which were designed for the traditional plastic card environment do not provide added value services such as User Interface support. To adapt these applications for the modern NFC environment, the application developer needs to implement some additional extensions into this cardlet. The Third Party Service Component (TPSC) and the Cardlet application (TPCA) coexist next to each other and realize the Third Party Service for the user. They run on the StoLPaN HOST resources, which makes them handset and platform agnostic.

4.2.4 Summary of the requirements

This section summarizes all the requirements necessary for a transparent application environment, which we believe is the key for NFC's success. We have also shown how we built our own flexible and transparent NFC wallet system (the StoLPaN HOST). This implementation will enable us to further analyse the environment requirements and at the same time create a tangible demo for the public.

This application environment concept, together with the application distribution principles explained in section 4.1, creates a complete description for an open NFC application environment. The advantages of the dynamic wallet approach can only be exploited if it is ready for rapid applications development. This requires open interfaces and libraries supported by an SDK for developers. The StoLPaN consortium is ready to cooperate on any further analysis of the related behaviours and requirements and to support any related standardization effort.

4.3 Demonstration of the StoLPaN solution in a smart shopping environment

So far in paragraphs 4.1 and 4.2 we presented the StoLPaN view of the utilization of the NFC technology and its contribution to the NFC market. In this section we will describe the retail demo application implemented by the consortium.

4.3.1 Overview of the retail industry scenario

In the past decades, the retail industry did not position itself as a great innovator when it comes to improving the customer's shopping experience through the implementation of new services and technologies. Instead, the most prevalent innovations were bigger packages for lower prices, for example in the range of products of hard discounters such as Aldi and Lidl. In the meanwhile, several technologies with the potential for significant innovations in the customer shopping process have matured and became available, e.g. mobile scanning barcode devices, new point-of-sale concepts, and radio frequency applications. Thanks to these technologies, retailers now have the opportunity to offer their customers additional services, such as self-scanning, self-checkout, information terminals, personal shopping assistants, and new, convenient methods of payment.

Studies (Benyó et. al. 2009, Wiechert et. al. 2009b) have revealed that the best thing a retailer can do to better serve its customers is to save their time. This includes the time that customers spend waiting at the checkout area and the time they spend waiting for a store employee to be available or to find a product that meets their needs. New checkout concepts and information devices can help retailers to shorten the lines at the checkouts and can help customers to become more independent from store personnel, for example through easy access to data on available products.

To enable mobile payment in retail commerce, support devices need to be designed and developed and the traditional business procedures need to be remodelled. The StoLPaN consortium is developing a mobile, contactless payment solution based on NFC mobile phones. With the help of these NFC mobile phones, customers will be able to pay offline via NFC, where compatible terminals are available, and over the air, where they are not. Asides the support for payment based on credit cards, debit cards, and an electronic purse, the project also includes the support of other concepts, such as e-tickets, loyalty cards, access ID, and e-prescriptions to be saved on the mobile phone.

The StoLPaN shopping process implements an individual information terminal combined with an individual POS, thus establishing a user friendly, efficient shopping environment.

The goal of the StoLPaN project is to create a pleasant shopping environment for the customers, while increasing efficiency of operation for the retail store operators. The core finding of the project, the basics of the StoLPaN shopping and payment process, is the personalization of the shopping experience, the delivery of personal services to someone's shopping cart and the removal of check-out and payment counters.

The new StoLPaN shopping and payment concept encompasses diverse functionality, including product information, loyalty programs, promotional programs, coupons, and payment. The smart shopping concept presented in this section has been developed in such a way that allows step-by-step migration from the traditional barcode based solutions to the new contactless, RFID based systems and services.

4.3.2 The StoLPaN shopping processes

As it was identified in our earlier studies (Benyó el.al. 2007, Wiechert et. al. 2009b), the various commercial sectors have significantly different shopping processes. It is however possible to identify some basic commonalities which can be summed up as a "generic customer shopping process" used across all retail sectors. This *generic customer shopping process* consist of the following steps:

1. choice of products,
2. registration of products,
3. payment,
4. security control (optionally including the deactivation of products),
5. procurement of product information (content, availability, pedigree, price),
6. general assistance (what, where, how, ...?),
7. use of the loyalty programme (collection and redemption of rewards).

Also, as it will be detailed in the following sections, we decided to split these steps into *core process* and *optional services*. While the core process consists of those steps indispensable for the completion of the shopping process, the optional services are not necessarily preconditions. The breakdown into these two areas is the following:

- Core process:
  - choice of products,
  - registration of products,
  - payment,
  - security control (optionally including the deactivation of products).
- Optional services:
  - procurement of product information (content, availability, pedigree, price),
  - general assistance (what, where, how, ...?),
  - use of the loyalty programme (collection and redemption of rewards).

This split up is important for the design of the StoLPaN Personal Shopping Assistant (PSA) prototype, because we want to enable people to be able to benefit of the full pallet of services enabled by NFC, but also reserving them the right to renounce to these services (e.g. use of a loyalty card) if they wish to do so. We plan for our solution to be just as easy to use, while giving the technology friendly use many more options.

The implementation of the StoLPaN solution will be possible with different Auto-ID solutions, namely barcodes and RFID tags. While the barcode solution is likely to be more popular with retailers, because of its lower cost compared to a tag based solution and the fact that almost all products are already equipped with barcodes, this position might change in the future. In fact, RFID based solutions offer many more potentialities, such as automatic check-outs, automatic inventory counts and the possibility to identify product instances instead of product types.

Our interviews with retailers have shown that they definitely appreciate the option of a partial or step-by-step implementation of NFC and RFID based solutions. The StoLPaN consortium will thus integrate this possibility. However, a partial implementation bears the risk of losing a part of the anticipated efficiency gains, because the necessary expenditures would result in a lesser return on investment.

The StoLPaN project is targeting a smooth migration path for the NFC/RFID technology into the retail environment as shown in Figure 5. In each of the different tasks covered by the StoLPaN solution the existing technical and business state-of-the-art implementation is considered as the base for the next step. Therefore the StoLPaN solution is taking a conservative investment saving approach.

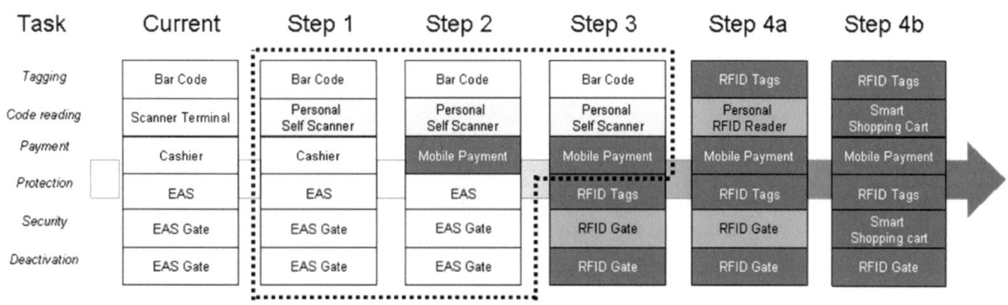

Fig. 5. Steps of migration from traditional shopping to shopping with the aid of applications using RFID/NFC technology

Taking into account the technical constraints as the limitation of the StoLPaN shopping experience, the following objectives are considered for the step-by-step migration path:

- delegate many functions to the customer's mobile handset to avoid the need for individual support devices;
- delegate many functions to the retail back office to avoid the need for expensive technology built into the smart shopping cart.

The first step is characterized by the introduction of a personal self-scanner, this could be a dedicated barcode scanning device, a personal shopping assistant or even the customer's mobile phone. The customer will be able to scan his items by himself and, before leaving, the payment procedure will be done at a fixed payment terminal which can be assisted by a cashier. In the following step the personal self-scanner will morph into a closed loop payment terminal enabling the customer to pay where and whenever he wants. Even in this step, barcode will be the main identification technology for objects in the store.

The step three can be seen as an intermediate step for moving from barcode to RFID tagging. As this approach will happen for high value goods first the security related tasks are mainly affected. Finally steps 4 a/b uses RFID tagged products as the base for providing the StoLPaN solution to the customer. Whereas the customer can still use a handheld device to scan its purchase (4a) and pay using the same method as developed in step 2, the final vision is a completely automatic checkout using a smart shopping cart being able to read all the tagged products inserted by the customer. In this step the cart will incorporate also the payment terminal functions by being wireless connected to the back office using the retailers WLAN network.

4.3.3 The StoLPaN support devices

The StoLPaN process is built on the use of various support devices. These modules enhance customer service as well as facilitate the use of new technology and enable the seamless migration of the StoLPaN services from a barcode based environment to one where the most of the products are carrying smart labels.

The StoLPaN shopping environment consists of multiple components:

- The smart shopping cart is the temporary storage area for products to be purchased.
- The on-board computer (Personal Shopping Assistant, PSA) provides the necessary user interface for customer interaction and facilitates the remote data exchange with the back office.
- The StoLPaN back office coordinates the front-end devices and provides for the data exchange and the integration with the legacy systems of the retail operation.

Let's see how these components have been implemented by the StoLPaN consortium.

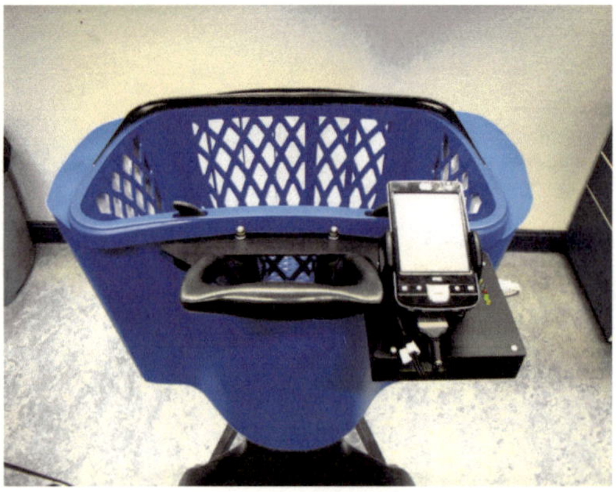

Fig. 6. Smart shopping cart

Smart shopping cart - It is a regular cart equipped with 3 pieces of UHF antennas, an RFID reader, a small computer (PDA) with display and battery (Figure 6).

It has a sophisticated antenna system that can read the RFID tags (EPC Class1 Gen2) on the products, found in the shopping cart in any position. The system consists of three switched high gain antennas, on the left and the right side, and at the bottom of the cart. At anytime only one antenna is active, while the others are operating as microwave absorbers. The cart is mounted with absorber in the front side to avoid reading tags outside the cart. The antenna system ensures that the whole volume of the cart can be read reliably at the 125mW radio frequency power level (European Standard Mode). Reading is reliable even with large number of tags. (The maximum number of tags to be handled can be adjusted in the collision algorithm). The UHF RFID reader reads the tags on the products inside the cart. We use the reader, according to the European standard, in 862-870MHz band, similarly the antenna system is designed according to the European regulations. The output power level can be programmed in the power range 9-24dBm at 50 Ohm. The system has an anti-collision algorithm for several RFID tags in the cart. The reader engine has a programmable serial interface for easy programmability from the local computer on the cart. The RFID reader has a serial to USB interface which connects it to the local on-board computer. From the local computer software the devices can be seen as serial devices and can be managed easily. The present program scans the cart with 500ms switching time at 12dBm radio frequency power level. The shopping cart has a high capacity battery. Using this battery the system can operate continuously more than 12 hours.

Personal Shopping Assistant (PSA) - The StoLPaN PSA will create a differentiated shopping experience and will increase customer loyalty as well as revenues. Nowadays retailer can no longer compete on price alone. In order to sustain and improve profitability in this highly competitive environment, retailers need to differentiate themselves from other stores, to strengthen customer loyalty and to increase overall sales. To survive in today's highly competitive environment, stores must achieve a new level of service excellence, eliminating long lines at the checkout counter and long waits for price and inventory checks. The StoLPaN PSA will provide a unique retail experience to promote customer loyalty as well as streamline everyday processes to maximize the productivity of retail associates, providing better control over labour costs while freeing up time to provide more personalized customer service. Customers can begin their enhanced shopping experience by simply swiping a loyalty card or by touching it with the mobile phone to unlock and activate the StoLPaN terminal. Afterwards, the customers are free to move throughout the store performing a wide variety of tasks – from scanning purchases to self-checkout and finally payment. Since the StoLPaN PSA can gather key data about customer's purchasing behaviour and decisions, the device enables the development of real-time push 1-to-1 promotions to shoppers. Wireless LAN connectivity delivers up-to-the-minute information on customers, while the StoLPaN architecture eases integration with current Point of Sale (POS) and Customer Relationship Management (CRM) systems, turning each customer visit into actionable business intelligence. In turn, shoppers benefit from more practical and valuable promotions that offer savings on regularly purchased products – increasing the likelihood of consumption.

From a technological point-of-view, the PSA is a general purpose PDA with a 4 inch 480x640 pixel display, with SD I/O port and WIFI interface, with a Windows Mobile operating system. (Presently an HP IPAQ214 is used, but any other types of small PCs could suit the purpose).

The computer serves the following functions:

- has a touch screen to allow the customer to select the various functions;
- has an NFC dongle (WD1010 connected to the SD i/o port) to communicate with the NFC handset or to read the contactless cards of the user;
- receives the antenna signals from the carts and converts them to a protocol manageable by the back office;
- communicates over secure WIFI with the back office;
- stores product information downloaded at the beginning from the back office to speed up response to the customer on specific queries;
- displays information received either through the NFC interface or through WIFI from the back office.

StoLPaN back office - The StoLPaN back office has its own business logic to manage the front-end shopping devices, as well as it provides the necessary interfaces for the legacy retail systems. It requests - using web-services - any information that is available and stored in the legacy. The back office is capable to communicate simultaneously with multiple legacy systems and to provide/request different data to/from the specified architectures.

The management of the individual virtual shopping cart is provided by the StoLPaN back office, as well as the operation of the integrated security architecture that checks whether products leaving the store are identical with the products paid for.

Smart security gates - The security gate is a shopping cart-wide, cart-length corridor with an UHF antenna system on the top, at a height of about 1.3 meter. Customers need to push the smart shopping cart through the corridor while they are passing through at its side. The antenna on the top of the gate can read the cart content and can retrieve the payment information of the respective cart from the back office. If the cart content matches the payment information the customer may leave the store without any further interruption, while if there is any discrepancy the security personal is alerted. In case of simpler architectures the security gate is replaced by a handheld security terminal to be carried by the security guards. Guards can make random spot checks on any of the carts leaving the store. The terminal provides the payment information on any of the carts they want to check which they can compare with the actual cart content.

4.3.4 Services

The StoLPaN smart shopping solution supports the following functionality:

1. **Loyalty sign-in -** Customers with loyalty cards or loyalty credentials in the NFC handset can sign in with the PSA. This greets the customer with his name and provides any relevant information (e.g. the number of bonus points available for use, personalized promotional offers or even a shopping list the customer has generated at home) the back office is providing as a response.
2. **Product pricing** - Upon placing the tagged products into the cart the on-board computer shows the running total of the purchase. There may be multiple prices shown, original price in one column and discounted price in another column.
3. **Product information -** When placing a product into the cart, it can be selected for further detailed information, for example detailed product description or the list of accompanying or related product, etc. Using the specific Query menu any products included in the store data base can be searched.
4. **Product location** - As part of the product information the exact location of a product in the store can be shown on the display of the PSA.
5. **E-coupons – discounts** - The PSA can read e-coupon information from vouchers or from the NFC handset. The PSA forwards the information to the back office and the response is the new reduced promotional price information or the rejection of the coupon including the specified reason.
6. **Payment** - The smart shopping cart with its PSA can substitute a cashier counter and can act as a POS terminal. There are three types of payment solutions supported by the StoLPaN implementation.
   - Cash payment: upon selecting the cash icon on the PSA, the invoice information is forwarded to a dedicated cashier desk and the customer is advised to proceed to that given counter. In this scenario products are not counted and scanned again by the cashier, payment is made based on the invoice generated and forwarded by the StoLPaN back office.
   - Card payment: upon selecting the bank card icon the customer is requested to present his bank card. The payment is processed by the retail back office as a Card-Not-Present transaction.
   - Loyalty payment: if the loyalty payment option is selected, then payment is made by using the pre-registered payment instrument of the customer. In this case the payment transaction is performed either as a regular card payment process, or using the available amount of loyalty points.

A future extension of the service can be the introduction of individual pricing. As smart tags on the products identify specific individual products and not just product categories, it is possible to price similar products differently on the base of various factors - like closeness of expiration date, damage of packaging, date of reception, etc.

4.3.5 Barcode and contactless

The original StoLPaN shopping process was developed for smart shopping operations, where all the products are tagged with smart RFID tags. However, as fully operational and completely smart retail operations are still a few years away, the solution has been extended to the traditional barcode based environment. In such an environment, the smart shopping cart does not have antennas, instead the PSA receives a built-in barcode reader. When a product is selected, the customer waves it in front of the reader and when the reading is successful a beep sounds. The process is similar with the loyalty sign-in and coupon redemption features. All the previously described services are available as well. At the back office level, the procedures are identical, no changes are necessary. Actually only the antennas on the cart and the smart security gate need to be added to the StoLPaN smart retail operation upon conclusion of a migration from the barcode based to RFID identification. All other features of the new StoLPaN shopping process can be continued without any modification and loss of investments.

5. Beyond the StoLPaN Project: future challenges for NFC-based services

The StoLPaN project ended in 2009, identifying four key topics for the future of NFC ecosystem (StoLPaN consortium, 2009b). The consortium is still working with Global Platform and NFC Forum to have the Project results endorsed by these standardization bodies.

Beyond the StoLPaN project, the authors have identified three major points that have to be considered for the mass adoption of NFC based applications and services. They are related to the evolution of devices and UICCs, the improvement of OTA communication capabilities, which make use of communication protocols such as Bearer Independent Protocol (BIP) with the overlaying Card Application Toolkit_Transport Protocol (CAT_TP), and finally the use of Smart Card Web Server (SCWS) technology for increasing SIM-based applications' capabilities.

The evolution of mobile devices includes the evolution of the UICC and related SIM logical module too. The capacities of the SIM, as well as the applications supported, improve and increase with the (U)SIM (Universal Subscriber Identity Module), which is used in 3G mobile phones. By increasing its capacity, the (U)SIM can host the Secure Element with the user's personal information along with the keys for data protection. In the (U)SIM the SE has a dedicated area for memory and logical elaboration. As we have already discussed in the paper, according to the Smart Card Alliance and Global Platform (Global Platform, 2006), the SE can be divided in different Security Domains (SD), which are separated and logically distinct domains controlled by different Service Providers.

As each SD can be dynamically managed via wireless networks, users can choose their favored services and personalize the carnet of applications on their mobile phone whenever they wish. This improves the service usability, while the user's satisfaction increases. Moreover, the mobile phone becomes a real multitasking object used to pay, to travel, to get discount coupons of the own preferred brands and to communicate with friends.

As we already mentioned, compared with smart card technology, NFC applications stored in the SE situated on the (U)SIM can be modified also after the issuance of the support thanks to OTA update and management service. In order to increase the amount of information exchangeable via wireless communication, OTA services can rely on new protocols besides SMS: the Bearer Independent Protocol and the overtopping layer named Card Application Toolkit_Transport Protocol. As a consequence of this improvement, it is possible not only to transmit a greater amount of data, but also to establish a more secure and reliable communication. More in detail, the BIP and the CAT_TP are able to open a channel for the transmission of data between the device, the OTA Server and the (U)SIM card. The communication channel can be opened either by the client or by the OTA Server, i.e. by the (U)SIM by means of a command to the host device or by the OTA Server by means of a SMS sent to the (U)SIM.

Finally, the future for mobile applications, even the NFC-based ones, is to use a web-compliant logic also for the user interface. The (U)SIM already offers a suitable space to host the Smart Card Web Server (SCWS), which is practically a web server stored locally on the UICC (Madlmayr et al., 2008). Through this Web Server the user can rapidly access to multimedia contents both static and dynamic. By using NFC in combination with a SCWS (now directly connected on the USIM) user can enjoy a richer, more consistent and more intuitive experience without paying any Internet connection fee, as he can benefit from local contents similar to the Internet ones. Moreover, since the MNO can update and manage the contents remotely, it can increase its offer to the end-user.

6. Conclusions

In this paper the authors presented the services, use cases and the future challenges for Near Field Communication, which is the most customer-oriented one among RFID technologies, as it can be described as the integration of an HF reader into the most popular personal device worldwide, i.e. the mobile phone.

After detailing NFC communication modes (card emulation, peer-to-peer and reader/writer modes) and related use cases such as payment, ticketing and information retrieval, the authors focused the attention on the standardization and interoperability within the NFC ecosystem that NFC based services need to achieve in order to reach mass adoption.

The authors presented the results of the research activities carried out by the StoLPaN consortium during a three-year Project co-funded by the European Commission.

The StoLPaN consortium has worked on overcoming interoperability issues, mainly dealing with application-level standardization, which has not been considered by standardization bodies yet. The consortium elaborated a procedure for dynamic card content management of Secure Elements placed in a mobile handset, identifying key and supporting roles within the NFC ecosystem. Moreover, the consortium has detailed the technical environment necessary for the dynamic management of NFC services, building a proof-of-concept prototype of the NFC wallet application based on a component structure. Finally, StoLPaN has demonstrated the effectiveness and the efficiency of the solution in a smart retail environment, tracing the way forward for the migration from traditional, barcode based, shopping to a smart shopping environment with the support of applications and services that use RFID and NFC technologies.

Beyond the results carried out during the three-year StoLPaN project, the authors have finally identified other three major points that have to be considered for the mass adoption of NFC-based services and applications. These are related to the evolution of the (U)SIM, the improvement of OTA protocols such as BIP and CAT_TP and to the migration to a web-compliant logic for the user interface making use of new technologies such as Smart Card Web Server.

7. Acknowledgment

The authors would like to thank their partners in the Framework of the IST-FP6 project StoLPaN (Store Logistic and Payment with NFC).

8. References

Benyó, B., Vilmos, A., Kovacs, K., Kutor, L., (2007) NFC Applications and Business Model of the Ecosystem. Proc. of the 16th IST Mobile and Wireless Communications Summit. Budapest, Hungary, 2007.07.01-2007.07.05., pp. 1469-1473. Paper 576.

Benyó, B., Vilmos, A., Fördős, G., Sódor, B., Kovács, L., (2009) The StoLPan View of the NFC Ecosystem. Proc. of WTS 2009, 8th Wireless Telecommunications Symposium. Prága, Csehország, 2009.04.22-2009.04.24., 5p., Paper 1569183809.

EPC 492-09, (2010), White Paper Mobile Payments, 1st Edition, Available from http://www.europeanpaymentscouncil.eu/knowledge_bank_detail.cfm?documents_id=402

ETSI TS 102 190 V1.1.1: Near Field Communication (NFC) IP-1; Interface and Protocol (NFCIP-1), (March 2003), Available from http://www.etsi.org

ETSI TS 102 622 V.7.5.0 : Smart Cards; UICC - Contactless Front-end (CLF) Interface; Host Controller Interface (HCI) (Release 7), (June 2009).

ETSI TS 102 613 V.7.7.0 : Smart Cards; UICC - Contactless Front-end (CLF) Interface; Part 1: Physical and data link layer characteristics (Release 7), (October, 2009).

GlobalPlatform, (2006), Card Specification Version 2.2, Available from http://www.globalplatform.org.

GSMA, (2007a), Mobile NFC services, Version 1.0, Available from http://gsmworld.com/documents/aa9310.pdf.

GSMA, (2007b), Pay-Buy-Mobile Business Opportunity Analysis, Version 1.0, Available from http://gsmworld.com/documents/gsma_pbm_wp.pdf.

GSMA (2007c), Mobile NFC technical guidelines, Version 2.0, Available from http://gsmworld.com/documents/gsma_nfc2_wp.pdf.

Innovision Research & Technology plc, (2007), Turning the NFC promise into profitable, everyday applications, In: *Near Field Communication in the real world – part I*, Available from http://www.nfcforum.org/resources/white_papers/Innovision_whitePaper1.pdf

ISO/IEC 14443 : Identification cards - Contactless integrated circuit(s) cards - Proximity cards (Part 1-4), Available from http://www.iso.org.

ISO/IEC 18092 (ECMA-340): Information technology - Telecommunications and information exchange between systems - Near Field Communication - Interface and Protocol (NFCIP-1), (First Edition, 2004.04.01), Available from http://www.iso.org

ISO/IEC 21481 (ECMA 352): Information technology - Telecommunications and information exchange between systems - Near Field Communication Interface and Protocol -2 (NFCIP-2), (January 2005), Available from http://www.iso.org

Kannainen, L., (2009). Global overview of commercial implementations and pilots of NFC payments during 2009, In : *Smart Card Technology International - globalsmart.com,* 08.11.2010, Available from http://www.mobeyforum.org

Madlmayr, G., Brandlberger, D., Langer, J., Scharinger, J., (2008), Evaluation of SmartCard Webserver as an Application Platform from a User's Perspective, *Proceedings of MoMM 2008.*

Mayes, K., Markantonakis, K., (Eds.). (2008). *Smart Cards, Tokens, Security and Applications,* Spinger, ISBN: 978-0-387-72197-2, New York.

Mobey Forum, (2010), Alternatives for Banks to Offer Secure Mobile Payments, Version 1.0.

NFC Forum, (2006), NFC Data Exchange Format (NDEF) Technical Specification, Version 1.0.

StoLPaN consortium, (2008a), Dynamic Management of multi-application Secure Elements, Public Whitepaper, Available from http://www.stolpan.com

StoLPaN consortium, (2008b), Dynamic NFC wallet, Public Whitepaper, Available from http://www.stolpan.com

StoLPaN consortium, (2009a), StoLPaN Smart Shopping, Public Deliverable, Available from http://www.stolpan.com

StoLPaN consortium, (2009b), NFC Application Distribution - Proposed Business Models, Public Deliverable, Available from http://www.stolpan.com

Wiechert, T., Thiesse, F., Schaller, A., & Fleisch, E., (2009a), NFC based Service Innovation in Retail: An explorative Study. In Proceedings of the 17th European Conference on Information Systems (ECIS)12, Verona, Italy, June 8-9 2009, p12., ECIS2009-0587.R1

Wiechert, T., Schaller, A., & Thiesse, F., (2009b), Near Field Communication Use in Retail Stores: Effects on the Customer Shopping Process. In Mobile und Ubiquitäre Informationssysteme - Entwicklung, Implementierung und Anwendung137-141. Bonn, Germany: Gesellschaft für Informatik e.V. (GI). - ISBN 978-3-88579-240-6.

16

RFID Applications in Cyber-Physical System

Nan Wu[1] and Xiangdong Li[2]
[1]*Nanjing University*
[2]*The City University of New York*
[1]*China*
[2]*US*

1. Introduction

A cyber-physical system (CPS) is a system which combines and coordinates the physical system and informatics or computational entities (including computation and communication) into a tight mode. Nowadays we can see the applications of cyber-physical system in the fields of aerospace, automotive, chemical processes, civil infrastructure, energy, healthcare, manufacturing, transportation, entertainment, and consumer appliances.

First, the typical feature of a cyber-physical system is the *combination*, the CPS is a system deeply combined with computing and physical system. Compared with the so-called embedded system, the percentage of the physical component involved in a CPS is higher than those in an embedded system (shown in Figure 1). In an embedded system, the main focus is on the computational elements, not on the link between the computational and physical elements. Second, unlike a traditional embedded system, usually a CPS is designed as a network of the interaction between the physical input and output, instead of being as a standalone device. The notion is tied to the concepts of robotics and sensor networks. The improvement of the link between computational and physical elements using the advances in science and engineering will boost the use of the cyber-physical systems. Several applications of the use of CPS are "the intervention (e.g., collision avoidance), precision (e.g., robotic surgery and nano-level manufacturing), operation in dangerous or inaccessible environments (e.g., search and rescue, firefighting, and deep-sea exploration), coordination (e.g., air traffic control, war fighting), efficiency (e.g., zero-net energy buildings), and augmentation of human capabilities (e.g., healthcare monitoring and delivery)" [1].

A Radio-frequency Identification (RFID) system is a typical cyber-physical system because of its mainly functional and physical components: (1) The computational element: although a passive RFID tag normally only contains the storage function, but the whole RFID system (mainly in a RFID tag reader) and the post-processing system have the computing and data-processing functions; (2) The controlling element: usually a RFID system is under the control of an inner micro- control-unit (MCU); (3) The communication element: in a RFID system, nearly all the information is exchanged via the wave of radio frequency (RF), the data and controlling flows are established via a 2-way RF communication. During the work process, the traditional RFID uses the electronic tags which are placed on the items to track their locations or descriptions. The RFID tags are tiny microchips that can, in some cases, be fabricated smaller than a pinhead or a grain of sand. The chip is attached to a tiny antenna which allows it to communicate and transmit information. Figure 2 is a blown up view of a simple RFID tag [2].

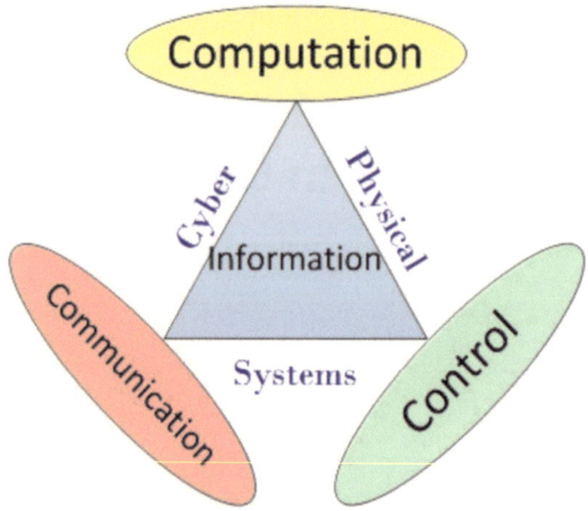

Fig. 1. Three main functional components in a cyber-physical system

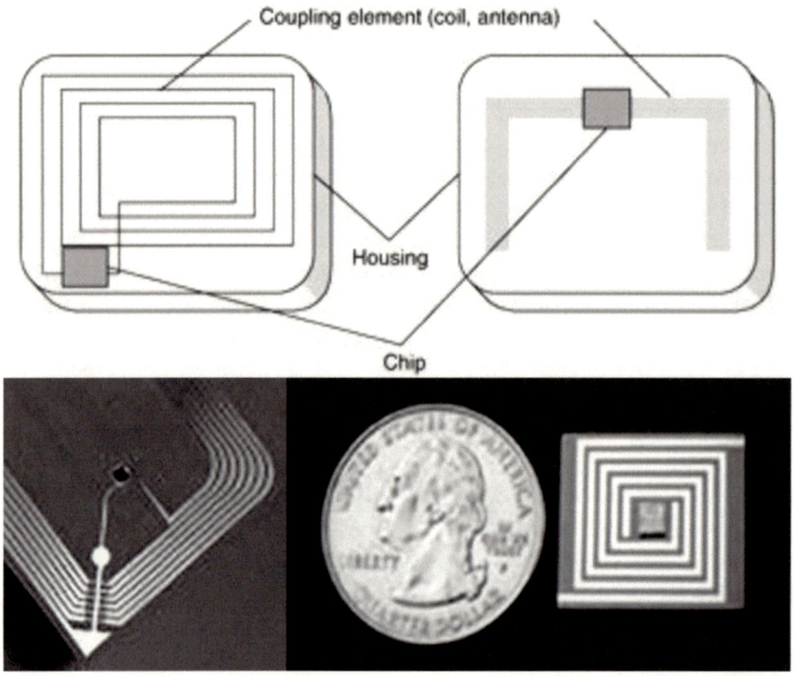

Fig. 2. The structure and outside view of RFID tags

Mostly the regular RFID systems for the civil use are classified into three types - the passive, the semi-passive, and the active RFID. A passive tag is dormant until it is triggered by a signal from a RFID reader. A passive tag does not have a built-in power supply, so it needs the radio frequency energy (electromagnetic wave) from the RFID reader. These tags are particularly popular in use because they can draw the power wirelessly, such that the size and price can be reduced much. Furthermore, these tags can be applied on almost everything because of the wide use of the wireless power supply. A semi-passive tag contains a small battery to function an inner timer or random access memory. However, the power supply does not actively communicate with a reader until it is requested. When it is requested, it uses the radio wave power to transmit the information to the reader, which is the same as that of a passive one. An active tag has a more powerful small power source (a battery or other changeable DC source) built-in. Unlike the semi-passive tags, it can actively communicate with the readers without the need of radio wave power.

The most common type of RFID tags used on the market is the passive type and the tags rely on the readers for the energy. A RFID reader usually has a Radio Frequency (RF) module that allows it to transmit and receive messages. It is also manufactured with additional interfaces (e.g. RS 232 or RS485) to allow the connection with the PC's, etc. Figure 3 shows a simple diagram of the communication between a RFID reader and a tag (or transponder). The "application" shown in the diagram is an enterprise network infrastructure.

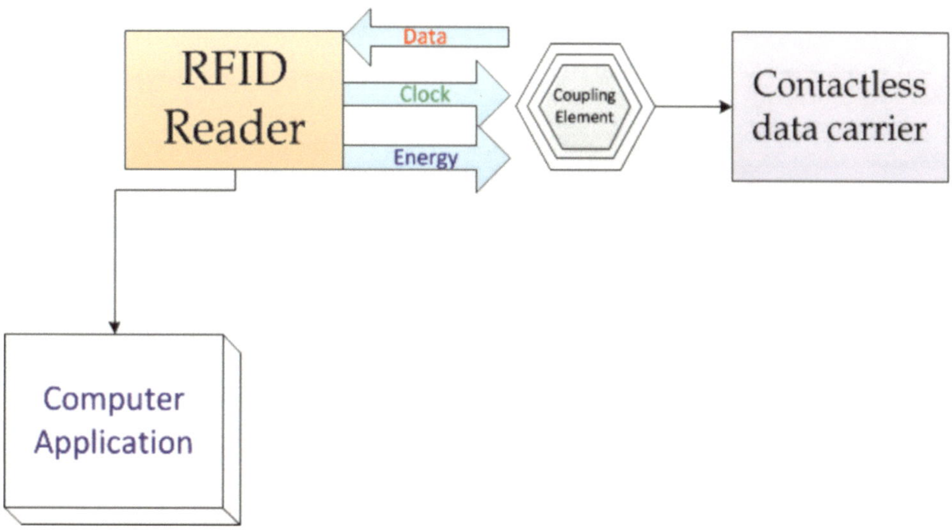

Fig. 3. A Simple passive RFID system diagram

In this chapter, we will study the mixture of a cyber-physical system using the RFID technology. As mentioned above, in a traditional embedded system with a built-in power supplier, using the passive RFID tags is subject to losing the processing ability without the RFID tag readers. To meet the requirements of CPS key application, it is necessary for the RFID tags to contain the batteries and operate the inner MCU and microchips. In the following sections, we will discuss the design on the key applications of the RFID system with the active mode [3].

2. Active RFID system

As discussed in the previous section, usually a passive tag holds a unique identification code or a number of 8 bytes in length, along with other small pieces of information. The active and passive tags are different based on the types of information they store. A common passive tag only stores the object identification information, whereas an active tag stores the object description and its transportation history, in addition to the identification information. A real active RFID tag is shown in Figures 4 and 5 [4, 8].

Fig. 4. A compacted active RFID tag

Fig. 5. An active RFID tag with a changeable antenna

To meet the requirements of the key application of cyber-physical system, we should analyze the applicability of a passive RFID system with details. Usually, the microchip in a passive RFID tag is sealed with a plastic cover statically and cannot be altered from its manufacture or configuration. But the information on the tags is able to be rewritten. There are three different core devices which are able to re-write the data into the RFID tags [6]: ①EEPROMs (electrically erasable programmable read-only memory) are most commonly used among these three. Usually an EEPROM memory capacity ranges from 16 bytes to 8 kilobytes. The disadvantage of using this device for the writing process is the high power consumption. ②FRAMs (ferromagnetic random access memory)'s reading power consumption is lower than that in the EEPROM. But the manufacturing problems in the past cause an impact on its market acceptance. The FRAMs have a similar limit in the memory capacity. ③SRAMs (static random access memory) are used especially in the microwave systems and have very high writing cycles. In order to retain the data, it needs an uninterrupted power supply, such an auxiliary battery or some other power sources should be equipped for the tags. This obviously limits its usefulness. The SRAMs memory capacity ranges from 256 bytes to 64 kilobytes. From the manufactory experience, a RFID tag can be read and written up to 10 billion times before its performance drops, so the future of this tag is optimistic.

As mentioned in section 1, the application of CPS inclines to "more computation power", the RFID system using passive tags shows several disadvantages when examined with the requirements of CPS applications.

- The processing ability of RFID tags is extremely based on the reader or the connected computers. The tag has a very weak computing ability, so a passive RFID tag is barely as an electrical ID container.
- A passive RFID tag is not able to take any kind of sensor to carry the environment data because of the lack of the driving circuits.
- Even in an active region of a passive RFID reader, the energy supply from the radio coupling of the electromagnetic coil is not sufficient for a more complex computation to function the RFID card's MCU.
- The two-way complex communication is subject to suffering more electromagnetic interference (EMI) during the communications between the card and the reader, plus the radio coupling interaction.

Due to these disadvantages, a passive RFID tag and its reader system cannot meet the requirements of CPS applications. Table 1 is shown in Yamada's research [5]:

Items	Passive RFID tag	Active RFID tag (Con.)	Active RFID tag (new)
Comm. Range	70cm/ 3m - 7m	more than 10m	around 10m
Battery life	(no battery)	around 1 year	around 1 year
Security	weak	N/A, or weak	strong
Cost	less than $1	less than $10	around $10
Application	distribution/ inventory control of goods.	tracking person (restricted area)	tracking person (no restriction)

Table 1. Classification of RFID tags

Apparently, an active RFID system can be described highly the likeness of wireless sensor, which has shown to be a successful and mature system. The largest deployment of the active RFID is done by US Department of Defense (DoD), the DoD uses the Savi active tags on each of its over a million shipping containers that travel outside of US.

However, different from a pure wireless sensor system, an active RFID system network is a kind of Ad-hoc network, that is, a heterogeneous network. From the communication protocol point of view, an active RFID reader and its corresponding tags can work with a one-to-many model (and vice versa): one tag can be coupled with many readers (the reader can be defined as a base station in the Ad-hoc model). So when designing the active RFID system protocols, we should consider the difference between the peer-to-peer model and one-to-many model (or many-to-one).

From the network topological structure point of view, a heterogeneous network is wireless based. It is a good carrier for the two-way wireless communication. Here, we define the RFID system used in a CPS system as the followings:

RFID application in CPS = active RFID system + wireless sensor + protocols + network collaborative mechanism

In the next section we will study a typical RFID application in CPS system.

3. A typical RFID application in CPS: a case study

This case study is about the use of an active RFID system which includes a few readers (as the base station) and many active tags (as the sensors) to build an active wireless positioning network, which is a pre-research one of our project [7].

The positioning based services for the geographic information are important in the civil applications, such as the travelling, geographic measurement, harbour operation, driving or logistics; as well as in the military, such as an emergency support or emergency logistics. Today, the global positioning system (GPS) is the most widely used and most well developed positioning system. A GPS receiver uses a high-precision referenced time from a low-orbit satellite to conduct the distance measurement, and it calculates the position by using the geometry methods. The GPS system provides a high positioning accuracy, an excellent timeliness and a strong anti-interference ability. The GPS has many advantages, but with a fatal weakness, that is, its positioning ability and performance are affected distinctly when the receiver is out of the region of the GPS satellite's signal. For example, in an application of the military emergency logistics, when the military vehicle is running in a tunnel or the soldiers are in a thick forest or in a construction, the GPS can not provide the robust positioning service. Especially in a situation such as the need for a rapid response, the loss of GPS performance may cause the possession lost or more casualties. To avoid this problem, the in-door positioning based service is needed for both the military and the civil applications. We study a kind of GPS-independent active positioning system, based on an active RFID system and the TOA (time of arrival) technology and related algorithms [7].

Based on the theory, the distributed node location service uses the referenced base stations (i.e. an active RFID reader, they have the absolute or relative positions of RFID tags) in a distributed network. The node location service is a highly potential core service in the location-based service when applied in a distributed scenario. It shows a great potential, especially when it is used for the positioning in the complicated or blocked indoor/outdoor environment, emergent logistics management, and disaster-relief emergent positioning, etc.

Currently, based on the positioned objects, the distributed node location service's algorithms and the systems can be categorized into a self-node positioning and a target-node positioning. Here we only focus on the self-node positioning. In the positioning technology, a node in the network is recognized as a beacon node or an unknown node based on whether the node is assigned or not assigned with the location information (relative location or absolute location via the GPS or other devices). As the unknown nodes gaining more relative information during the process, therefore, in order to reduce the overall networking loads and the communication cost, the number of the beacon nodes should be limited.

We consider the unknown nodes in the network as the sensors with some special functions (e.g. a function of measuring the distance) and the beacon nodes as the base stations, such that the network with a specific topological structure is a heterogeneous wireless sensor network (WSN). Generally, a well-designed WSN mainly contains the following units:

- Transmission units (including the distance sensors and A/D modules);
- Processing units (including the MCU and embedded software system);
- Communication units (including the radio frequency modules).

We could see clearly that these requirements can be well met based on the active RFID system. In this section we address some key issues on the range based positioning service and study a novel model of the node-location service based on the aforementioned CPS model. The theoretical analysis shows that this model can provide a good and stable positioning accuracy and a strong robustness for the scenes where the network topology structure or the node's surrounding environment varies. In addition, it also benefits the re-organizing of the network and the corresponding the active RFID tags' surviving time.

When we combine the node-location service with the proposed model, a new positioning service mode is formulated, that is the wireless sensor network positioning technology. Compared with the traditional wireless positioning technology, it provides new features below:

- Large scale;
- Low hardware resource requirement;
- Non-centralized Ad-hoc network;
- Low energy cost;
- Self-organizing;
- Dynamic topology;
- High positioning accuracy;
- Dynamic positioning supported;
- Communication and positioning;

The accuracy of the range-free positioning service and its convergence rate highly depends on the estimated accuracy on the network's average jumping distance. When the anisotropy or the topological structure of this RFID system's organized WSN becomes complicate, the performance of the algorithm will be significantly weaken. Therefore, compared with the range-free positioning algorithm, the range-based positioning algorithm has such advantages as a better accuracy and a shorter response time. However, on the other hand, it also holds some disadvantages, such as a higher requirement for the hardware and the more power cost. To balance the advantages and the prices, we choose a TOA (Time-of-Arrival)-based distance measurement technology under the WSN as the prototype model.

According to the discussion above, we realize that the in the pursuit of the highly realizable accuracy of the positioning sacrifices the fast response rate and narrow bandwidth with less power cost. An optimization should be reached. We propose a method to simulate and calculate (or optimize) this point by using a statistical model. A lower bound of the positioning accuracy based on the TOA and WSN methods is also discussed in this section.

The algorithm of the TOA costs less hardware resource in the applications, it is helpful to enhance the reliability and robustness of the model. Using the TOA distance estimation, we can apply the trilateration method for the positioning, which is shown in Figure 6.

Considering the mixture of cyber-system and physical system; the aforementioned theory can be designed as algorithms and be coded in MCU of the active RFID tags.

Figure 7 shows the experimental results of the proposed method. The detailed experimental platform is described below.

We physically implemented the specified positioning equipment using the proposed model. The equipment includes an IEEE 802.15.4a chirp spread spectrum (CSS) system, its time-domain and frequency-domain characteristics are shown in the Fig 8. The receiving sensitivity of beacon nodes and unknown nodes is greater than -97dBm. They work in a duplex mode, cooperating with a gain antenna and a related operating system, and the experiment shows a good signal-to-noise ratio (SNR). The average loaded RF power is only 1mW, which could significantly ensure the surviving time of the unknown nodes.

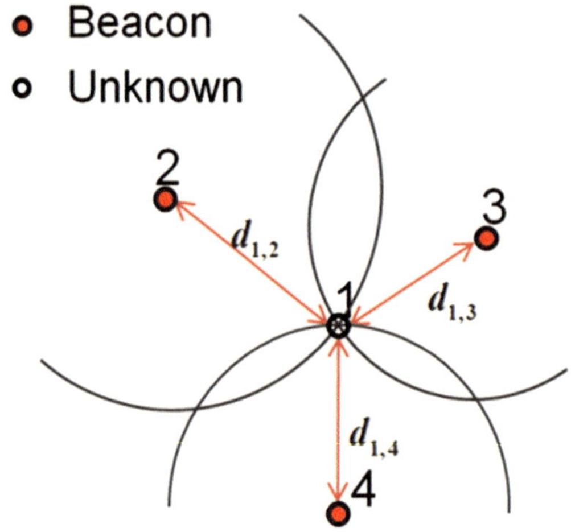

Fig. 6. The estimation of the position of an unknown node by using the trilateration method

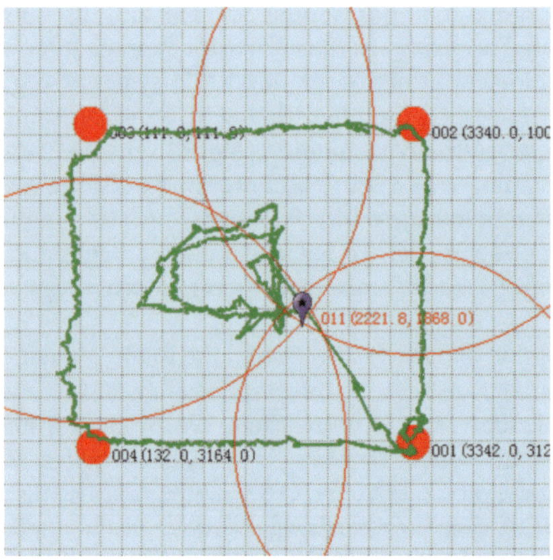

Fig. 7. The experimental results of the bacon-based active position system

The model system has a strong ability of the anti-multi-path-interference and anti-human-interference. In the modulation, the pulse resolution is adjustable in order to be adapted in different application environments. The networking protocol is uncomplicated and reliable; it can be also added with a 128-bit hardware encryption, which can effectively prevent the interference from outside and the disclosure of the location information.

Based on our prototype of the design, we have made the experimental product. Figure 9 shows the experiment platform which realizes the algorithm and the active RFID system.

We improved this product with a compact size and low-power consumption. The key parameters of the tag are listed in Table 2.

Other product's entities and the related software interface are listed in following Figures 10, 11 and 12.

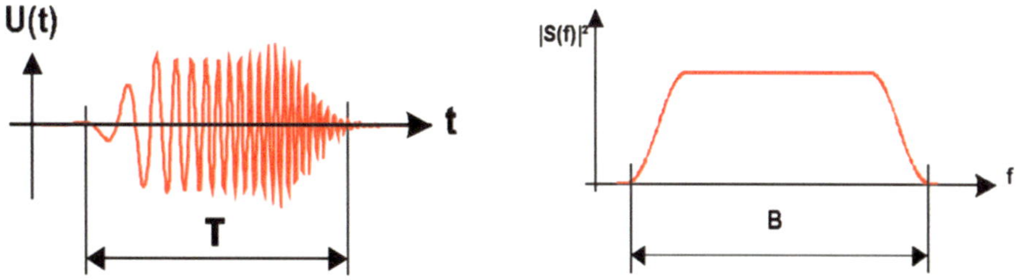

Fig. 8. The time-domain and frequency-domain characteristics of the CSS signal

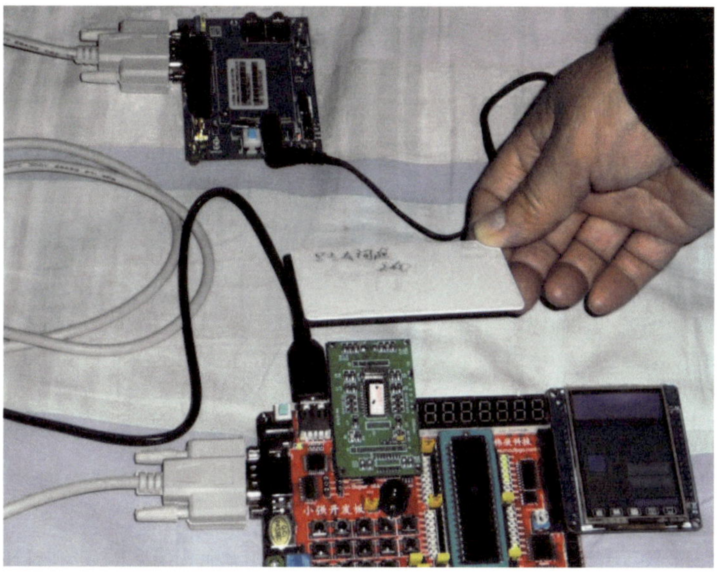

Fig. 9. Experimental platform using RFID reader, tag and wireless transmitting system

Dimensions:	35×14×2mm	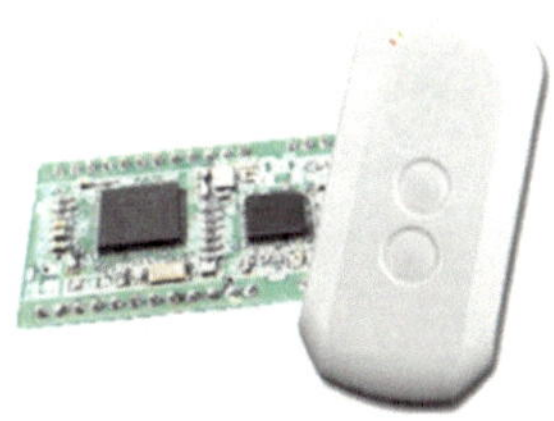 Picture: Chip module and integration module
Power supply:	2.7V	
Energy consumption:	0.15W	
Weight:	30 g	
Radio frequency:	2412～2484M	
Antenna type:	Built-in（Ceramic or micro-strip）	
Installation:	OEM module，SMD	
Interface:	TTL, SPI	
Level of protection:	IP65	

Table 2. The compact-sized positioning active RFID tag

Fig. 10. The indoor positioning sensor and active RFID positioning tag

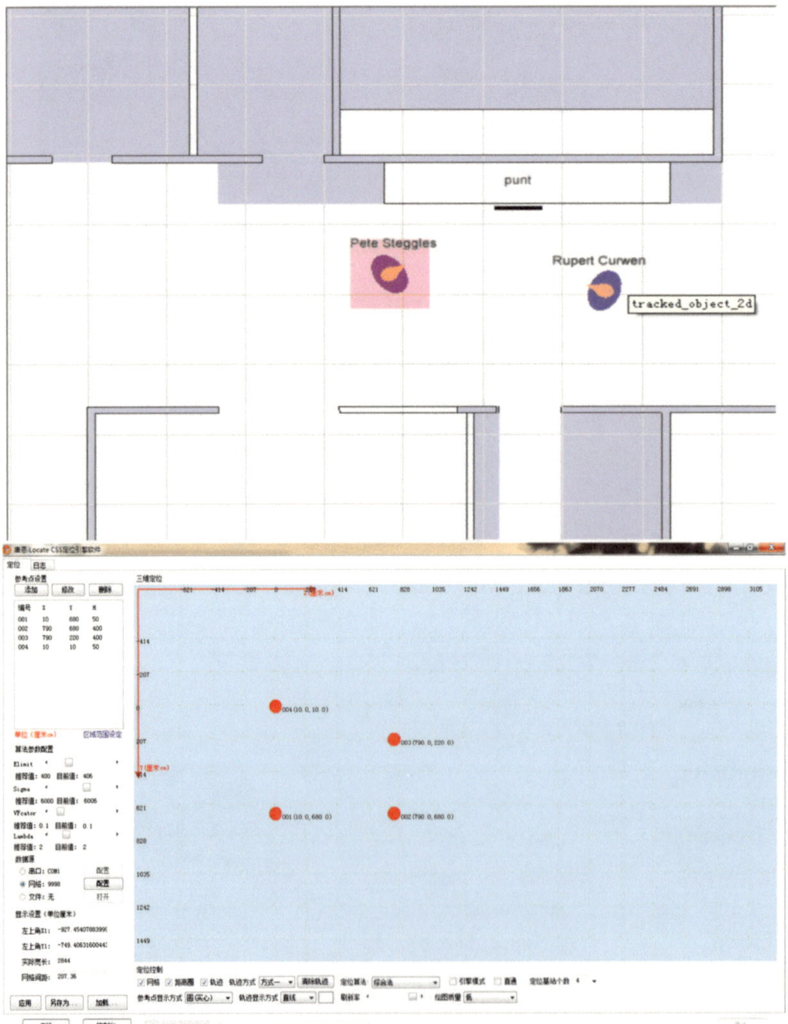

Fig. 11. The related software running on the reader's computer (in 2D mode, red spots indicate the tags' relative position and the blue shadow area is the error space in XY-plane)

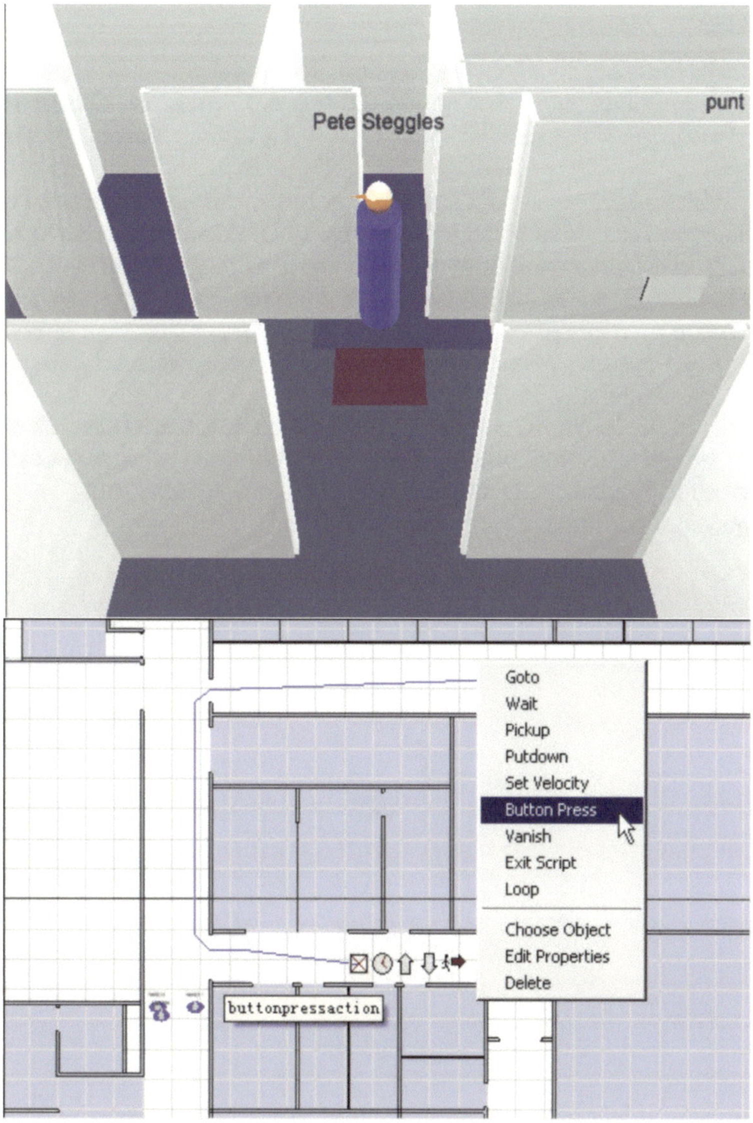

Fig. 12. The related software running on the reader's computer (in 3D and live-action modes. In the left, the yellow spot means the tags' relative position, the purplish red shadow area is the error space in XY-plane, and the blue cylinder is the height in Z-axial)

4. Summary

In conclusion, the active RFID system has shown the gain of a great potential for building a highly-mixed system of information and the physical devices. In this chapter, we compare the RFID system with a traditional wireless sensor network system and discuss the applicability of the type of RFID systems. We propose and study the design idea, methodology, product and experimental results of an active RFID based relative positioning system.

5. References

[1] http://en.wikipedia.org/wiki/Cyber-Physical_Systems

[2] Sztipanovits, J.; Stankovic, J. & Corman, D. *Industry - Academy Collaboration in Cyber Physical System (CPS) Research White Paper V.1* ,Georage Mason University, May 19, 2009

[3] Jalasto, M. *Mobile phone applications for radio frequency identification systems*, Proc. Reserach Seminar on Telecommunications Bussiness II, Luukkainen S. Eds., April, 2005

[4] http://www.scienceprog.com/how-does-rfid-tag-technology-works/

[5] Yamada, I., Shiotsu, S., Itasaki, A., et al. *Secure active RFID tag system*, Ubicomp2005 Workshop

[6] Rhodes, J. *RFID System Protocols and Standards Overview*, Research Report CPET 384 - Wide Area Networks, 2007

[7] Lai, X.; Li, J. ; Li, X ; & Wu, N. *A novel model of node location service based on wireless sensor networks and statistical method*, Proc. SPIE Wireless Sensing, Localization, and Processing, V, Sohail A. Dianat; Michael D. Zoltowski Eds., 2010

[8] http://tronweb.super-nova.co.jp/tronnews05-4.html

17

SAW Transponder – RFID for Extreme Conditions

Alfred Binder, Gudrun Bruckner and René Fachberger
Carinthian Tech Research AG
Austria

1. Introduction

Harsh or hazardous environments, e.g. continuous furnaces, process chambers, rotating or moving objects, require a robust wireless passive transponder technology for sensor and RFID applications. The transponders' operating temperature often exceeds 200°C in these applications and is way above the thermal limit of CMOS devices. Surface acoustic wave (SAW) devices are excellent candidates for high temperature applications as their operation has been shown at temperatures of 1000°C (Hornsteiner, 1997). With the use of an RF (radio frequency) antenna SAW devices can be interrogated passively and wirelessly.
The main advantage of surface acoustic wave sensors is their outstanding thermal stability specially compared to semiconductors. The sensors utilize the piezo-effect that creates so-called surface acoustic waves by means of a transducer structure on the surface of the sensor. Metallization gratings, so called reflectors are used to supply the device with a unique identification code (ID), achieved by pulse position coding (Reindl, 1998). This allows using the device as a high temperature stable RFID transponder (radio frequency identification). Depending on temperature or mechanical strain the surface acoustic waves are also affected. These changes on the surface acoustic waves can be used to implement an additional sensor functionality. In this way pressure sensors in combination with temperature sensing have been demonstrated [Pohl,1997; Kalinin, 2004].
In ideal application fields of SAW based RFID systems environmental conditions like high temperature or high doses of γ-radiation exist. Successful application examples are the automatic identification of pressure sensors, vehicle identification in paint shops and several tagging tasks in the steel industry. Often the temperature information contained in the response signal gives valuable additional process information of the tagged goods.
This chapter gives an overview of SAW based RFID transponders made for extreme conditions like temperatures up to 400°C or cryogenic temperatures down to -196°C. Their function principle and system performance is explained and pertinent application examples are given.

2. Principle of operation

A wireless surface acoustic wave based RFID system essentially comprises a reader unit emitting and receiving radio waves and a SAW reflective delay line attached to an antenna, building the transponder (Figure 1). For data acquisition, the impulse response of the SAW transponder is analysed by a digital signal processor. The response signal contains a pattern of reflectors, which resembles for instance a binary, decimal or hexadecimal code. Utilizing the natural sensitivity of the piezoelectric substrate crystal, e.g. on temperature or strain, the SAW tag can operate as a sensor.

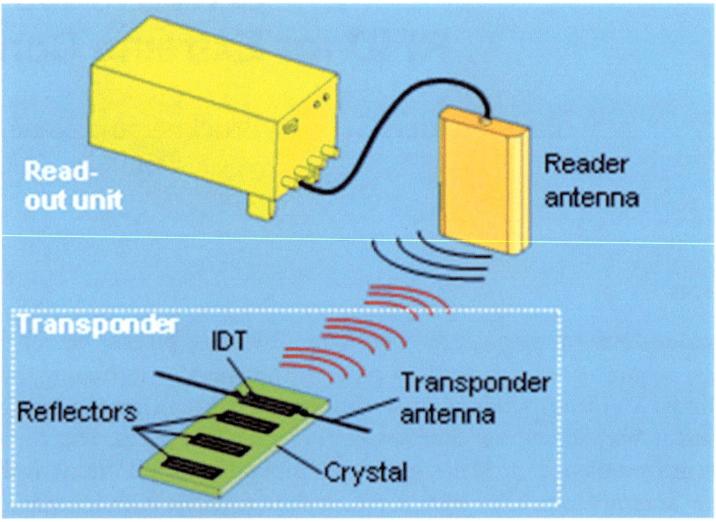

Fig. 1. SAW transponder interrogation setup.

The SAW RFID system is suited for high operating temperatures as it is purely based on piezoelectricity and therefore fully passive. It makes use of the piezoelectric-substrate lithium niobate. The operating principle of the system is as follows:

A high-frequency electromagnetic (EM) interrogation signal is picked up by the antenna of the passive SAW device and conducted to a transducer. The interdigital transducer (IDT) converts the received signal into a surface acoustic wave (SAW) by the converse piezoelectric effect. The SAW propagates towards reflectors distributed in a characteristic barcode-like pattern and is partially reflected at each reflector. The acoustic wave packets returning to the IDT are reconverted into electrical signals by the IDT and sent back to the request unit by the antenna. This response contains information about the number and location of reflectors as well as the propagation and reflection properties of the SAW. It is evaluated by the interrogation unit to extract the desired information.

In a particular design example eight reflectors are used to supply the SAW device with the unique identification code (ID) and a temperature sensing functionality. The first response pulse should have an adequate time delay towards the interrogation pulse to avoid environmental electromagnetic reflections and echoes corrupting an early sensor response. A practicable value for this delay time is 1.0 µs. In Figure 2, a typical impulse response of the designed SAW tag is shown. The SAW's edge reflection (at 0.4 µs) and crosstalk signal rests can be seen in the time between 0 and 1 µs. Then, the tag's eight response pulses rise clearly out of the surrounding noise level (about 1 µs up to 2.25 µs). The first and last pulse take the function of start- and stop-bit and are used for compensation of temperature changes, the second pulse is additionally taken for temperature measurement. Via pulse position coding, (Reindl, 1998) the other pulses are used to encode a unique ID comparable to an ID stored in a microprocessor's ROM.

While wireless interrogation can be achieved at any readout frequency, there is only a distinct number of radio frequency (RF) bands which are free for industrial-scientific-medical (ISM) applications. Here, the ISM band from 2.4 GHz to 2.4835 GHz proved to be most suitable as it has an adequate bandwidth (83.5 MHz) and an almost worldwide geographical licence. At the same time it allows a read out at a distance of several meters. At this frequency, the RF wavelength is about 13.5 cm, thus permitting the usage of simple and small antennas, e.g. dipoles, slot- or patch antennas, favorably for the transponder part.

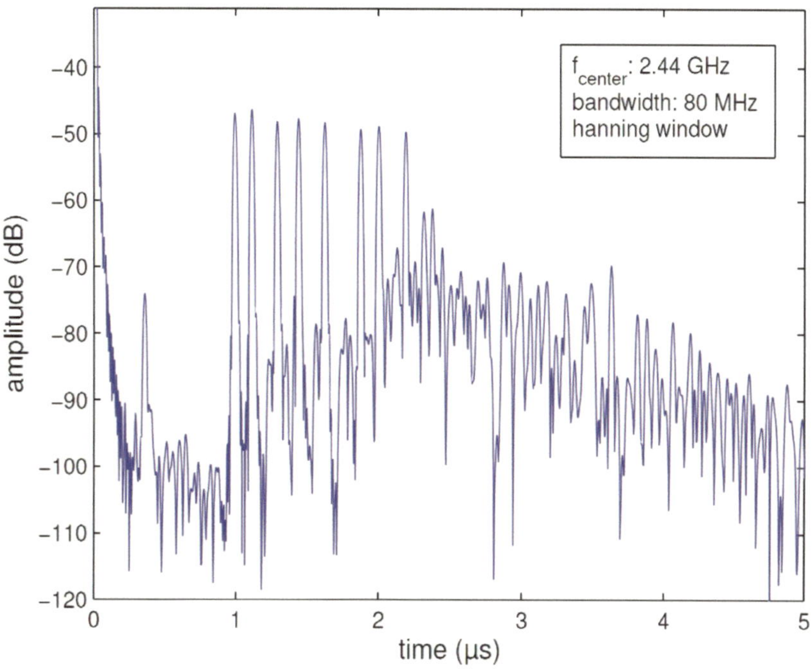

Fig. 2. Impulse response of a SAW transponder with eight reflectors.

2.1 Reader systems for SAW transponder

Reader systems for SAW transponders usually utilize the continuous wave radar principles. Impulse radars could be considered but are inferior in cost and are not efficient in terms of feeding the electromagnetic energy in the transponder. A set of three radar types are investigated: First, a frequency modulated continuous wave (FMCW), second, a frequency stepped (FSCW) and third, a switched frequency stepped (S-FSCW) radar. All three realized types generate a RF ramp within the ISM band at 2.4 GHz. The FMCW radar is equipped with a fast direct digital synthesis (DDS) based frequency synthesizer that provides fast frequency sweeps of 100 µs duration (Figure 3). The DDS works with a frequency pre-distortion to combat non-linear frequency chirps as reported in (Scheiblhofer, 2006). The Tx and Rx paths have separate antennas to achieve better signal isolation. The front-end collects during one frequency sweep 1024 data points. Data averaging is performed by repeated frequency sweeps over the whole bandwidth.

The FSCW radar generates the frequency ramp with a phase locked loop (PLL) based synthesizer (Figure 4). The synthesizer is significantly slower than the DDS providing sweep

durations of 100 ms. During one frequency sweep the radar collects 636 data points. Contrary to the FMCW the measurements are taken on discrete frequency steps.
The S-FSCW radar front-end is additionally equipped with Tx- and TRx switches (Figure 5). The switches are accurately synchronized to the signal response of the SAW delay line (Figure 2). The method yields in a significant reduction of environmental echoes and noise (Stelzer, 2004). The radar collects 636 data points during one sweep. The FSCW and the S-FSCW radar can average data either on a single frequency step or alike the FMCW over the whole sweep.

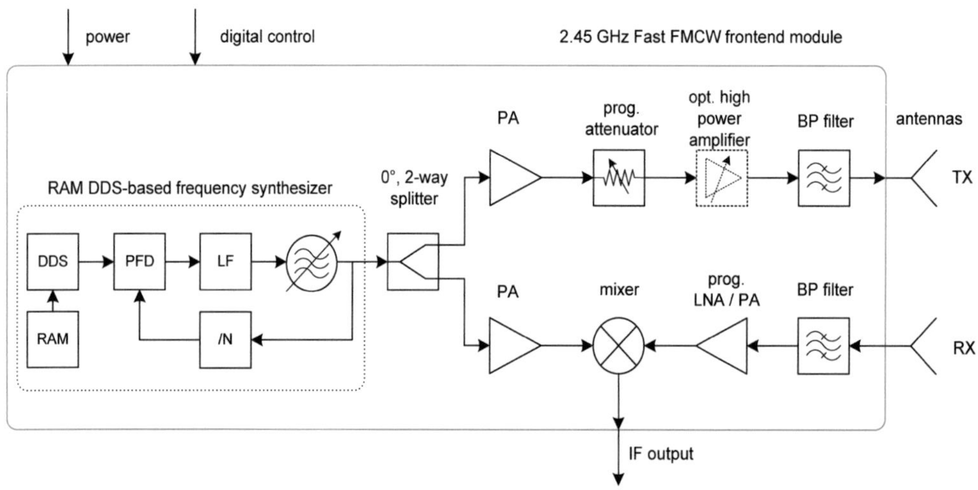

Fig. 3. Block diagram of a FMCW radar front end.

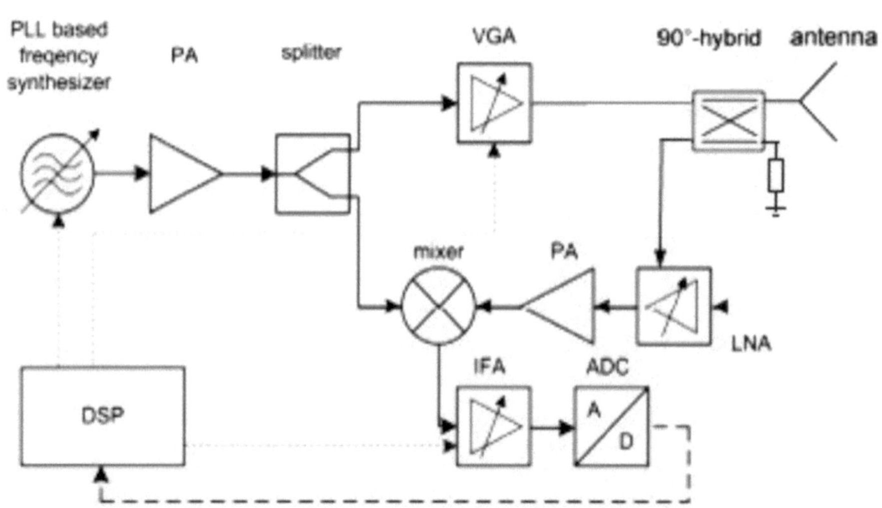

Fig. 4. Block diagram of a FSCW radar front end.

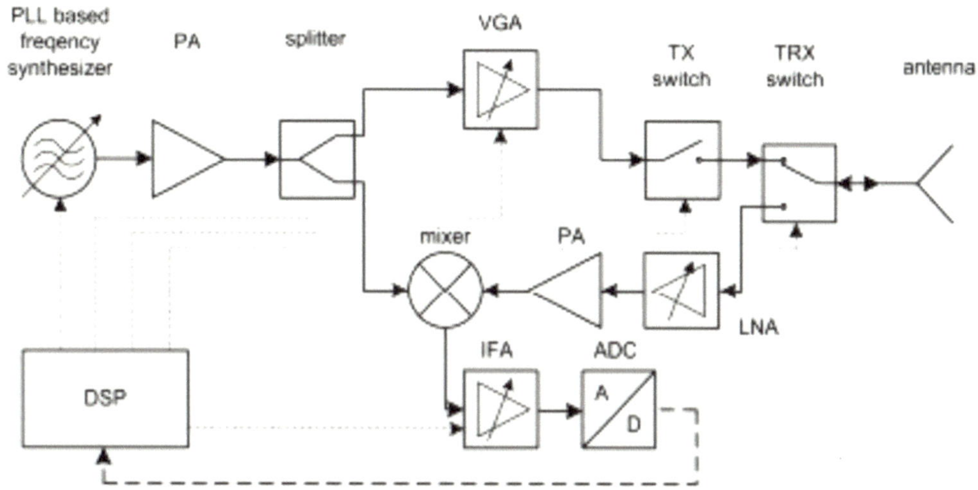

Fig. 5. Block diagram of an S-FSCW radar front end.

2.2 Package for SAW transponder

An important aspect for the transponder is the development of an appropriate packaging, which is functional at high temperatures (HT). A metallic housing with glass feed-throughs, withstanding temperatures up to 400°C is shown in Figure 6. The sensor tags' fixation inside the housing is done by a polyimide glue and its electrical interconnection by wire bonds. It is essential that the packaging is hermetically sealed. This is best done by resistance welding the lid to the socket. The weld is robust against high temperature and high temperature gradients. A complete transponder tag, using a monopole antenna welded to the pins is the simplest form of tag-antenna integration. This monopole tag is fully functional and does not need any additional hardware or any power supply. It can be injection molded into various plastics or ceramics depending on target applications. For optimized read range the connector pins can be welded to a stainless steel slot antenna. The interconnection between tag housing and antenna is done by laser welding, making this RFID transponder system HT resistant well beyond 400°C. It can easily be screwed on metallic objects via the integrated assembly units. By the design of this transponder, a metallic surface acts as a

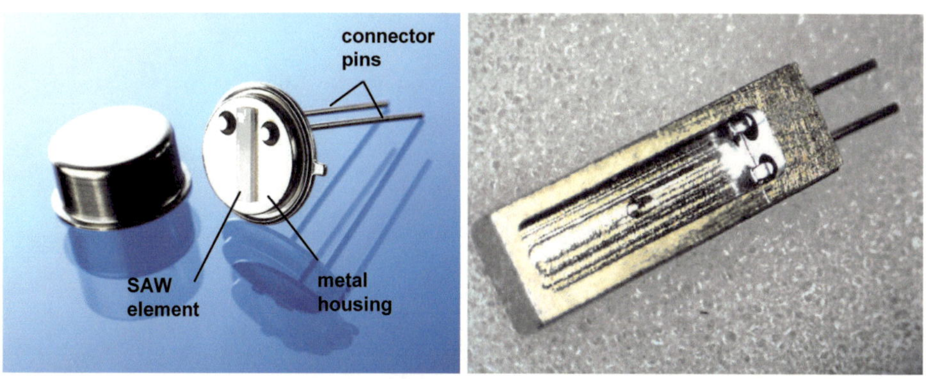

Fig. 6. Transponder housing TO39 (left) and custom KOVAR® housing (right).

reflector, increasing the antenna gain. This in turn doubles the operable readout range. An alternative package is shown in Figure 6 (right). This KOVAR® package is optimized for thermal conductivity to the sensing SAW element, thus increasing the temperature dynamics by a factor 5.

3. System performance

The robustness of SAW transponder technology was proven with various temperature tests. SAW transponders are not only stable at high temperatures; they even can be read out operationally. In case the transponder is read out at high temperature a reduced read-out distance has to be taken into account in the system design.

3.1 Durability

The durability of the packaged SAW devices has been tested by thermal aging (Fachberger, 2008). Several tags of each type of metallization were stored in an oven at temperatures of 300°C and 350°C. Measurements were made in an air-conditioned cabinet at 22°C. The devices were placed in a test jig equipped with lateral interfering spring pins and measured with a network analyzer (NWA). In some cases the contact pins were oxidized on the surface; these pins had to be cleaned with a blade to achieve a good electrical contact to the spring pins. For each measured device the weakest peak amplitude was recorded. The rejection criterion was defined to be a decrease of more than 3 dB in the peak amplitude (measured from the initial level). That is the failed devices were still operating but were significantly degraded.

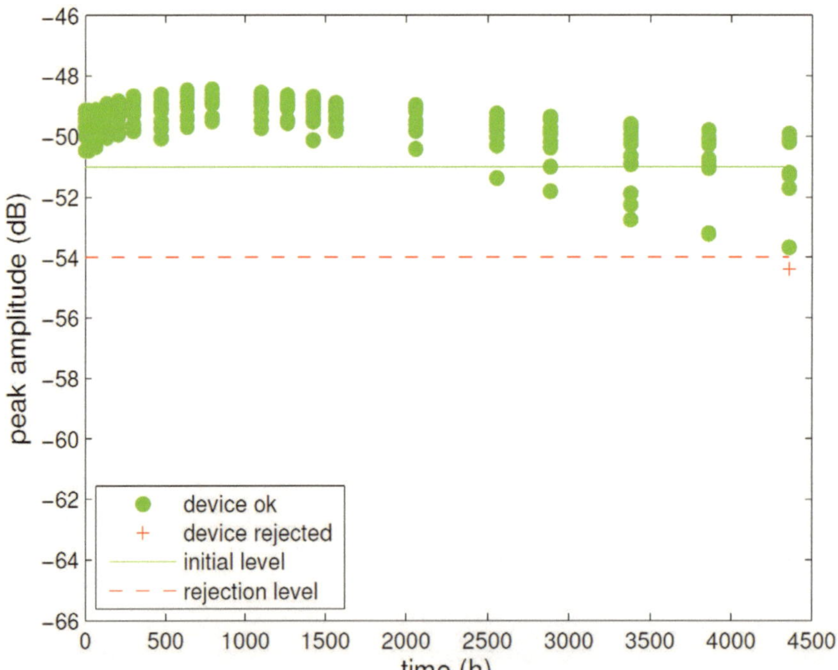

Fig. 7. Annealing at 300°C. Peak amplitudes over annealing time

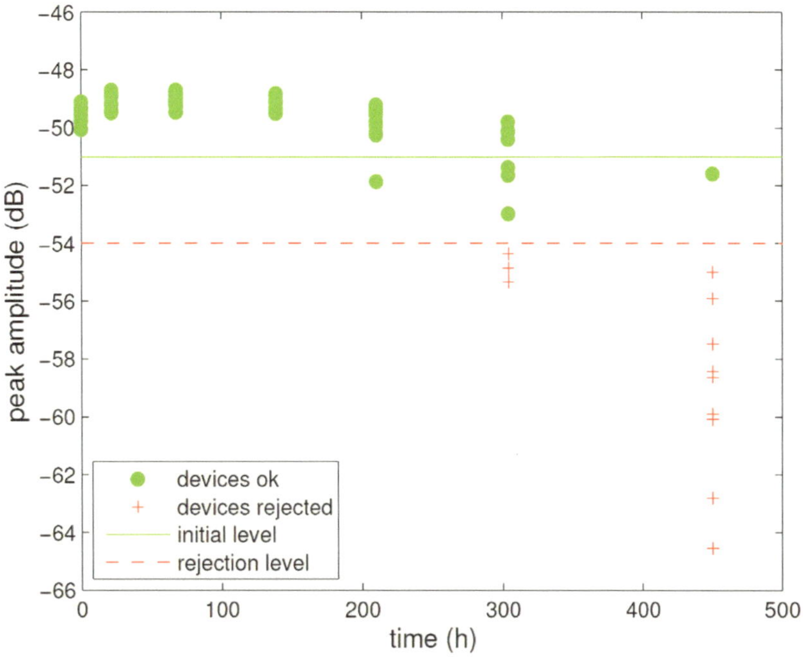

Fig. 8. Annealing at 350°C. Peak amplitudes over annealing time

First tests using a standard Al/Ti metallization showed very poor behavior at 300°C annealing. Some of the devices showed a run-in effect, where the peak amplitude dropped below the rejection limit in the first 20 h at 300°C and recovered after further heating. After aging for 450 h at 300°C all devices were rejected. Subsequently an Al/Ti sandwich metallization was developed.

The Al/Ti sandwich devices also showed a run-in effect for the first 1000 hours at 300°C during which the signal level actually increased. Figure 7 shows the amplitude over time at 300°C. The rejection criterion was exceeded after 4350 hours (more than 6 months) for one of ten devices. Aging at 350°C produces a similar behavior; run-in with increasing signal level is observed within the first 50 hours (Figure 8). After 300 hours three of ten devices dropped below the rejection limit. Interpolating the overall trend, we estimate a lifetime of 4000 hours at 300°C and 250 hours at 350°C. According to the Arrhenius equation, a lifetime of 15 hours can be estimated at a temperature load of 400°C (assuming that no further reactions are activated).

To check the effect of temperature changes, cycling tests were carried out. A batch of 15 packaged Al/Ti sandwich devices was placed in a preheated oven at 240°C. Every 15 minutes air cooling was turned on or off. Cycles between 30°C and 230°C were achieved in this way. The devices were measured according to the procedure described in the previous section.

Even after 5600 cycles none of the device exceeded the 3 dB rejection level (Figure 9). Some deviations in the peak amplitude are present, however, as no trend can be observed, these are presumably artifacts of the measurement (e.g. variations of the electrical contact resistance between spring pins and contact pins due to oxidation of the contacts' metal surface).

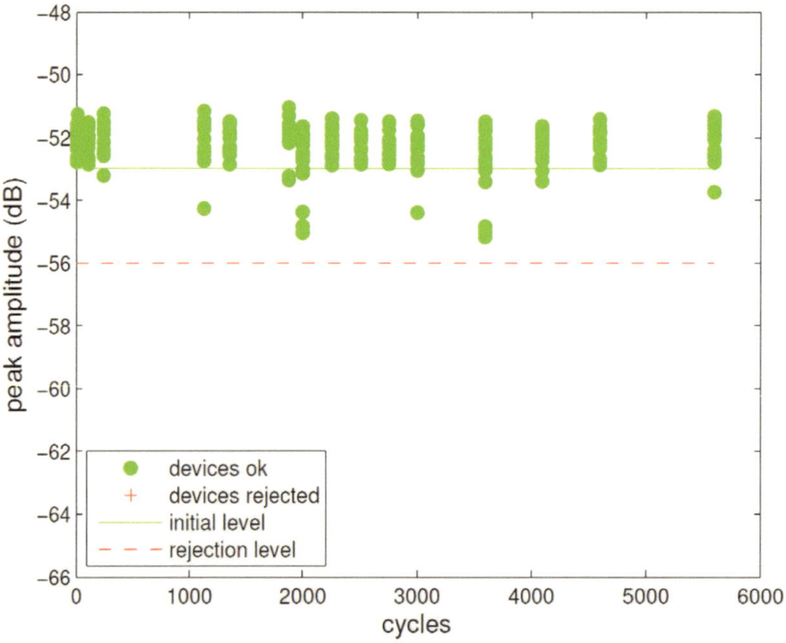

Fig. 9. Cycling between 30°C and 230°C. Peak amplitudes over cycles.

3.2 Read out range

The read out range can be a compulsory system specification especially in harsh environment where antennas cannot be placed arbitrarily. The achievable range of typical SAW transponders was measured in various benchmark configurations. Depending on antenna gain, output power and noise reduction via averaging the read range results in 5 m and above. All measurements were carried out with a FSCW (frequency stepped continuous wave) reader and a SAW transponder with a slot antenna at 25°C and a radiated power of 10mW EIRP. The antenna gain, the antenna configuration and the averaging settings were varied.

The results of the distance measurements are summarized in Figure 10. The readout distance is defined as the distance where the signal power sinks below 80 % of the reference signal, where 1x 9 dBi or 1x 18 dBi means that one reader antenna per channel with an antenna gain of 9 dBi or 18 dBi is used, 2x 9 dBi or 2x 18 dBi means double antenna per channel mode. In addition, some measurements have been performed using the averaging ability of the system. This facility increases the readout range but at the cost of readout speed. With a two antenna system using 18 dBi each and an averaging factor of 8 (8x av.), a maximum readout range of about 6.5 m has been achieved.

Figure 11 shows the readout range between RT and 300°C using a 9 dBi antenna in single channel mode, for a single shot measurement. The measured read out range at RT was taken as reference distance. At 300°C, the readout range decreases to 30 % of the original range at RT. Due to physical effects, the attenuation of the transponder signal increases with operating temperature. In the temperature range between RT and 300°C, the loss is almost linearly 0.05 dB/µs °C. Roughly half of this value, 0.02 dB/µs°C [13], can be ascribed to the change of the acoustic propagation attenuation of the crystals with temperature. The other half of the attenuation can be referred to the temperature dependent frequency shift of the transducers and the transfer function of the transponder antenna relative to the fixed ISM band.

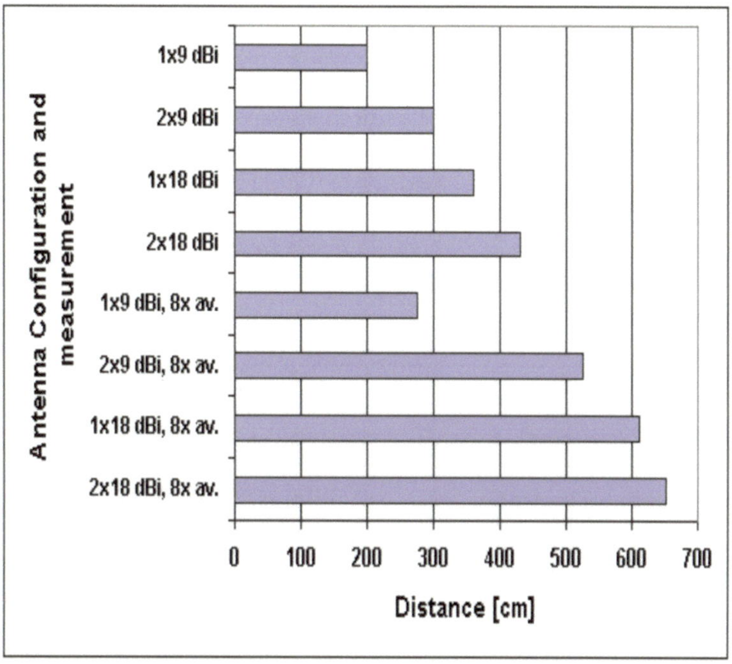

Fig. 10. Read range for various antenna configurations and averaging factors.

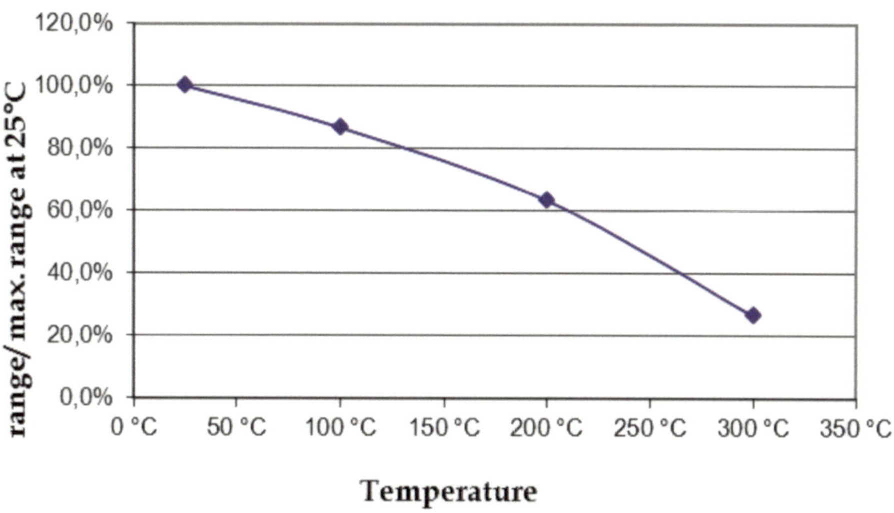

Fig. 11. Read range depending on transponder temperature.

The described tags work also as temperature sensors. It is possible to measure temperature with at least 0.6°C of accuracy, depending on calibration sets and evaluation algorithm. This theme is discussed in detail in (Fachberger, 2006) and is omitted here because this chapter focuses on the RFID-applications of the tags.

3.3 Exposition to gamma radiation

To examine the robustness of the tags further, the devices where exposed to strong gamma ray doses to investigate their behavior in sterilization chambers e.g. those for food or medicine sterilization. The dose started with 6 kGy (Gray, 1 Gray = 1 Joule per kg mass of absorbed radiation) and was increased up to 100 kGy so that every tested tag had to withstand a summarized radiation of more than 500 kGy. Table 3 lists some ranges for specified radiation doses. During that test, no degradation was observed. The tags can be used in environments with high doses of gamma radiation, e.g. food or medical sterilization chambers.

Radiation dose	Application / Effect
0.1 mGy	Flight in stratosphere
1.2 - 2 mGy	Natural Radioactivity
4.5 - 5 Gy	Human lethal dose
40 - 70 Gy	Tumor treatment
Up to 10 kGy	Sterilization of food
Up to 50 kGy	Sterilization of medicals and surgery instruments

Table 1. Table of radiation dose and specific applications

4. Application examples

In practice the application SAW transponders comes in discussion literally when alternative technologies fail, i.e. do not survive the process conditions. SAW transponders do not offer memory and work with proprietary reader systems and above all they are in the price range of active tags. Still the cost-benefit analysis works out for certain processes. The key factors are: cost of tagged investments, cost of lost tracks or lost assignment of assets, follow-up costs of safety hazards and cost savings through process optimization using RFID.

4.1 Automotive paint shop

Modern automotive paint lines are fully automated. However, the automatic assignment of the car body to the process parameters (shape, paint color) is rather difficult as heating cycles with maximum temperatures of up to 250°C are involved. Conventional RFID-tags, and even conventional SAW-tags cannot stand these conditions for long term. Through the optimization of the SAW-metallization and a customized metal antenna it was possible to reach the required lifetime of the tag. In combination with a solder free packaging and assembly technique a very robust and stable SAW-transponder was achieved. The cycling test showed, that after 3000 cycles from room temperature to 220°C and back, with a hold time of 30 min for each temperature level, the drop out quote was well below 10%. The drop out was defined as a degradation of the signal amplitude of 3 dB compared to the initial amplitude values, the same definition used for the durability tests described in chapter 3.1. In this case a different SAW metallization based on Al and Cu was used. In this way a higher readout range could be achieved, however the temperature stability of these tags is slightly lower than of those presented in chapter 3.1.

Fig. 12. SAW transponder used in automotive paint shops (Source: Baumer Ident).

4.2 Identification of slag vessels

SAW transponders are increasingly investigated by steel producing companies. Due to the high temperatures of handling assets like steel and slag vessels the advantage of SAW transponders comes in place. The cost ration of tagged item per tag remains small and the revenues through a well-documented track process are high. The example shows the tagging of slag vessels from tapping to the heap.

Recycling processes of slag coming from the converter and stored in special slag ladles strongly depends on the slag composition. A correlation of the casting process and the slag ladle is recommended for high quality post-processing of the slag. For an automatic transport logistics a SAW transponder for identification was placed directly on the ladle. The transponder had to withstand temperatures of up to 350°C and heavy mechanical shocks during emptying of the ladle. An exemplary system configuration is shown in Figure 13a. In spite of the metallic surrounding and coexisting WLAN, a readout distance of 4 m had been achieved. The tags were mounted inside the transport ears to have more protection against slag splashes and collisions with the transport hook (Figure 13b). Figure 13c and 13d show the rough environment around the slag tapping and the emptying of ladles on the heap. Slag vessels have also been tagged while being transported on the crane as shown in figure 14.

Fig. 13. (a) System setup for interrogation, (b) mounting of the tag, (c) area of slag tapping, (d) slag vessel emptying on the heap

Fig. 14. Slag vessel on the transport crane.

4.3 Identification of slide gate plates

Another example from the steel industry shows the application of SAW transponders for tagging of slide gate plates and monitoring their temperature behavior during the cast. A RFID sensor system to track slide gate plates and furthermore to continuously monitor the temperature during steel casting was developed. The sensor system should help to optimize the casting process and as a consequence reduce costs by improving slide gate plate logistics, maximizing individual plate usage, and minimize unscheduled casting interruptions.

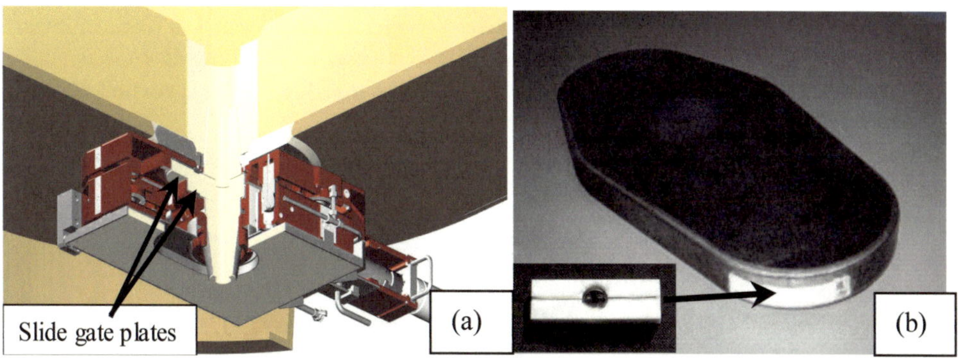

Fig. 15. (a) Slide gate mechanics, (b) implementation of a SAW transponder in the plate

Slide gates, as shown in Figure 15a, are large valves, which are used to regulate the flow of liquid steel (Fachberger, 2010). The flow is controlled via the overlap of two holes, one in each of the two refractory plates. During casting these ceramic plates are exposed to high temperatures of more than 1500°C, high mechanical loads of several tons of liquid steel, and chemical attack by e.g. alloys. Depending on the steel grade, the throttling, and the casting time the plates have to be replaced before a critical degree of wear is reached. SAW transponders are mounted on the outer radius to limit the heat exposure (Figure 15b).

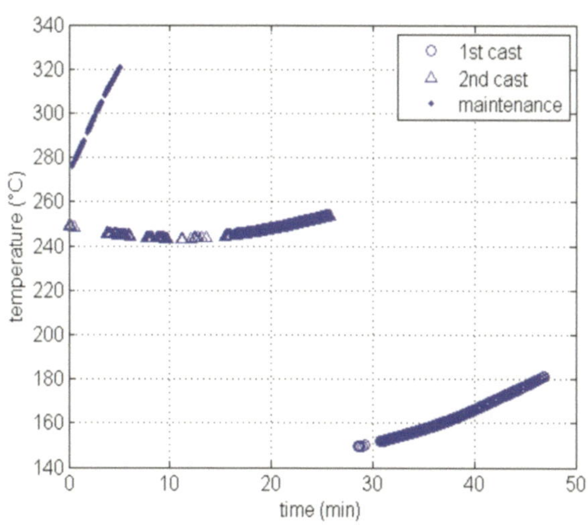

Fig. 16. Temperature monitoring of slide gate plates during casts and maintenance; the usage of plates (casting time, replacement frequency) is monitored throughout the casts via antennae inside the slide gate mechanics and further data transfer to the control stand.

4.4 Identification of automotive pressure sensors

Identifying individual sensors is often desired, in particular when sensors are frequently replaced or recalibrated, as e.g. in test blocks for combustion engines. Here a commercially available pressure sensor from AVL, which is typically mounted close to the combustion chamber of an engine, has to be identified, and therefore, was equipped with an identification tag (Figure 17). The aim of the identification (ID) is to automatically assign correct calibration data. With an ID-tag placed inside the pressure sensor housing instead of a standard RFID tag being attached to the end of the sensor cable, as in existing solutions, cables can be disconnected to reduce set-up time without the need of carefully reconnecting sensors to assigned connectors. This reduces set-up time as well as prevents from incorrect measurements caused by wiring errors and the use of incorrect calibration data (Bruckner, 2003).

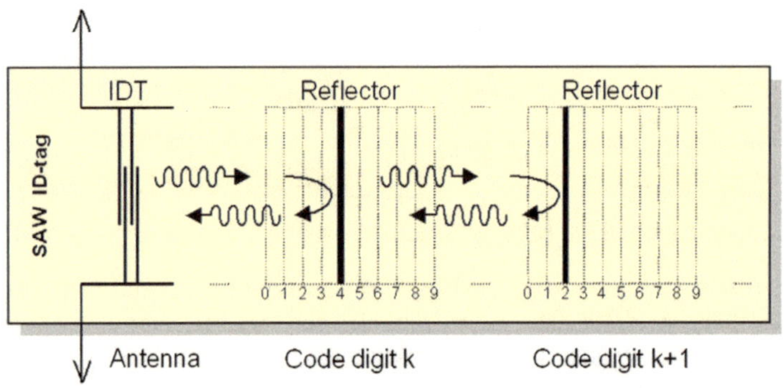

Fig. 17. Principle of the RFID coding using pulse position coding with a decimal basis.

The concept of the presented identification system and the integration in existing sensor systems is sketched in Figure 18 below. In the given example an AVL pressure sensor operating under extreme environmental conditions (pressure, temperature, vibration, shock) is connected to a sensor-evaluation unit. The identification system may only use the existing shielded connection without direct ohmic connection, not destroying the high isolation

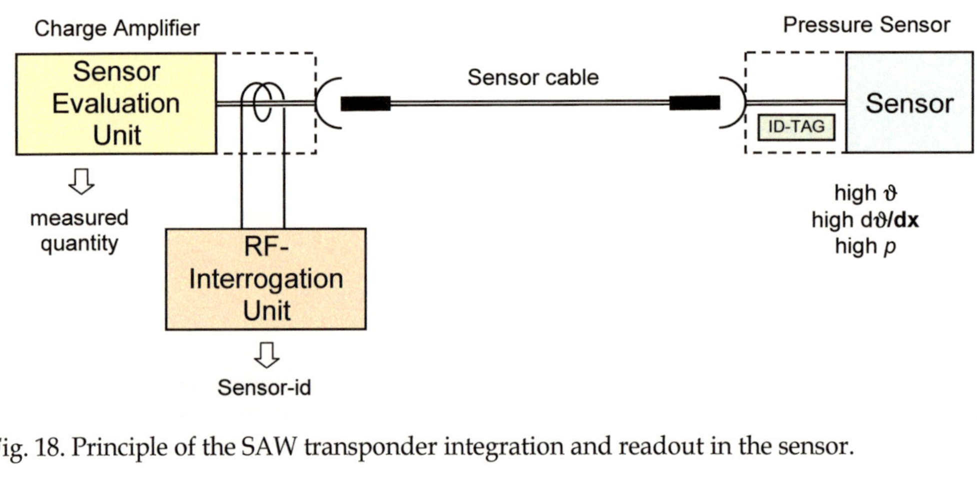

Fig. 18. Principle of the SAW transponder integration and readout in the sensor.

resistance necessary for correct charge amplifier operation. Thus the standard sensor cable remains unaffected. The RF-interrogation unit is coupled to the signal line. Inside of the pressure sensor the SAW-ID tag is coupled via an antenna structure directly on chip to the signal line to preserve high resistance. The interrogation unit identifies any sensor connected to the evaluation unit and can provide additional sensor information like calibration data or sensor age from the database.
The stability of the assembly shown in figure 19 was tested up to 3500 g. Also further rigid tests referring to the temperature stability up to 400°C and temperature gradients up to 70°C along the length of the SAW ID tag were performed. At least up to 400°C no trace for the impact of pyroelectricity on the metallization was observed.

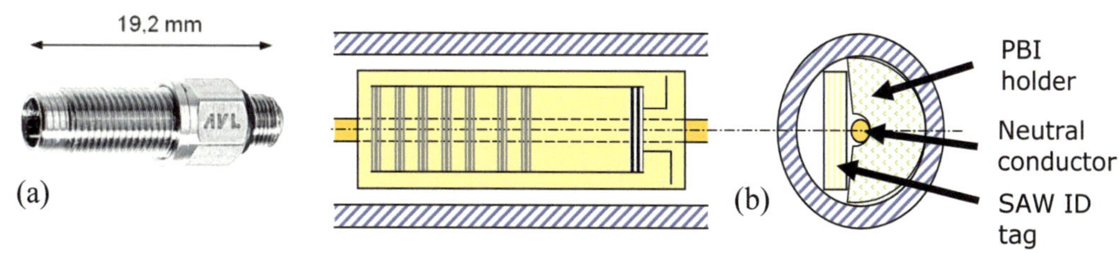

Fig. 19. A standard pressure sensor of AVL Type GM12D (M5*0,5) (a) and schematic arrangement of the SAW ID-tag mounted inside of the sealed pressure sensor (b).

5. Conclusion

In this chapter the operation principle of SAW transponders was discussed for RFID applications. The SAW transponder systems are to be considered for harsh environment processes. This has been demonstrated for various applications in steel and automotive industries.
Future work in research and development deals with the increase of temperature stability of transponders. This includes the stabilization of metallization films, substrate and packaging technology. A unique RFID code on sensors has its advantage for automatic calibration of individual sensors. Thus SAW based pressure and strain sensors are under development for wireless high temperature applications.

6. Acknowledgment

The authors would like to thank the industrial co-operation partners RHI AG, AVL LIST GmbH and HESCON. The work presented was partly funded by the Austrian COMET program operated by FFG Austria.

7. References

Bruckner, G; et.al. (2003). A high-temperature stable SAW identification tag for a pressure sensor and a low cost interrogation unit. IEEE Sensors 2003.
Fachberger, R ; Bruckner, G.; Hauser, R.; Reindl, L. (2006).Wireless SAW based high-temperature measurement systems. Proc. IEEE Frequency Control Symposium, pp. 358-367.

Fachberger, R.; Bruckner, G. and Bardong, J. (2008). Durability of SAW transponders for wireless sensing in harsh environments. Proc. of IEEE Sensors Conference, pp. 811-814.

Fachberger, R.; Erlacher, A. (2010). Applications of Wireless SAW Sensing in the Steel Industry. Proc. Eurosensors XXIV, September 5-8.

Hauser, R.; et.al.(2004). A wireless SAW-based temperature sensor for harsh environment. Sensors, 2004, Proceedings of IEEE, pp. 860-863.

Hornsteiner, J.; Born, E. and Riha, E. (1997). Langasite for high temperature surface acoustic wave applications. Physica Status Solidi A, vol. 163, p. R3-R4.

Kalinin, V.(2004). Passive wireless strain and temperature sensors based on SAW devices. Radio and Wireless Conference, 2004 IEEE, 187 – 190.

Kalinin, V.; Lohr, R.; Leigh, A. and Bown, G. (2007).Application of Passive SAW Resonant Sensors to Contactless Measurement of the Output Engine Torque in Passenger Cars. Frequency Control Symposium, 2007 Joint with the 21st European Frequency and Time Forum. IEEE International, pp. 499-504.

Pereira da Cunha, M.; Moonlight, T.; Lad, R.; Bernhardt, G. and Frankel, D. J. (2007). Enabling Very High Temperature Acoustic Wave Devices for Sensor & Frequency Control Applications," 2007 IEEE Ultrasonics Symposium, pp. 2107-2110.

Pohl, A.; Ostermayer, G.; Reindl, L.; Seifert, F. (1997). Monitoring the tire pressure at cars using passive SAW sensors. Ultrasonics Symposium, 1997. Proceedings., 1997 IEEE, 471 - 474 vol.1.

Reindl, L.; Ostertag, T.; Ruile, W.; Ruppel, C.C.W.; Lauper, A.; Bachtiger, R. and Ernst, H. (1994). Hybrid SAW-device for a European train control system. Ultrasonics Symposium, 1994. Proceedings., 1994 IEEE, vol. 1, pp. 175-179.

Reindl, L.; Scholl, G.; Ostertag, T.; Scherr, H.; Wolff U. and Schmidt, F.(1998). Theory and application of passive SAW radio transponders as sensors. Ultrasonics, Ferroelectrics and Frequency Control, IEEE Transactions on, vol. 45, no. 5, pp. 1281-1292.

Scheiblhofer, S.; Schuster, S.; Stelzer, A. (2006). Signal Model and Linearization for Nonlinear Chirps in FMCW Radar SAW-ID Tag Request. IEEE Transactions on Microwave Theory and Techniques, Vol. 54, NO. 4, pp.1477-1483.

Stelzer, A.; Scheiblhofer, S. and Schuster, S. (2004). S-FSCW Radar based SAW Sensor Interrogator for Highly Accurate Temperature Monitoring. Proc. Asia Pacific Microwave Conference APMC, (New Delhi, India), p. 751.

18

Internetworking Objects with RFID

Rune Hylsberg Jacobsen, Qi Zhang, and Thomas Skjødebjerg Toftegaard
Aarhus School of Engineering, Aarhus University
Denmark

1. Introduction

The Internet of Things refers to the networked interconnection of everyday objects. Everyday objects, such as cars, coffee cups, refrigerators, bathtubs, and more advanced, loosely coupled, computer resources and information services will be in interaction range of each others and will communicate with one another. The Internet of Things has the potential to be used by billions of independent devices co-operating in large or small combinations, and in shared or separated federations. It is going to be based on information about objects in the physical world and their respective surroundings. This information will be provided by "the things", as they obtain and reveal information through RFID, wireless sensors and communication devices embedded in systems or worn by users. Through unique addressing schemes these things are able to be networked with each other on a global scale and to cooperate with neighbors and remote systems to reach common goals.
During the last few years an increasing number of conferences, workshops, research projects and coordinated actions on a global as well as European level shape the current understanding of the important topics of RFID and Future Internet including Internet of Things. Buckley (2006) summarized recent trends in Radio Frequency Identification (RFID) integration with Internet of Thing. The coordinated action CE RFID in Europe has published a Final report on RFID and its applications. In the report edited by Wiebking et al. (2008), a comprehensive summary of RFID and its applications are provided.
In a recent publication, Khoo (2010) reviews current RFID technology, its usage, and the necessary development required for RFID technology to enable the Internet of Things. Atzori et al. (2010) describes how the basic idea is to have the pervasive presence around us by using a variety of things or objects such as RFID tags, sensors, actuators, mobile phones etc.
The vision of an Internet of Things powered by next generation RFID has many potential advantages. It offers new industrial opportunities for the Information Communication Technology (ICT) market, and enable a breakthrough improvement in process efficiency and product/service quality in several application scenarios, such as environmental monitoring, e-health, intelligent transportation systems, military, and industrial plant monitoring. Moreover, it increases the usefulness of the Internet to the majority of citizens, who are interested in getting physical support to their daily needs.
RFID devices and systems are showing significant potentials in applications from manufacturing, security, logistics, airline baggage management to postal tracking. The technology enables an organization to re-engineer its business processes and to increase the efficiency that results in lower costs and higher effectiveness. Manufacturers and distributors deploy RFID to handle the logistical overload that results from the large increase in global sales from electronic commerce or to improve the efficiency of an enterprise supply chain.

While current deployment of RFID technology is focusing on use cases for object tracking and object monitoring, the integration with wireless sensor network (WSN) technology adds another dimension. The integration of RFID and WSN allows RFID tags and readers to form networks in order to implement complex functions where the communication of one tag and one reader is insufficient. The networks can be further enriched by the integration of sensors. One of these functions could be the range enhancement by distributing messages over multiple network nodes. Static network nodes could also locate each other as well as locate nodes moving within the network. By taking a holistic approach to RFID/WSN in the Internet of Things we move from connection of objects to the networking of objects.
This chapter discusses the RFID/WSN technology in a networking perspective. We outline the development needed to integrate RFID systems with the Internet of Thing and look at the evolution from today's connection of objects to the future networking of objects.

2. Internetworking scenarios

It can be observed that the Internet of Things should be considered as part of the overall Internet of the future, which is likely to be remarkably different from the Internet we use today. Fig. 1 illustrates this principle. A wide-spread interconnection of everyday objects to the Internet adds another "onion ring" to the communication infrastructure. As we move from

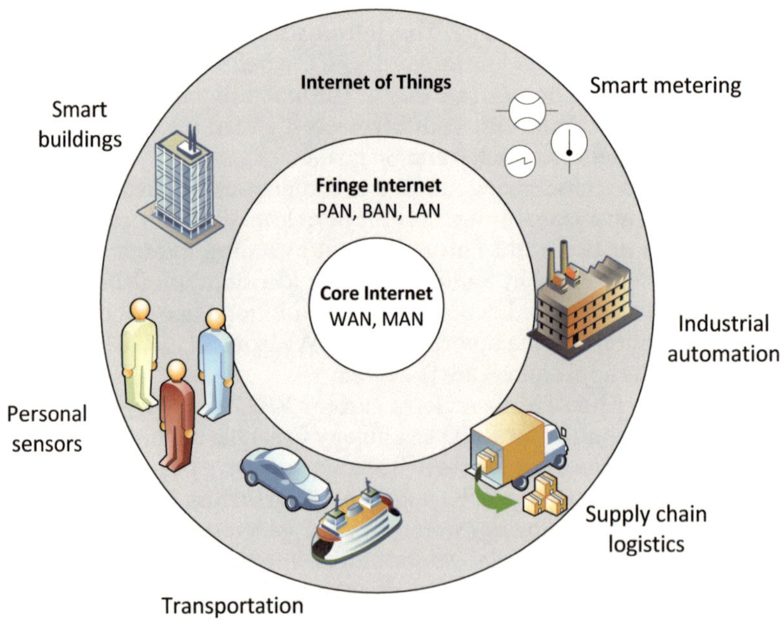

Fig. 1. Interconnecting objects to the Internet adds an outer "onion ring" to the communication infrastructure.

the core of the Internet with its high capacity routers to the outer network edges, i.e. the fringe Internet, where different local networks and access networks such as personal area networks (PAN), body area networks (BAN), local area networks (LAN) we gradually get closer to the physical objects in our surroundings.
The integration of RFID and WSN technology into the infrastructure adds new possible usages of RFID technology. Mitrokotsa & Douligeris (2010) describe how integrated RFID

sensor systems essentially allow two new categories of usage: First, integrated *RFID sensor-tag* will allow the tracking of sensor data of an object through-out its life-cycle. This might be very important for the transportation and storage of hazardous goods (e.g. chemicals, nuclear waste etc.), and medical samples that e.g. must stay within some temperature interval during transportation. Another possible use is to track the usage of a mechanical system known to be prone to failures due to fatigue built up over time such as a weapon system. These usage scenarios represent a further enhancement of the object tracking and object monitoring applications. When tags are brought into proximity of the reader an asynchronous data transfer can occur. Thus connecting the physical objects to the Internet. Second, integrated *RFID sensor-reader* systems will add the wireless networking dimension to the RFID system thereby introducing enhancements such as mobility support, naming and addressing, resiliency, end-to-end architectures, networking security etc. This allows portable readers to be connected to the Internet of Things whereby data can be readily accessed, processed and distributed over the Internet.

Fig. 2 and 3 illustrate two different network architectures for the integration of RFIDs and wireless sensor nodes. The integration of wireless sensing nodes with RFID tags allow devices

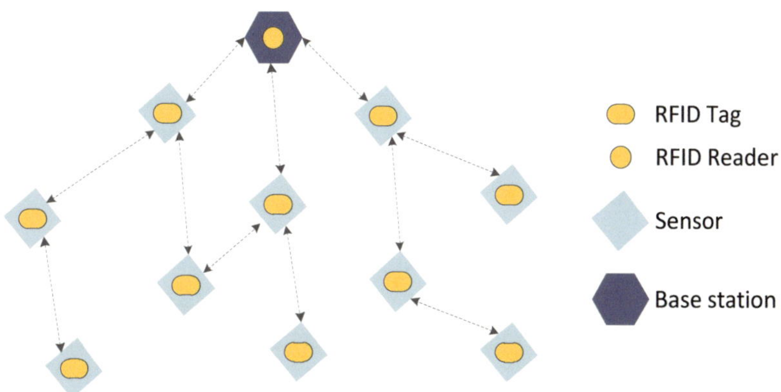

Fig. 2. RFID sensor-tag network architecture. (Adapted from Mitrokotsa & Douligeris (2010)).

to communicate with each other as well as with other wireless devices. The main feature of such integrated device is that the RFID sensor-tags can collect data related to the conditions around them and transmit and share these data with each other. The network of the integrated sensor-tags is able to communicate with a wider network, such as an enterprise network and/or the Internet, via base stations.

Another possible strategy of integrating RFID systems with WSNs is by integrating RFID readers with sensor nodes as shown in Fig. 3. Zhang & Wang (2006) labeled this integrated RFID sensor/reader node a "smart node" with the interpretation of "smart" meaning an autonomous physical/digital objects augmented with sensing, processing, and network capabilities. Smart nodes are able to relay information and to be configured as relay nodes or routers of a WSN. Likewise the RFID sensor-tags, smart nodes are able to communicate with each other by creating an ad hoc communication network. From an architecture point of view this integrated network, is similar to the hierarchical clustering-based two-tiered WSN.

RFID and WSN are key enablers to realize the Internet of Things scenario described above. On the other hand cost will be the key driver for the evolution. The main argument for bringing the WSN into the discussion is to offer connected mobility for relatively small and power/resource limited devices as an integral part of the Internet of Things. The necessity

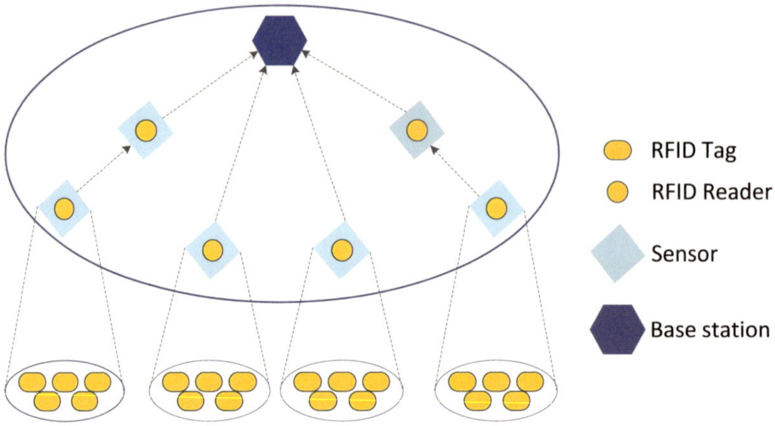

Fig. 3. RFID sensor-reader network architecture. (Adapted from Mitrokotsa & Douligeris (2010)).

for RFID is basically the same. However with a factor in increased volume of 1000 the constraints are even stronger. Especially the cost of the nodes becomes very critical. Given the potential ultra-low cost of RFID objects, as shown in e.g. Lakafosis et al. (2010) we can reach a completely new layer in the Internet of Things. Therefore the combination of the two will give us a technology with extended capabilities, scalability and of course portability while still being able to control the cost.

3. Technologies for identification, sensing and communication

In this section we introduce the essential technologies for identification, sensing and communication in the Internet of Things. We do not provide for an in-depth presentation of all relevant topics but merely focus on the technological aspects that are the most significant ones for an internetworking scenario. In the following we will address RFID system components, WSN technology as well as infrastructure aspects of the Internet of Things.

3.1 RFID systems

Several reviews and surveys of RFID technology have been published in the literature such as as the articles by Floerkemeier & Sarma (2008) and Krishna & Husalc (2007). Essentially, an RFID system is composed of a number of tags coupled with one or more readers that are connected to an ICT infrastructure. RFID tags (transponders) fall into two general categories, active and passive RFIDs, depending on their source of electrical power. RFID tags are typically of very small size and of very low cost. Passive tags harvest the energy required for transmitting their Identification (ID) from the query signal transmitted by a RFID reader (interrogator) in the proximity and their lifetime is not limited by the battery duration. An RFID reader communicates with one or more RFID tags via electromagnetic radio frequency fields. The radio frequency band used for RFID range from low frequency (LF), via high frequency (HF) up to ultra high frequencies (UHF). In fact, this signal generates a current into the tag antenna by induction. The current is utilized to supply the microchip which will transmit the tag ID. Usually, the antenna gain i.e. the power of the signal received by the reader divided by the power of the signal transmitted by the reader, of such systems is very low. Thanks to the highly directional antennas utilized by the RFID readers, tags ID can be correctly received within a radio range that can be on the order of few meters. At least the reader reads tag ID. Furthermore, it may read auxiliary data from tags or write data to tags that support additional data memory (read only, read/write). The transmission of an RFID system is subjected to the same radio wave impairments as any other wireless communication systems.

Other RFID tags get power supplied by batteries. In this case we can distinguish between semi-passive and active RFID tags. For semi-passive RFID tags batteries are used to power the microchip while receiving the signal from the reader. Like in the passive RFID tags, the radio is powered with the energy harvested by the reader signal. In contrast, active RFID tags use the battery power for the transmission of the signals as well. Obviously the radio coverage is higher for active tags compared to the semi-passive and passive tags.

A typical RFID reader (interrogator) is comprised of a radio module, a central processing unit (CPU), a network interface, and general input/output pins. The CPU can be a low-end microcontroller or an advanced embedded microprocessor with significant computing resources. RFID readers do not require line-of-sight access to read the tag and the read range of RFID is larger than that of a bar code reader. Tags can store more data than bar codes and readers can communicate with multiple RFID tags simultaneously. Because of this capability, an RFID reader can capture the contents of an entire shipment as it is loaded into a warehouse or shipping container.

By using RFID it is possible to give each object, e.g. each product in a grocery store, its own unique object ID. There are several different standardized schemes for identifier encoding format. The unique object ID must have a global scope that is capable of identifying all objects uniquely and acts as a pointer to information stored about the object and the functionalities of the tag somewhere over the network. In general, the identification will be a number that contains information about the tags ID format, the organization issuing the tag, the class of the objects as well as serial number information.

3.2 Wireless sensor network technology

Several books and research papers exist on wireless sensor network (WSN) technology and applications such as e.g. Karl & Willig (2005). WSNs bring about key enabling technologies for the Internet of Things. Wireless sensor technologies allow objects to provide information about their environment and context, whereas smart technologies allow everyday objects to "think and interact".

WSNs have evolved from the idea that small wireless devices distributed over large geographical areas can be used to sense, collect, process, and distribute information from the physical environment. An essential building block of a wireless sensor is the microcontroller. The processor core can be 8-, 16- or 32-bit based but the CPU performance is not by itself that critical as a wireless sensor network is not expected to process large amount of data. WSN devices run with a low duty-cycle alternating between sleep and active mode. The active period of operation can be shorter with a more efficient CPU. The devices are unable to communicate during the sleep periods and in most scenarios WSN devices spend a large part of their time in a sleep mode to save energy and cannot communicate. This is absolutely anomalous for internetworking devices in today's Internet.

A WSN typically connects the physical environment to real-world applications, e.g., wireless sensors. Different wireless protocols have evolved for personal area networks and sensor networks as e.g. Z-wave and Zigbee with its IEEE 802.15.4 radios and several standards for wireless communication exist today. Until recently the perception has been that a full-fledged Internet Protocol (IP) communication stack was too large and complex to implement in small devices. However, a new and appealing wireless standard for interconnecting wireless sensor

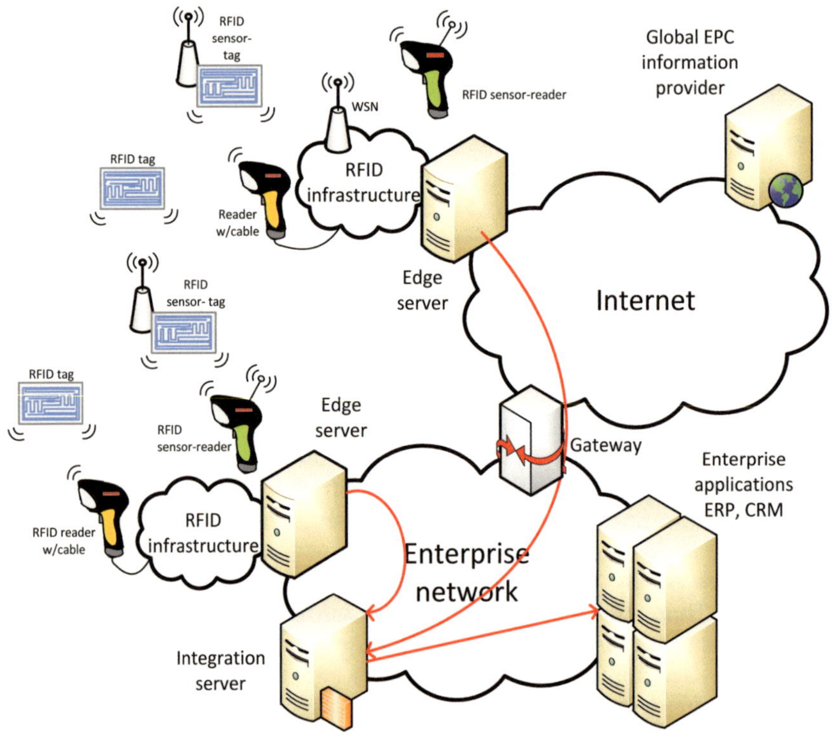

Fig. 4. RFID network scenario.

networks is the IEEE 802.15.4 standard. In this particular case it seems that through a wise Internet protocol adaptation, IEEE 802.15.4 devices can be incorporated into the Internet architecture. This allows us to rely on already adopted schemes for forwarding, routing, addressing etc.

WSNs can potentially consist of a very high number of sensing nodes communicating in a wireless multi-hop infrastructure. The number of nodes usually reports their sensing data to a small number (in most cases, only one) of special nodes called sinks.

3.3 Network reference model

From a networking perspective, an RFID system consists of several components that communicate. Typically an RFID system is built as an enterprise system that integrates RFID with enterprise legacy systems over a common ICT infrastructure. Together with existing enterprise systems, a RFID network system is built that may interact and communicate with other networks (e.g. business to business) as well. Fig. 4 shows a possible RFID scenario. Via a wired or wireless interface, the reader connects to an RFID edge server. This edge server adapts and co-ordinates the data transfer from a number of readers to enterprise resource planning systems (ERP), such as integration and/or control servers. RFID middleware running on the edge server helps to convert usually proprietary and incompatible interfaces between readers and enterprise systems.

Issues related to how to represent, store, interconnect, search, and organize information generated by the Internet of Things will become very challenging.

4. Integrations aspects of RFID/WSN in the Internet of things

Upon interconnecting objects to the Internet a number of central questions can be raised. How will the Internet architecture evolve when a large scale of limited devices represented by objects get globally connected? What is the essential protocols to use and what needs further development. How to provide application and service interoperability? And how can the security and privacy issues be handled? In this section we will discuss these aspects in more details.

4.1 Internet architecture evolution

The integration of RFID sensor networks in the Internet of Things adds further heterogeneity to the networks. We are working towards an evolved architectural model for the Internet of Things that supports a loosely coupled, decentralized system of smart objects. In contrast to simple RFID tags, smart objects carry chunks of application logic that let them interact more "intelligently" with human users.

The Internet of Things will include an incredibly high number of nodes, each of which will produce content that should be retrievable by any authorized user regardless of her/his position. To make a universal communication system there is a need for globally accepted methods of identifying how each object is attached to a network. This requires effective addressing schemes (and policies) by which objects can identify themselves, locate other objects and discover the communication path between them. Due to the rapid depletion of IPv4 addresses and its short address length (32-bit) it is clear that other addressing schemes than the IPv4 addressing scheme should be used. In this context IPv6 addressing has been proposed. IPv6 uses 128-bit addresses and therefore, it is possible to define on the order of 10^{38} addresses, which should be enough to identify any object which is worth to be addressed. Accordingly, we may think to assign an IPv6 address to all the things included in the network. Since RFID tags use 64 or 96-bit identifiers ways to associate RFID identifiers with network addresses can be inserted. One such method that has been proposed is the recent integration

	EPC™	**IPv6**
Scope	Global	Global
Namespace depth	3	3
Naming authority	EPCglobal	IANA
Identifying objects	All physical	All network interfaces
Length	64 or 96 bits	128 bits
Identifies through	Information pointers	Routing address
Identifier assignment	Permanet	Temporary

Table 1. Comparison between RFID EPC™identification and IPv6 addressing schemes.

of RFID tags into IPv6 networks. Table 1 compares the addressing schemes for RFID and IPv6 devices.

As an example for the 96-bit EPC™identification scheme the space for a company is 60 bits with 24-bit Object Class and 36-bit Serial Number. The standardization body EPCglobal assigns the General Manager Number. A single IPv6 subnet can map this entire space. With the integration, the RFID Object Class and Serial Number become the IPv6 Interface ID. This is illustrated in Fig. 5. So, each RFID tag can be addressable in the IPv6 network. The IPv6 prefix defines the scope of reach.

Another issue is the way in which addresses unknown to the requester are obtained. A name service is needed to map a reference to an address and a description of a specific object and

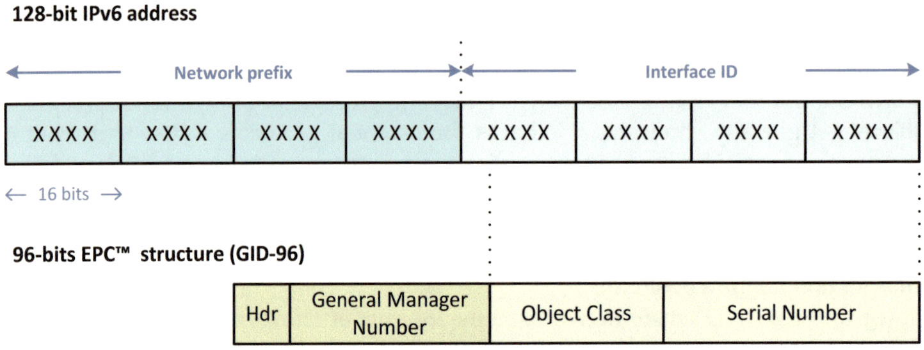

Fig. 5. IPv6-RFID address mapping with EPC^{TM}GID-96.

the related identifier, and vice versa. In today's Internet any host address is identified by querying appropriate domain name servers (DNS) that provide the IP address of a host from a certain input name. In the Internet of Things, communications are likely to occur between (or with) objects instead of hosts. Therefore, the concept of an Object Name Service (ONS) must be introduced, which associates a reference to a description of the specific object and the related RFID tag identifier.

Another promising usage for RFID in the Internet of Things is the potential support for mobility. Recently, Papapostolou & Chaouchi (2009) demonstrated the RFID-assisted IP mobility by using topology information provided by an RFID system to predict the next point of attachment of an RFID-enabled mobile node. There are several proposals for objects addressing but none for mobility support in the Internet of Things scenario, where scalability and adaptability to heterogeneous technologies represent crucial problems. The Internet of Things presents a further challenge that mobile objects may need to re-register their presence on different name servers as a consequence of moving.

4.2 Protocols

The OSI seven-layer model has conditioned a whole generation of telecommunications and information technology protocols. The basic concept of separating functionalities in layers according to clearly separated interfaces through protocols has proven to be powerful for large system designs. For resource limited devices or objects this approach is now showing its limitations. Protocols typically used in the Internet today need hundreds and more of kilobytes of program code to run but this is exceedingly too large for even device object with modest computing resources. Lighter protocols and lighter implementations that compress the explicit protocol layers into a single communications module are now required in the Internet of Things. The protocol header overhead introduced in each layer is a severe limitation to the effective data throughput of narrow-band wireless links. Therefore, existing data communication protocols may be inappropriate for the small objects of the Internet of Things.

New alternative cross-layer based protocols need to be re-engineered in order to cope with the changes that the connecting of objects bring. Stateful protocols as e.g. TCP cannot be used efficiently for the end-to-end transmission control in the Internet of Things. Furthermore, TCP requires excessive buffering to be implemented in objects and its connection setup and congestion control mechanisms may be useless. So far, no complete solutions have been proposed to solve this issue for the Internet of Things and therefore, research contributions

are required. There are several proposals for objects addressing but none for mobility support in the Internet of Things scenario.
Finally with the resource limited nodes and objects and the combination of RFID and WSN the Internet of Things is bound to deal with both an RFID protocol stack as well as a protocol stack for WSN e.g. IEEE 802.15.4 together with a higher layer internetworking protocol stack such as the 6LoWPAN stack.

4.3 Service-oriented architectures

Application and service interoperability is a key aspect for the success of the Internet of Things. In today's service architectures, a middleware layer that translates different data formats and protocols are typically implemented. User interfaces like web services offer necessary interaction and application control. However, according to Wiebking et al. (2008) conventional middleware is inappropriate for handling the range of devices needed for the pervasive internetworking of everyday objects.
Architectures proposed in the recent years for the Internet of Things often follow the service-oriented architecture (SOA) approach. The adoption of the SOA principle allows a decomposition of complex and monolithic systems into applications consisting of an ecosystem of simple and well-defined components that interplay. The use of common interfaces and standard protocols gives a horizontal view of a (potentially) globally distributed enterprise system. Advantages of the SOA approach are recognized in most studies on middleware solutions for Internet of Things. The development of business processes enabled by the SOA is the result of the process of designing workflows of coordinated services, which eventually are associated with objects actions. Furthermore, these processes can be directly linked to the business logic of the enterprise. This facilitates the interaction among the parts of an enterprise and allows for reducing the time necessary to adapt itself to the changes imposed by the market evolution.
An SOA approach does not impose a specific technology for the service implementation and hence allows for software and hardware reuse and can cope with a large degree of heterogeneity. Fig. 6 shows a simplified service architecture for an RFID enriched Internet of Things. The proposed solutions face essentially the same problems of abstracting the devices functionalities and communications capabilities, providing a common set of services and an

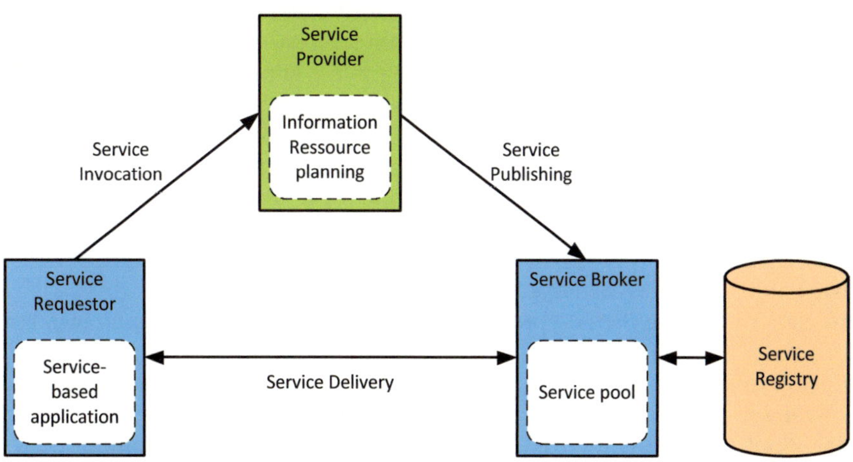

Fig. 6. Service architecture for RFID sensor-network system.

environment for service composition. These problems are further strengthened by the lack of resources available in the RFID/WSN devices. However, for most devices foreseen to be connected with objects in the Internet of Things, the SOA framework becomes impractical to be used because of its demands for computing resources.

Shelby (2010) describes the enabling for web services in contained devices such as WSN. In the described approach a RESTful service architecture based on light-weighted protocols and schemes are introduced to allow a transparent web service to become a reality. Functions that are not needed are omitted, redundant information compressed and service interoperability can be achieved.

4.4 Security and privacy

In general, good security depends on a holistic system-oriented view. Weis et al. (2004) reviews the security aspects related to RFID. From a networking point of view the security threats in an RFID empowered Internet of Things are much similar to that of wireless ad hoc and sensor networks. The wireless and distributed nature of the networks increases the spectrum of potential security threats. The threat model is further stressed by the resource constraints of the RFID/WSN devices. One major challenge in securing RFID tags is a shortage of computational resources within the tag. RFID sensor devices tend to be prone to failure, for example due to battery depletion. The lack of resources prevents intensive security approaches from being deployed. Standard cryptographic techniques require more resources than that is available in most low cost RFID devices. Therefore, manufacturers are looking at more light-weighted encryption schemes, but often with the trade-off in form of a weaker security. A viable security approach should adapt small code size, low power operation, low complexity, and small bandwidth across all nodes in the sensor network.

However, RFID also brings in new perspectives and challenges into the protection of systems, goods and other assets. End-to-end protection in the Internet of Things require confidentiality and integrity protection. This can be provided at the application, transport, network, and at the link layer. Daou et al. (2008) as well as Sharif & Potdar (2008) outline several of these aspects.

Authentication is difficult in the Internet of Things as it requires appropriate authentication infrastructures that will not be available in Internet of Things scenarios. Also the protection from man-in-the-middle attacks is a big challenge for the system design.

Data integrity is usually ensured by protecting data with passwords. However, the password lengths supported by Internet of Things technologies are in most cases too short to provide a strong level of protection.

The network used to share product data between trading partners i.e. EPCglobal Network, by design, is also susceptible to denial of service (DoS) attacks. Using similar mechanism with DNS in resolving EPC™data requests, the ONS root servers become vulnerable to DoS attacks. Any organization planning to implement RFID technology based on EPCglobal Network may discover that the EPCglobal Network infrastructure inherits security weaknesses similar to the weaknesses of DNS.

A second class of defense uses cryptography to prevent tag cloning. Some tags use a form of "rolling code" scheme, wherein the tag identifier information changes after each scan, thus reducing the usefulness of observed responses. More sophisticated devices engage in challenge-response authentication scheme where the tag interacts with the reader. In these protocols, secret tag information is never sent over the insecure communication channel between tag and reader in accordance with a well-defined protocol scheme. Rather, the reader issues a challenge to the tag, which responds with a result computed using a cryptographic

circuit keyed with some secret value. Such protocols may be based on symmetric or public key cryptography. From a networking point of view it is less evident if and how public key infrastructure can be adopted by RFID/WSN devices in the Internet of Things.
A primary security and privacy concern comes from the illicit tracking of RFID tags. Tags, which are readable, pose a risk to both personal location privacy and corporate/military security. Indeed, unseen by users, embedded RFID tags in our personal devices, clothes, and groceries can unknowingly be triggered to reply with their information. Hence, a lot of private information about a person can be collected without the person being aware. The control on the diffusion of all such information is impossible with current techniques. Potentially, this enables a surveillance mechanism that would pervade large parts of our lives. Privacy organizations have expressed concerns for the context of ongoing efforts to embed RFID tags in consumer products. Thus the essential question to address is how to provide user control over their own privacy for them to build trust in the systems. For RFID technology to be a successful part of the Internet of Things public entities need to be made aware that the pervasive networking concepts pose new challenges in terms of personal privacy.
The Internet of Things must be reliable and robust in the face of device malfunction, abnormal traffic loads and traffic patterns and malicious attack. It should safeguard policies regarding ownership of information and authority to access devices, giving due respect to people's rights of privacy.

5. The road ahead for Internetworking of objects

Regarding the future of RFID technologies, with a time horizon between medium-term (5-10 years) and long-term (10-20 years), it is obviously difficult to see where vision reaches beyond what is realistic. What seems clear today is that we are witnessing a paradigm shift from the "identification of objects at a distance" to the more challenging "communication between objects". This implies that besides the next generation of RFID technology there must be a scalable, efficient, reliable, secure and trustworthy infrastructure able to internetwork all involved objects. Technological issues relating to laws of physics must clearly be addressed.
In the European Union, as well as other places around the globe, Future Internet and Internet of Things has been a key strategic challenge for research and technological development. Wiebking et al. (2008) presents a focused roadmap for the Internet of Things that provides a forecast for the evolutions on medium-term and long-term. Among other things the roadmap addresses standardization efforts, technological trends, basic research, and interoperability aspects. Fig. 7 summarizes the road ahead for the evolution of Internet of Things based on a large scale of interconnected objects.

5.1 Standardization

RFID technology efforts towards standardization are focusing on principal areas such as RFID frequency spectrum usage and reader(s)-tags communication protocols, and data formats for tags and labels. The major standardization bodies dealing with RFID systems are EPCglobal, the European Telecommunications Standards Institute (ETSI), and the International Organization for Standardization (ISO). With respect to the Internet of Things, ETSI has started the Machine-to-Machine (M2M) Technical Committee to conduct standardization activities relevant to M2M systems and sensor networks. The objectives of the ETSI M2M committee include the development and maintenance of an end-to-end architecture for M2M based on internetworking standards. This seems to be a wise choice due to the immediate strengthening of the standardization efforts by including sensor network

	Now	Before 2015	Beyond 2015
Vision	• Connecting objects	• Networked objects	• Intelligent objects
Use	• RFID adoption in logistics and retail • Interoperable frameworks	• Increased interoperability • Industry specific deployments	• Unified network that connects, people and things • Integrated industries
Technology trends	• Smaller and cheaper tages and sensors • Smart multi-band antennas • Higher frequency tags • Miniaturized, embedded readers • Low power chipsets • Reduced energy consumption • Network security • Ad hoc sensor networs • Protocols for distributed processing	• Increasing memory and sensing capacities • Extended range and transmission speed of tag-reader communication • Improved energy management • Better batteries • Interoperability protocols and frequencies • Fault tolerant protocols • Ad hoc hybrid networks • Communication in harsh environments	• Cheaper materials • Executable tags • Intelligent tags • Autonomous tags • New materials • Energy harvesting • Intelligent device coorporation • Global internetworked applications • Self-adaptive systems • Distributed memory and processing
Standards	• RFID security and privacy • Radio frequency usage	• Sector specific standards (IETF, ISO ...)	• Interaction standards

Fig. 7. Roadmap for the extrapolation of current technology trends and research topics towards a RFID-enabled Internet of Things. (Adapted from Wiebking et al. (2008)).

integration, naming, addressing, location, QoS, security, charging, management, application, and hardware interfaces for related fields.

As for the Internet Engineering Task Force (IETF) standardiztion activities related to the Internet of Things, it is worth noting that recently the IPv6 over low-power wireless personal area networks (6LoWPAN) IETF group was formed. The 6LoWPAN working group is defining a set of protocols that can be used to integrate sensor nodes into IPv6 networks. Essential protocols composing the 6LoWPAN architecture have already been specified and commercial products that implement the 6LoWPAN protocol stack have been released. Another relevant IETF Working Group is named Routing Over Low power and Lossy networks (ROLL). The working group is currently designing the RPL routing protocol for routing in WSNs – a draft standard which have already got a wide acceptance and a large community support behind it. This will be the basis for routing over low-power and lossy networks including 6LoWPAN. More recently a working group Constrained RESTful Environment (CoRE) formed with the objective to look at the support of RESTful environments for constrained devices such as wireless sensors. This is the key focus of the IETF CoRE working group.

What is also worth pointing out in these standardization areas is the tight collaboration on standards integration as well as the collaboration with other world-wide Interest Groups and Alliances such as IP in Smart Objects (IPSO) Alliance and the ZigBee Alliance. It seems that the whole industry is willing to cooperate on achieving the Internet of Things.

Although there are several standardization efforts to support the integration of heterogeneous networks, a comprehensive framework lack and in a broader perspective for the real-world integration of all sorts of networked contact-less devices there will be a need for substantial progress in the field.

5.2 Technology trends

In terms of technology evolution the current trends towards smaller, more powerful, and more efficient devices is expected to continue. In WSN, energy consumption is of highest priority and the RF communication design blocks consume the most energy. Wireless sensor network designers strive to reduce the power consumption of the blocks in general.
For the CPU part it is likely that it will approximate the evolution expressed by Moore's law, i.e. doubling of capacity each 18-24 months. The improvement for WSN is likely to be used to reduce size and power consumptions instead of increasing capacity and speed. The use of energy harvesting is an important aspect of RFID/WSN devices. With a combination of energy efficient protocols and energy harvesting methods, the optimal solution for achieving autonomous and long-lasting RFID/WSNs can be reached.
Power management plays a significant role in prolonging node life time. The support of advanced power management schemes needs further research and it needs to be taken from a device-level to a network-level. The IEEE 802.15.4 standard defines only a limited set of power management mechanisms for devices. However, most commercial implementations and industrial standards built on IEEE 802.15.4 seem to deviate from the defined power management mechanisms. Efficient protocol support is also needed for the internetwork based WSN and the ongoing work of the relevant IETF working groups is heading in this direction. This includes protocol optimization for smart devices.
Although movements can have severe impact on the received signal strength a global optimum for the network could still be achieved in some cases. While some protocols already exist that take care of the link layer and networking layer, this area still has a lot of open research issues. More specific link layer protocols need to be developed that take into account the movement of the nodes, in addition to the development of low power features such as an adaptive duty cycle for lowering the idle listening and for adapting to the dynamics of the network. A security framework adapted to internetworked objects in the Internet of Things has to be sufficiently light-weighted to meet the constraints of the RFID/WSN devices. On the other hand it also needs to be capable of providing the in-depth security required for the RFID applications.

5.3 Interoperability

Interoperability issues are also very important because RFID tags increasingly travel across a large number of different geographical and organizational environments, together with the object which they identify, thereby imposing new technical requirements such as multi-protocol, multi-frequency integrated circuits and appropriate antenna solutions for tags. For systems, such as in a supply chain applications, where multiple entities have the ability to access RFID tag related information that is shared across geographic or organizational boundaries, there are issues which need to be addressed through research and development. Not all issues can be addressed by the RFID hardware or middleware or similar technological advancements. They include notably look-up services for efficient data retrieval; business models for data sharing among multiple partners (selective data retrieval, access rights); support for distributed decision-making further than just data sharing; networked RFID systems; interoperability requirements and standards; and network security (access authorization, data encryption, standards).

5.4 Research

The ensuring research targets include the hardware aspects (tags, readers, and embedded systems), the software/system aspects and the networking aspects.

The RFID devices themselves need more capabilities to broaden the range of applications. They need to acquire larger memory, local intelligence, encryption and security features, extended functionalities such as integrated sensors, and much more. To support this functionality, new breakthroughs in battery technology are needed, in particular to enable more energy, less space (or printing of the tag), and more reliability than ever before. Lakafosis et al. (2010) demonstrate prototypes that uses inkjet printed RFIDs integrated with wireless sensors.

Today almost all conventional RFID devices contain a silicon-based microchip. The potential in low cost RFID is split between chip-based technologies and "chip-less" tags. These chip-less tags can still be interrogated through a brick wall and hold data; although more primitive in performance than silicon-based chip tags, they hold the potential of much lower production costs and other advantages that will become clear as the technology matures. Further miniaturization of the tag antenna and more efficient and reliable antenna connecting technologies are seen as another priority before mass introduction is affordable.

Research does not only apply to the RFID tag and/or the reader themselves, but also to the information systems which process the RFID events. Using RFID events within enterprise applications, such as Enterprise Resource Planning (ERP) or Customer Relationship Management (CRM), require new RFID middleware and reorientation of these business applications. Research on RFID software is needed to ensure data security, integrity and quality in large networks. It is also needed to provide solutions enabling a reduction of counterfeit.

The Internet of Things will generate data traffic with patterns that are expected to be significantly different from those observed in the today's Internet. Accordingly, it will also be necessary to define new Quality of Service (QoS) requirements and related support schemes.

6. Related work

Technically the combination of wireless sensor network and RFID gives rise to a number of challenges e.g. for the networking. We need to figure out how to evolve the Internet architecture to handle the novel user scenarios. How can service interoperability be ensured? How can we ensure security and privacy and what are the protocols to use in the system? On top of that, seen from a RFID perspective, we argue that to gain the full potential it is necessary to bring the classic scenario of RFID tags "being connected" to a scenario where we actually have networked RFID objects.

The combination of RFID and wireless sensor networks has been studied in a great range of applications, e.g. from healthcare to transportation/logistics and smart environments (home, office, plant). Mitsugi et al. (2007) argues how medication errors such as outdated treatments orders, inaccurate medical records, and increased costs can be avoided with the use of an integrated RFID sensor network. In the healthcare domain the integration of RFID and wireless sensor networks includes real-time monitoring of temperature, blood pressure measurements, heartbeat rate, heartbeat rate variability and pH value.

Bacheldor (2007) reports that the Ghent University hospital in Belgium has implemented an RFID-based real-time locating system to provide nurses and other caregivers with a patient's location in the event of an emergency. The implemented integrated RFID-sensor network detects when a patient is having cardiac distress and sends to the caregivers an alert indicating

the patient's location. In the proposed prototype Aero Scout T2 active Wi-Fi tags are used, which transmit the tags' unique IDs to the hospitals Wi-Fi network.
Besides a large amount of issues that needs to be address to have a successful internetworking of objects in the Internet of Things with RFID there are also a number of applications that have big potentials for the future. By embedding transponders in everyday object used by individuals, such as books, payment cards, and personal identification we will find new ways to improve our daily lives. As an example, Meingast et al. (2007) discusses the electronic passport that has been investigated in the US.

7. Conclusion

The Internet of Things era represents a gradual evolution from ICT around us to ICT on us. Many challenging issues still need to be addressed and both technological as well as social knots have to be untied before the Internet of Things idea can be widely accepted. The current trend of integrating RFID and WSN seems to be a natural step towards a Internet of Things that provides internetworking opportunities for objects but also allows objects to become smarter and interact more "intelligently" with humans. Generally, it can be concluded that the trend towards an even larger population of connected intelligent objects is irreversible, because the economic value of a system of objects and devices is directly related to the fact that objects are "networked".
Due to the large volume of objects in the Internet of Things cost is a major issue. By introducing RFID technology to internetwork objects a lower system cost compared to wireless sensor network technology can be achieved.
RFID technology is a key enabler for the transition from today's scene of connected objects to the scene of networked objects of the future. For an efficient and smooth transition a number of research issues need to be addressed. In this chapter, we have discussed important aspects of RFID/WSN technology in the Internet of Things with emphasis on what is being done and what are the issues that require further research.

8. References

Atzori, L., Iera, A. & Morabito, G. (2010). The internet of things: A survey, *Computer Networks* 54(15): 2787–2805.
URL: *http://dx.doi.org/10.1016/j.comnet.2010.05.010*

Bacheldor, B. (2007). Belgium hospital combines rfid, sensor to monitor heart patients.
URL: *http://www.rfidjournal.com/article/articleview/ 3120/1/1*

Buckley, J. (2006). *The Internet of Things: From RFID to the Next-Generation Pervasive Networked Systems*, Auerbach Publications 2008. Final report of conference organized by DG information society and media, networks and communication technologies directorate.

Daou, H., Kayssi, A. & Chehab, A. (2008). Rfid security protocols, *Innovations in Information Technology, 2008. IIT 2008. International Conference on*, p. 593.
URL: *http://dx.doi.org/10.1109/INNOVATIONS.2008.4781675*

Floerkemeier, C. & Sarma, S. (2008). An overview of rfid system interfaces and reader protocols, *RFID, 2008 IEEE International Conference on*, p. 232.
URL: *http://dx.doi.org/10.1109/RFID.2008.4519372*

Karl, H. & Willig, A. (2005). *Protocols and Architectures for Wireless Sensor Networks*, John Wiley & Sons.

Khoo, B. (2010). Rfid- from tracking to the internet of things: A review of developments, *Green Computing and Communications (GreenCom), 2010 IEEE/ACM Int'l Conference on Int'l Conference on Cyber, Physical and Social Computing (CPSCom)*, pp. 533–538. URL: *http://dx.doi.org/10.1109/GreenCom-CPSCom.2010.22*

Krishna, P. & Husalc, D. (2007). Rfid infrastructure, *Communications Magazine, IEEE* 45(9): 4. URL: *http://dx.doi.org/10.1109/MCOM.2007.4342872*

Lakafosis, V., Rida, A., Vyas, R., Yang, L., Nikolaou, S. & Tentzeris, M. M. (2010). Progress towards the first wireless sensor networks consisting of inkjet-printed, paper-based rfid-enabled sensor tags, *Proceedings of the IEEE* 98(9): 1601. URL: *http://dx.doi.org/10.1109/JPROC.2010.2049622*

Meingast, M., King, J. & Mulligan, D. K. (2007). Embedded rfid and everyday things: A case study of the security and privacy risks of the u.s. e-passport, *RFID, 2007. IEEE International Conference on*, p. 7.

Mitrokotsa, A. & Douligeris, C. (2010). *Integrated RFID and Sensor Networks: Architectures and Applications*, RFID and Sensor Networks: Architectures, Protocols, Security and Integrations, Auerbach Publications, CRC Press, Taylor and Francis Group, chapter Chapter 18, pp. 511–535.

Mitsugi, J., Inaba, T., Patkai, B., Theodorou, L., Sung, J., Lopez, T. S., Kim, D., McFarlane, D., Hada, H., Kawakita, Y., Osaka, K. & Nakamura, O. (2007). Architecture development for sensor integration in the epcglobal network, *Technical Report WPSWNET-018*, Auto-ID Labs. White paper. URL: *http://autoidlabs.mit.edu/CS/files/folders/whitepapers/entry3012.aspx*

Papapostolou, A. & Chaouchi, H. (2009). Rfid-assisted movement detection improvement in ip mobility, *Proceedings of the 3rd international conference on New technologies, mobility and security*, NTMS'09, IEEE Press, Piscataway, NJ, USA, pp. 378–382. URL: *http://dx.doi.org/10.1109/NTMS.2009.5384701*

Sharif, A. & Potdar, V. (2008). A critical analysis of rfid security protocols, *Advanced Information Networking and Applications - Workshops, 2008. AINAW 2008. 22nd International Conference on*, p. 1357. URL: *http://dx.doi.org/10.1109/WAINA.2008.212*

Shelby, Z. (2010). Embedded web services, *Wireless Communications, IEEE* 17(6): 52. URL: *http://dx.doi.org/10.1109/MWC.2010.5675778*

Weis, S., Sarma, S., Rivest, R. & Engels, D. (2004). *Security and Privacy Aspects of Low-Cost Radio Frequency Identification Systems*, Vol. 2802 of *Security in Pervasive Computing*, Springer Berlin / Heidelberg, pp. 50–59. URL: *http://dx.doi.org/10.1007/978-3-540-39881-3_18*

Wiebking, L., Metz, G., Korpela, M., Nikkanen, M. & Penttilä, K. (2008). A roadmap for rfid applications and technologie, *Technical report*. Final report of the coordination action CE RFID. URL: *http://www.rfid-in-action.eu/public/results/roadmap*

Zhang, L. & Wang, Z. (2006). Integration of rfid into wireless sensor networks: Architectures, opportunities and challenging problems, *Grid and Cooperative Computing Workshops, 2006. GCCW '06. Fifth International Conference on*, p. 463. URL: *http://dx.doi.org/10.1109/GCCW.2006.58*

19

Applying RFID Technology to Improve User Interaction in Novel Environments

Elena de la Guía, María D. Lozano and Víctor M.R. Penichet
University of Castilla - La Mancha
Spain

1. Introduction

At present our surrounding environment is constantly changing, the impact of new technologies and the information age is spreading dizzily in environments that were previously unthinkable. In order to take advantage of new technological developments we need solutions adapted to the rhythm of everyday life.

The interaction with applications has changed. Nowadays we aim to find new scenarios where the interaction between user and computer needs to be improved. We keep advancing, moving closer and closer to the idea of ubiquitous computing . All objects will be connected and the computer user will be involved. Thus, we will be closer to the natural way in which people interact with the environment. To achieve this type of scenarios we make use of the different advantages offered by mobile technologies, communication Wi-Fi (IEEE 802.11) and identification (RFID).

In this paper we describe the development and implementation of three different case studies, actually implementing the concept of context-awareness, location awareness and Internet of Things . In the first case, we describe GUIMUININ (Wireless Intelligent Museum Guides). This is a context-awareness project aim to improve the user experience in museums, so that users can see their location on any of the floors of the building and receive multimedia information on the museum item they select. In the second case, the project RCAR (Robotic Context Awareness by RFID) creates an environment sensitive to the location of a robot, which develops a tracking system and high resolution scanning for indoor or outdoor environments.The third case is the project Co-Interactive Table (Collaborative Interactive Table) that presents a digitalized table with RFID technology which executes collaborative tasks in a meeting room, face-to-face or remotely.

This article has six sections. Section 2 describes a RFID system and defines some concepts: context-awareness, Ubiquitous Computing, the Internet of Things and the types of interaction from user to new systems depending on the level attention. Section 4 presents the general infrastructure of RFID systems. In section 5, we describe our RFID systems and present their advantages and disadvantages. Finally, conclusions are set out in Section 6.

2. Related works

Technological developments in the miniaturization of microprocessors have opened up new possibilities for user services through the manipulation of information in their environment.

The major developments related to the computer field herald a new era where the systems should be adapted and integrated into the daily lives of people, occupying second place. This would ensure that the individual does not explicitly interact with the computer but in a passive way and implicitly dealing only with its target. This paradigm is defined by the term "Ubiquitous Computing" given by Mark Weiser from Xerox research center in Palo Alto in 1988 [2].

Weiser sees the technology as a means to an end based on the current user-computer interaction which he considers inadequate. The computer is a complex device and requires much concentration, thus distracting the attention of the user from the actual final task and he argues that the barriers between people and computers will disappear as we know them today, trying to provide ordinary physical objects with switching communication capacity, thus creating a large network of interconnected devices. The main objective is that the computer is hidden from the user, interacting with the user implicitly.

According to Bill Schilit [3] with the arrival of Ubiquitous Computing we will find a new way to interact with the system, the execution environments are constantly changing due to a very important factor is context. The context is defined by Dey as "any information that can be used to characterize the situation of an entity. An entity may be a person, place or object considered relevant to the interaction between a user and an application, including the user application [7]. Another concept closely related to the theory that explains Weiser is The Internet of Things, also known as the Internet of Objects, which refers to the networked interconnection of everyday objects.

Supporting these new environments requires the identification and communication technology which enables the user to work transparently. In the next section we briefly describe RFID technology. Specifically the Internet of Things is partly inspired by the success of this technology, which is now widely used for tracking objects, people, and animals. RFID system architecture is marked by a sharp dichotomy of simple tags and an extensive infrastructure of networked RFID readers. This approach optimally supports tracking physical objects within well-defined confines (such as warehouses) but limits the sensing capabilities and deployment flexibility that more challenging application scenarios require.

2.1 RFID technology

This section presents a detailed description of RFID technology. It is used to implement interactive, collaborative and context-awareness scenarios.

RFID (Radio Frequency Identification) is a system for storing and remotely retrieving data. This technology allows the identification of the object from the distance with no contact. An RFID system consists of:

- Reader or transceiver, which transmits request signals to the tags and receives the answers to these requests, it is a receiver / transmitter radio device. It needs one or more antennas to transmit the RF signal generated and receive the response. Readers may have an integrated antenna in their own hardware but it is not a requirement. Readers can range from card-sized PCM / CIA to fit a PDA to having a considerable size. Its main function is to communicate with the tags and facilitate the transfer of data to a control system.
- The RFID tag, label or transponder in the field of electronics, is an essential component of the RFID system, because of its operating scheme, it is capable of receiving and

transmitting signals, but these signals are only transmitted as a response to a request from a transceptor. The tag is a small chip or integrated circuit, adapted to a radio frequency antenna that enables communication via radio. According to their feeding mode, tags can be divided into: passive, obtaining the transmission power from the reader; active, using their own battery, and semi-active or semi-passive, using a battery to activate the chip circuitry but the energy to generate communication is that received from the reader´s radio waves, as passive ones. The most common type is passive tags, allowing the transponder device to work without its own power supply, making it cheaper, smaller and with an unlimited life cycle. As disadvantage, they present the distance limitation for identification.

- RNC (Reader Network Controller). This component is necessary to control the information received from the tags transforming it into useful information.
- Consumers. These are applications based on data received from RFID tags and will offer one service or another.

The operation of an RFID system is as follows: readers emit a magnetic, electric or electromagnetic field exciting the labels, they respond with the information they contain (unique id) via radio waves. When the reader receives the information, it is transmitted to the RNC that is responsible for pre-processing the data which will finally reach the customer.

2.2 The new interaction style with RFID

The new scenarios are generating new ways of the interaction between the user and the computer. The scenario where the user uses the mouse and keyboard to get a service has been replaced by others scenarios where the computation is implicit and user-directed. The new systems can provide information from the context which is captured by the system or by simple natural gestures. According to Ricardo Tesoriero in [1], the attention required from the user to use the new systems can be divided into two levels.

2.2.1 Lower level of attention

The user does not need to focus on the task to execute it. The system anticipates him/ her to provide a service. This is possible thanks to the context-aware applications. Some context-aware systems that use RFID technology are detailed next.

Then, it describes the context-aware systems developed with RFID technology to improve the cultural environment. eXspot is an RFID device evaluated in the Exploratorium museum in San Francisco (USA). Visitors have RFID tags and carry automatic cameras which take pictures depending on the preferences offered previously. The identification is used in a kiosk to view the captured images, creating custom web pages automatically [4,5,6,7].

Matthias Lampe [11] built some models for smart box application, among which we find the following: an application displays a smart medicine cabinet which prevents the problems of medication by monitoring the use of medicines [8]. In this application, the bottles are equipped with RFID tags. The temperature of medicines is monitored constantly to prevent damaging substances. The Cabinet monitors if there are drugs that should not be combined to prevent dangerous situations. A variety of similar works focused on medical applications can be found in [9] and in the next chapter [10].These systems use RFID technology.

Smart Tool Box contains different tools equipped with RFID tags and a toolbox with an RFID reader included. The toolkit sends different warnings according to different situations

to workers in the workplace. It also monitors the time period that the tools have been used. It is designed for working environments such as maintenance air critical. [11].
Context-aware systems have also been implemented to improve house environment. An example developed for cooking is the RFIDChef [12] , a device equipped with RFID which read everyday products tagged with RFID. A suggestion is offered according to products, different dishes to be prepared depending on the products available.
Transnote [13] is a system based in RFID built to improve classroom teaching. It stores and shared notes default, using a PDA with an RFID reader, which are sent by students during the lesson.

2.2.2 Medium level of attention

It requires more attention from the user. Distinguishing between a low or medium level of attention from the user is difficult. This level includes the environments that contain digitized objects but a simple action is required, such as, a natural gesture or closer the mobile device to the object.
The advantage is that the user has total control about the functions executed by the system, because the functions will not be executed unless the user performs a natural gesture.
This level requires physical mobile interaction. It is an interaction paradigm that allows physical objects to be increasingly augmented and associated with digital information and on the other hand, mobile devices can provide increasing capabilities to ubiquitously acquire and process this information. This level uses mobile devices to extract information from augmented physical objects and to apply it for a more intuitive and convenient interaction with associated services. This approach optimally supports tracking physical objects within well-defined confines (such as warehouses) but limits the sensing capabilities and deployment flexibility that more challenging application scenarios require. Next some RFID systems using physical mobile interaction in different environments are described. They are also called the Internet of Things. There is also a description of a RFID system at this level.
Libraries where the books are digitized and can be electronically browsed by physically Mobile and RFID tags [14]. Scenarios and smart objects at airport [15]. The concept of smart packaging is real thanks to technologies like RFID [16] . The definition of AID (Appliance Interaction Device) is an environment in RFID Internet of Things.
The home, intelligent office and other scenarios specific additions to the environment using RFID tags [18-29]. Multiple platforms and tools for the development of physical interfaces including the Phidgets based on this technology [30,31]. Papier-mâché using RFID, computer vision with codes bars to create tangibles interfaces [32]. Sports equipment increased electronically with RFID and ubiquitous technologies [33]. In addition to these systems we can find more examples where the objects are digitalized by RFID to approach the future Internet of Things.

3. RFID systems architecture

The hardware and software architecture used to design and implement the RFID system adaptable to new context-awareness and collaborative scenarios is:

3.1 Hardware architecture

The general infrastructure is shown in Figure 1. It consists of physical objects that incorporate identification technology such as RFID, QR codes, Bidi codes, etc. The device in the client side includes a reader and a controller that is responsible for processing information received by the physical object and transform it into useful information, such as an XML message that is sent to the server, which will process the message and will trigger an action, such as the generation of user interfaces or the information requested at that time. To notify the customer with web services, the network technology is used, connecting the two components, the client and the server.

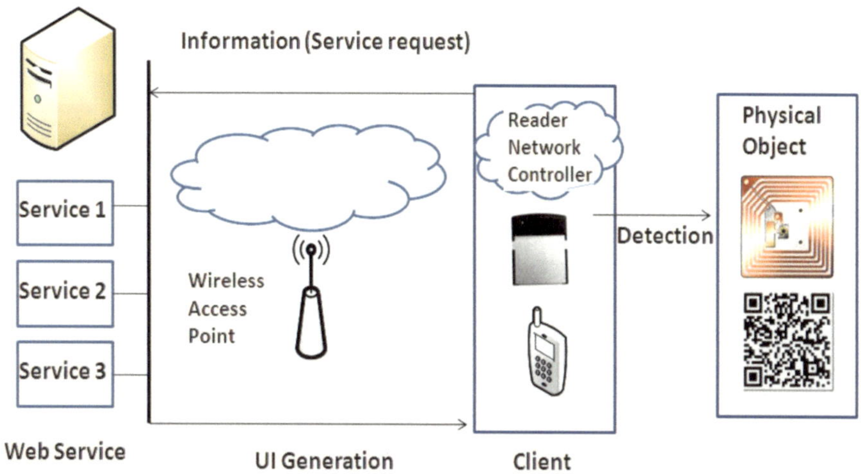

Fig. 1. Hardware architecture

To support the new systems we have used the following hardware devices:

- RFID-tags embedded in objects.
- RFID reader integrated into mobile devices or other object.
- RFID-Network Controller (RNC - RFID Network Controller) which is responsible for providing the necessary data to client applications.
- Server: it is a computer with functionality for hosting Web Services and Database.
- The communication network is responsible for connecting all the devices in the RFID system; the technology used in the systems is WiFi (IEEE 802.11).

3.2 Software architecture

We have used the Model-View-Controller to model the entities and actions that can execute in the described scenarios. It is a software architecture pattern that separates data from an application, user interface and control logic into three different components.

The general scheme covers several kinds of systems. These systems can be divided in two groups: The context-awareness system or location-awareness, which use the ID RFID to identify context or location information or digitalized objects that contain an id RFID associated to a service, it can generate an interface or execute a specific function.

The systems built are highly distributed. Figure 2 shows the three parts: Controller, View and Model.

Controller. This part is on the client side and consists of the following entities: **RFIDReader**, which is the RFID reader. It will inform the **Context Model** that an Identifier from any tag has been detected. The **ContextModel** will transmit the identifier to the server through a proxy, represented by the **ServerProxy** class. Such identifier is processed by the server running the Web Service . There is only one identifier and it is associated with the class command which is referred to a command or service. The server component contains the **WebService**, it is a software component that communicates with other applications by coding the XML message and sending this message via standard Internet protocols such as HTTP (Hypertext Transfer Protocol). Web Services are a set of protocols and standards used to exchange data between applications in order to offer services. They facilitate interoperability and offer automated services to be invoked, causing the generation of user interfaces automatically, thus allowing the user consistency and transparency in using the technology. There are two possibilities depending on the request made. The first case when it is only necessary to view some information, we just need to send the identifier of the RFID tag. The second case when the function requested is more complex, as the generation of a new interface, sharing files, etc., it is also necessary to pass the corresponding parameter.

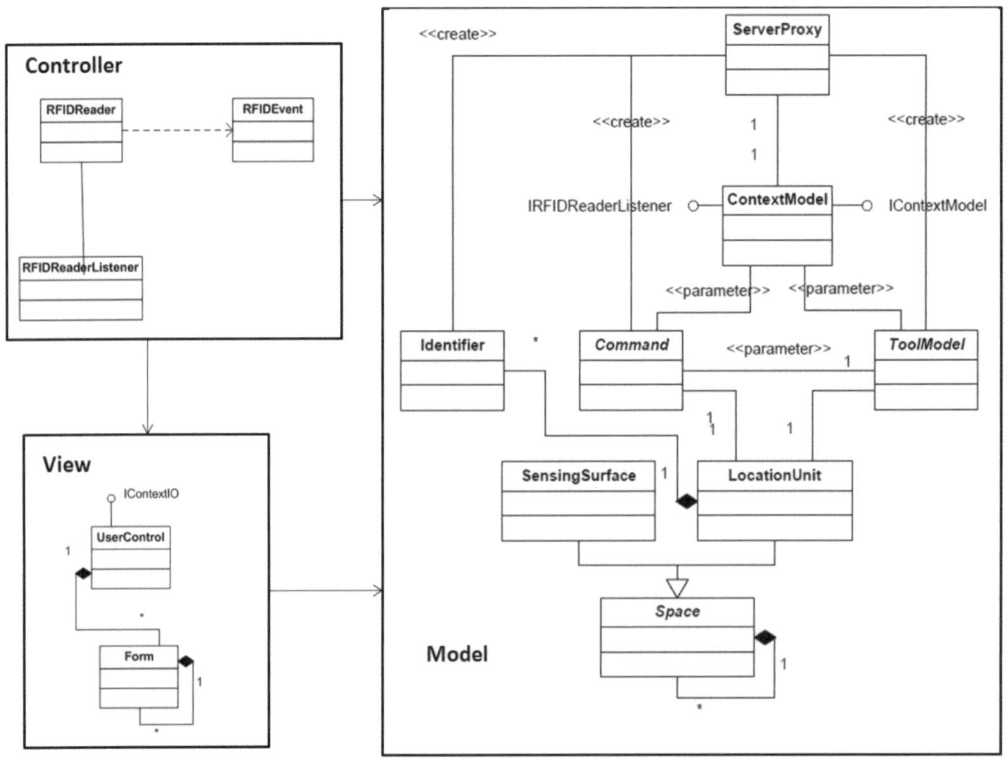

Fig. 2. Software architecture

4. Case studies developed by using RFID technology

In this section we describe three systems built in the University of Castilla-La Mancha (Albacete). The main objective is to take advantage of RFID technology to built systems that improve the user experience.

4.1 A context-aware system in cultural environments (GUIMUININM)

Modern museums offer visitors devices to guide them and help them to enjoy their visit. Often, these electronic guides provide visitors with audio information about the pieces exhibited in the museum. These devices require a higher level of attention from users; we attempt to address this problem by converting the environment in a context-aware place. Thus, we make the users' interaction with the system invisible, allowing them to enjoy the experience in the cultural environments.

GUIMUININ (Wireless Intelligent museum guides) is a context-aware project aimed to improving the user experience in museums in which the system may know his/her location on any of the floors of the building and retrieve multimedia information about the museum pieces near the user.

We have used technology based on mobile devices and RFID to implement the system. An art object or piece displayed along with extra information. Active RFID tags to identify a showcase and passive RFID tags to identify a piece in the showcase. A positioning subsystem is responsible for giving the mobile devices an identification to locate the device according to a relative or absolute position. On the other hand, the mobile device is able to detect automatically the position of the user in the museum and retrieve the correct information according to the user location in the museum.

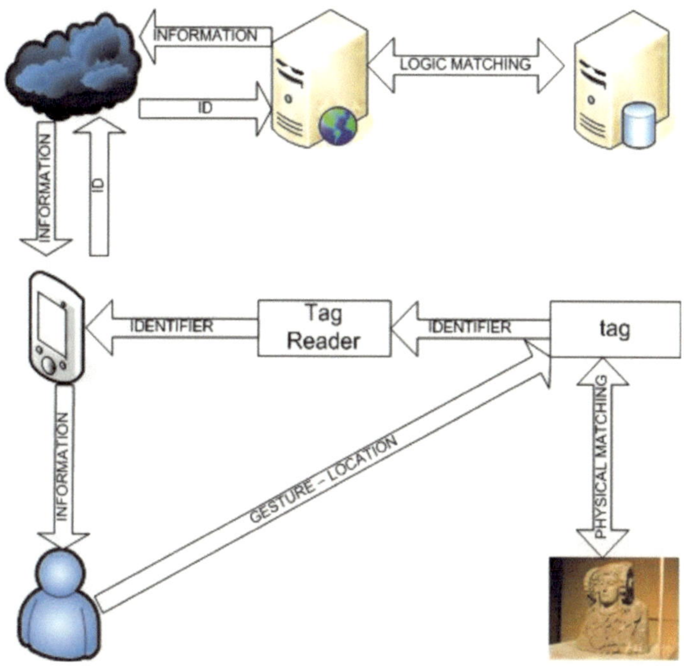

Fig. 3. Active(location) and passive (gesture) RFID scheme. The RFID reader is installed on the mobile devices.

4.2 System based on location awareness (RCAR)

The location of the entities (user, robots...) is necessary to build context-aware systems. Currently the development of systems for outdoor location seems to be solved by Global Positioning Systems (GPS). However, this technology is not enough in indoor environments, where the satellite signal undergoes a total attenuation, or rather where the accuracy and precision are very low. To solve this problem we have built the system RCAR (Robot context awareness by RFID).

RCAR is an indoor tracking system. It is capable of locating and track autonomous entities inside buildings. To achieve this goal we propose awareness surfaces using passive RFID tags.

Sensitized surfaces are physical surfaces where an autonomous system can be located (floors, walls, ceilings, tables, etc.). This surface consists of tracking units, which have at least one single identifier associated. For the identification code of a paging unit, physically we have to assign an identification code to a physical area and implement a mechanism to get that ID.

A tag represents a paging unit and a mesh of tags a sensitized surface. The mechanism used to obtain information from the tags is a passive RFID reader which is integrated into the autonomous entity. A robotNXT is chosen to incorporate the PDA and RFID reader. The PDA is connected via Wi-Fi (IEEE 802.11) to a wireless service control. The location manager is responsible for mapping the identifier to the corresponding physical surfaces.

Figure 4 shows a schematic representation of the location system, Lego Mindstorm NXT robot built with PDA and the surface used to sensitize with RFID.

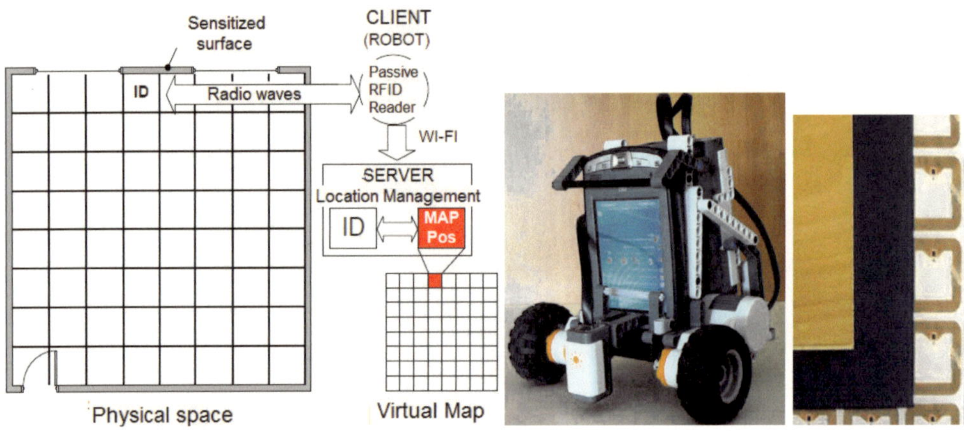

Fig. 4. Schematic representation of the location system. Lego Mindstorm NXT robot built with PDA and sensitized with RFID.

4.3 The co-interactive table: RFID system to improve collaborative meetings

Most activities performed in our life present a high degree of collaboration. One of the critical points of this process is the lack of information which may lead to waste users' time, thus increasing the feeling of frustration in the tasks. Some distributed collaborative applications may allow users to perform some tasks as videoconferences, chats, e-mail, etc., but in some other situations it is necessary to perform such tasks face-to-face to obtain better results.

The Co-Interactive Table is a system designed and built to facilitate collaborative tasks necessary in any meeting, such as sharing information and files among the participants, using simple, natural and intuitive gestures.

We have used technology based on mobile devices and RFID to implement the system, which is composed of several panels (one per user) forming the interactive table. A projector connected to a PC updates instantly important information such as ideas, notes and specific user information. The Co-Interactive Table client application runs on mobile devices with RFID reader. The system can recognize and offer the service required by the user with a gesture as simple and natural as placing the mobile device near the interactive table to select the desired action.

The interactive table is composed of different panels. The size of each panel is 210x297 mm, the same as a din-A4 paper. Its interface shows the operations that a user may perform in the meeting. These operations are graphically represented by attractive and intuitive metaphors. RFID tags that provide the functionality to the panels are hidden under the external interface

Each user has a panel and a mobile device with an RFID reader. Each mobile device has the application client running in it. This application shows the information necessary to use the interactive table. All the devices are connected by a Wi-Fi access point to the server that stores the important data, files and methods necessary for the operation of the system. In addition, the room has a projector connected to a PC that shows the users connected at each moment and the post-it notes that they have sent during the meeting through their panels. It can also support remote meetings where users are distributed in different geographical locations connected to the same server.

Fig. 5. The image on the left hand side shows a meeting room that incorporates the "Co-Interactive system". The image on the right shows the natural gesture required to run a function by interactive panels and the tags RFID integrated under the interface.

As an example of use, figure 5 shows a specific scenario in which four users participate in a meeting using the Co-Interactive Table. Each participant has a panel with all the functionalities available depicted on it and a mobile device with an integrated RFID reader. The interactive panel (See figure 6) shows visual metaphors that represent a specific operation. The available functions are the following:

- Log in: It suggests the user to log in. At this moment the system panel is associated with the user.
- Transfer File: This function allows users to share files with the participants in the meetings, this way you can share and save information easily in a few seconds. Depending on the metaphor selected, the file can be shared with a particular user or with all the users in the meeting.

- View User Information and Files: This metaphor shows the users´ academic and professional information. This function facilitates communication in meetings where people do not know each other previously. In addition they can see the files uploaded by other users.
- Returns to the main screen: This function returns to the main screen.
- View my files: This metaphor shows the files received during the meeting.
- Select user: Selecting a user is always necessary to carry out a collaborative activity. In our case it will be necessary to transfer and view files and information from another user.
- Exit: The user logs out. This function removes the association between the panel and the user.

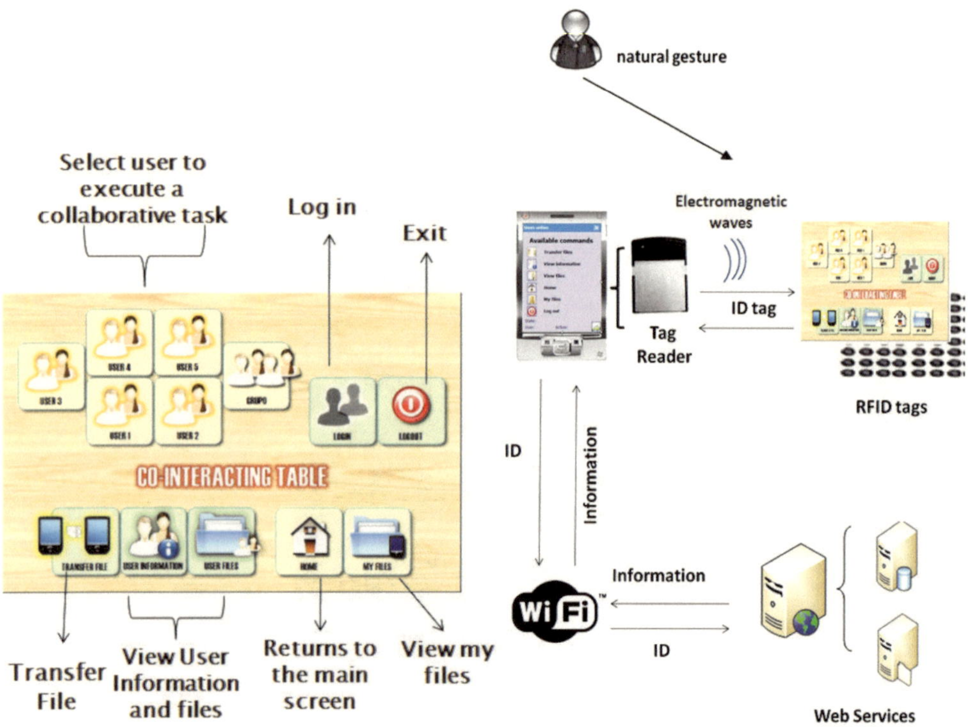

Fig. 6. User interface interactive panel: it shows the functions offered by the intelligent object and scheme showing the operation of the panels incorporating RFID technology with the mobile device and Web Services.

The system architecture is client-server, where the processing capacity is shared between both. Its operation is very simple: the client, which in our case is the application running on the mobile device, makes requests for a service from the server which provides the response. As shown in Figure 6, from the client side we can see the scanned physical object (interactive panel) that communicates with the mobile device using RFID technology. Both the mobile device and the PC are responsible for implementing the services offered by the server via Wi-Fi wireless technology.

The server runs on a PC that is connected to an access point via Wi-Fi. Its main objective is to provide the shared resources and the services required to the customer.

The server content has been divided into two parts: one is responsible for hosting the files from the meeting and the other one stores the database.

The hardware components used to build the system are the following:

- The interactive panel is a lightweight and portable rectangular object of 210x297 mm. Its interface contains visual metaphors that indicate the functions that can be carried out using the panel. Under the interface RFID tags are hidden, as seen in the right image of Figure 5, thus providing functionality to panels. Altogether the object is made up by 48 tags.
- RFID-Tags are under the interactive panel interface. The tags used are passive, i.e. no external battery is required to operate.
- RFID Readers contain CompactFlash connectivity. The maximum reading range is 10 meters and can accommodate multi-tag (anti-collision), i.e. they can read multiple tags at a time without causing any conflict or loss of information read. It is easy to connect to computers and PDA handheld readers.
- The Access Point, also called WAP, (Wireless Access Point), is a terminal that connects devices to form a wireless network, thus allowing communication among them. It is always ready to serve new customers we serve and its main function is to receive, store and transmit information to the corresponding device. Its use has allowed us to create a high performance safe and reliable wireless LAN.
- The Webcam is a small digital camera connected to a computer which can capture images and transmit them through the network to users who are located elsewhere; therefore, meetings can be held remotely, because the interactive panel functionality is not lost despite users not being in the same room. In our case it is optional; it would only be used in specific cases.
- Mobile devices, also known as handheld, PDA, Palmtop or just handheld devices are small, with some processing capabilities, permanently connected to the network. This is the device the user will use to interact with the panel.
- The Projector: Its function is to receive a video signal and project the corresponding image on a projection screen. In our case it is used to project the image of an application whose function is to show users present at the meeting and the important information. It is connected to the server and is constantly updated.
- The PC: Its function is to run the application to be displayed on the projector. The server is contained in the PC, as well as all its resources and optionally the web cam which is incorporated can be connected when wishing to hold a meeting remotely. The PC is always connected to the access point to provide the services required by the user.

The software components that allow the operation of all the functions required are:

- The "controller." It is located on the mobile device and its main function is to control communication between the mobile device and the interactive table, supported with RFID technology. For communication to be perfect the component is continuously waiting for the execution of an event as the ID reading of the RFID tags contained in the panel and then it will send the information to the web server to run the specific operation.
- The "web Service". Its function is to receive the identifier read by the RFID reader and check it with the database in order to return the corresponding data to the client running on the mobile device.
- The "co-it". It runs on the mobile device and its main objective is to display the information corresponding to the function selected in the interactive table. This component communicates via Wi-Fi with the Web service.

- The "post-it system". It runs on a PC and its function is to verify if there is a new user or event in the meeting, as sending a "post-it type" note to the public folder, and subsequently inform all users of the meeting by displaying the updated information on the projector. The "post-it" system incorporated allows people to expose their ideas easily, facilitating the exchange of information at the meeting.

The communication process is as follows: (See Figure 6). The user makes a gesture as simple and intuitive as approaching the device to the metaphor selected; this is the event that triggers the next action:

The RFID reader incorporated on the mobile device emits electromagnetic waves that excite the RFID tags. The reader will read a single ID, and then sends it to the RNC (Controller) which is responsible for pre-processing data to be sent to the server. Using wireless technology the reader sends the ID and the panel ID to the server, the ID is then mapped to the database for the action to be executed. This information is sent via Wi-Fi to the client´s mobile device, which is responsible for implementing one window or another depending on such information, thus triggering the interface automatic change without requiring the use of a pen.

5. Benefits and drawbacks - lessons learned

In this section we will discuss the advantages offered by the integration of RFID technology in the new scenarios.

The benefits about RFID tags are that they can be hidden allowing users to work in a natural way without being aware of them.

The execution of functions is easier, this depends on the context or on simple natural gestures. In this way this is actually the definition given by Weiser on Ubiquitous Computing, which describes a technology that runs implicitly to the user, ensuring that the user has complete control over the application without concentrating on it.

Low cost deployment as mobile devices will incorporate this technology in the short term, passive RFID tags are very inexpensive, thus offering very cheap and affordable systems.

Greater flexibility in the dynamic content of the system. The application can update their information quickly and effectively, thanks to the existing interconnection network. This type of systems offer the possibility to work in different physical environments, presenting fewer limitations than other systems such as infrared or ultrasound, which suffer greater attenuation level to consistent materials. Tags or passive tags do not require additional batteries, thus providing easy maintenance and high resistance.

Thanks to this technology the implementation of new interfaces can be developed for any mobile device, allowing the system usability and user-friendly interaction, thus improving user satisfaction.

Some possible limitations are: It requires connectivity to another network interconnection. The server needs to contain all the data from RFID tags, so in very complex systems we can find a lot of data, which might be difficult to manage.

6. Conclusions

The computer human interaction currently undergoing is not yet as natural as desired. The user must pay too much attention to the computer.The main objective is for users to focus on their tasks and not on the system performance. This type of systems, implicit by the user, also called context-awareness or the Internet of Things are difficult to develop. The use of embedded technologies into the objects is required to implement them, in addition to the network wireless and server requirements to manage all the information. In this chapter we present three projects, one of them is focused on improving the user visit the cultural environments, the system captures the user's context information and sends back information on works of art that are near the user at that moment. Another project has focused on improving indoor tracking systems for objects that have been sensitized with RFID tags. The last project improves collaborative tasks carried out at the meetings. In order to facilitate such tasks, we have digitized panels. RFID technology has been used because of the advantages it offers. We can see that RFID is a technology with far more profits than previously thought, which has moved from being the star product identification, to be able to scan simple objects and scenarios, providing intelligent environments where information is readily available. It facilitates human interaction with the environment through mobile devices and overcomes the limitations of mobile phones by providing a new type of interface that is easily adaptable.

7. Acknowledgements

This research has been partially supported by the Spanish CDTI research project CENIT-2008-1019, the CICYT TIN2008-06596-C02-0 project and the regional projects with reference PAI06-0093-8836 and PII2C09-0185-1030.

8. References

[1] Tesoriero, R., Gallud. J. A., Lozano, M. D. and Penichet, V.M. R. (2009). Supporting New Ways of Interaction in Cultural Environments, *Human-Computer Interaction*, Inaki Maurtua (Ed.), ISBN: 978-953-307-022-3, InTech, Available from: http://www.intechopen.com/articles/show/title/supporting-new-ways-of-interaction-in-cultural-environments

[2] Weiser, M. (1991) The Computer for the 21st Century, Sci Amer., *ISSN*: 1064-4326

[3] Schilit, B., Adams, N., Want, R. (1994). Context-aware Computing Applications.1994 First Workshop on Mobile Computing Systems and Applications, ISBN: 978-0-7695-3451-0, Santa Cruz, CA, USA.

[4] Sherry, H., Holly, F., RFID enhances visitors' museum experience at the Exploratorium, Communications of the ACM, ISSN:0001-0782, v.48 n.9, September 2005 - 240.

[5] Fleck, M., Frid, M., Kindberg, T., O'Brien-Strain, E., Rajani, R., and Spasojevic, M. (2002) Rememberer: A tool for capturing museum visits, ISBN:3-540-44267-7 Tech. Rep. HPL-2002-187, Hewlett Packard Laboratories, July 17 2002.

[6] Fleck, M., Frid, M., Kindberg, T., O'Brien-Strain, E., Rajani, R., and Spasojevic, M. From informing to remembering: Ubiquitous systems in interactive museums, *IEEE Pervasive Computing*, vol. 1, no. 2, pp. 13–21, 2002. ISSN:1536-1268

[7] Hsi, S. (2004).I-guides in progress: Two prototype applications for museum educators andvisitors using wireless technologies to support informal science learning, in *WMTE'04: Proceedings of the 2nd IEEE International Workshop on Wireless and Mobile*

Technologies in Education (WMTE'04), ISBN:0-7695-1989-X (Washington, DC, USA), p. 187, IEEE ComputerSociety, 2004.

[8] Wan, D. (1999). Magic medicine cabinet: A situated portal for consumer healthcare, *Lecture Notes in Computer Science*, vol. 1707, pp. 352–355, 1999. ISBN:3-540-66550-1.

[9] Fuhrer, P., Guinard, D. (2006). Building a Smart Hospital using RFID technologies: Use Cases and Implementation October 12 - 13, 2006, Fribourg, Switzerland and will be published in its proceedings in the Lecture Notes in Informatics. ISBN: 978-0-7695-3336-0.

[10] Benelli, G., Parrino, S. and Pozzebon A. (2010). RFID Applications for Sanitary Environments, Sustainable Radio Frequency Identification Solutions, Cristina Turcu (Ed.), ISBN: 978-953-7619-74-9, InTech, Available from: http://www.intechopen.com/articles/show/title/rfid-applications-for-sanitary-environments

[11] Floerkemeier, C., Lampe, M., and Schoch, T. (2003). The Smart Box Concept for Ubiquitous Computing Environments. Presented at Smart Objects Conference (Grenoble, 2003).

[12] Langheinrich, M., Mattern, F., Romer, K., Vogt, H. (2001). First steps towards an event based infrastructure for smart things, Dec. 15 2001.

[13] Miura, M., Kunifuji, S., Shizuki, B. and Tanaka, J. (2005). Airtransnote: Augmented classrooms with digital pen devices and RFID tags, " in *WMTE*, pp. 56–58, IEEE Computer Society, 2005. ISBN: 0-7695-2385-4.

[14] Salcedo, R. M., Ogata, H. and Yano, Y. (2005).Towards a new digital library infrastructure with RFID for mobile ELearning, in *WMTE*, pp. 261–263, IEEE Computer Society, 2005.ISBN: 0-7695-2385-4.

[15] McCoy, T., Bullock, R. J. and Brennan, P. V. (2005). Rfid for airport security and efficiency, in *Proc. IEE Seminar on (Ref Signal Processing Solutions for Homeland Security No. 2005/11108)*, pp. 9 pp.–, 2005., ISBN: 0-86341-564-4.

[16] Pala, Z. and Inanc, N.(2007). Smart parking applications using rfid technology, in *Proc. 1st Annual RFID Eurasia*, pp. 1–3, Sept. 5–6, 2007. ISBN: 978-975-01566-0-1.

[17] Roduner, C., Langheinrich, M., Floerkemeier, C. and Schwarzentrub, B. (2007). Operating appliances with mobile phones - strengths and limits of a universal interaction device, "in *Pervasive Computing, 5th International Conference, PERVASIVE 2007, Toronto, Canada, May 13-16, 2007, Proceedings* (A. LaMarca, M. Langheinrich, and K. N. Truong, eds.), vol. 4480 of *Lecture Notes in Computer Science*, pp. 198–215, Springer, 2007. ISBN: 978-3-540-72036-2.

[18] Jung, K. K., Son, D. S. and Eom, K. H. (2009). Rfid footwear and floor system, in *Proc. WRI World Congress on Computer Science and Information Engineering*, vol. 3, pp. 72–75, Mar. 2009. ISBN:978-0-7695-3507-4.

[19] Ha, S., Jung, H. and. Oh, Y. (2006).Method to analyze user behavior in home environment, *Personal Ubiquitous Comput.*, vol. 10, no. 2-3, pp. 110–121, 2006. ISSN:1617-4909.

[20] Barrett, R. and Maglio, P. P. (1998).Informative things: how to attach information to the real world, " in *UIST '98: Proceedings of the 11th annual ACM symposium on User interface software and technology*, (New York, NY, USA), pp. 81–88, ACM, 1998. ISBN:1-58113-034-1.

[21] Want, R., Fishkin, K. P., Gujar, A. and Harrison, B. L. (1999).Bridging physical and virtual worlds with electronic tags, " in *CHI '99: Proceedings of the SIGCHI conference on Human factors in computing systems*, (New York, NY, USA), pp. 370–377, ACM, 1999. ISBN:0-201-48559-1.

[22] Kindberg, T., Barton, J., Morgan, J., Becker, G., Caswell, D., Debaty, P., Gopal, G., Frid, V. Krishnan, M., Morris, H., Schettino, J. and Serra, B. (2000). People, places, things: Web presence for the real world, " Tech. Rep. HPL-2000-16, Hewlett Packard Laboratories, Feb. 13 2000. ISSN:1383-469X.

[23] Ljungstrand, P., Redstr¨om, J. and Holmquist, L. E. (2000). Webstickers: using physical tokens to access, manage and share bookmarks to the web, " in *Designing Augmented RealityEnvironments*, pp. 23–31, 2000.

[24] Rekimoto, J., Ayatsuka, Y. and Hayashi, K. (1998). Augment-able reality: Situated communication through physical and digital spaces, in *ISWC*, pp. 68–75, 1998. ISBN:0-8186-9074-7.

[25] Rohs, M. and Bohn, J. (2003).Entry points into a smart campus environment - overview of the ETHOC system, " in *ICDCS Workshops*, p. 260, IEEE Computer Society, 2003. ISBN:0-7695-1921-0.

[26] Smith, M. A., Davenport, D., Hwa, H. and Turner, T. (2004). Object auras: a mobile retail and product annotation system, " in *Proceedings of the 5th ACM conference on Electronic commerce (EC-04)*, (New York), pp. 240–241, ACM Press, May 1-8 2004. ISBN:1-58113-771-0.

[27] Riekki, J., Salminen, T. and Alak¨arpp¨a, I. (2006). Requesting Pervasive Services by Touching RFID Tags, *IEEE Pervasive Computing*, vol. 5, no. 1, pp. 40–46, 2006. ISSN:1536-1268.

[28] Ailisto, H., Pohjanheimo, L., V¨alkkynen, P., Str¨ommer, E., Tuomisto, T. and Korhonen, I. (2006).Bridging the physical and virtual worlds by local connectivity-based physical selection, " *Personal and Ubiquitous Computing*, vol. 10, no. 6, pp. 333–344, 2006. ISSN:1617-4909.

[29] Wagner, D., Pintaric, T., Ledermann, F. and Schmalstieg, D. (2005). Towards massively multiuser augmented reality on handheld devices, " in *Pervasive Computing, Third InternationalConference, PERVASIVE 2005, Munich, Germany, May 8-13, 2005, Proceedings*(H.-W. Gellersen, R. Want, and A. Schmidt, eds.), vol. 3468 of *Lecture Notes in Computer Science*, pp. 208–219, Springer, 2005.ISBN: 3-540-26008-0.

[30] Marquardt, N. and Greenberg, S. (2007) Distributed physical interfaces with shared phidgets, " in *Tangible and Embedded Interaction* (B. Ullmer and A. Schmidt, eds.), pp. 13–20, ACM, 2007. ISBN: 978-1-59593-619-6.

[31] Greenberg, Saul and Fitchett, Chester, (2001). Phidgets: easy development of physical interfaces through physical widgets, in *Proceedings of the ACM Symposium on User Interface Software and Technology*, Papers: Tactile user interface, pp. 209–218, 2001. ISBN:1-58113-438-X.

[32] Klemmer, S. R., J. Li, J. Lin, Landay, and J. A., (2004) Papier-mache: toolkit support for tangible input, in *Proceedings of ACM CHI 2004 Conference on Human Factors in Computing Systems*, vol. 1, pp. 399–406. ISBN 1-58113-702-8, Vienna, April, 2004.

[33] M. Kranz, Spiessl, W. and Schmidt, A. (2007). Designing ubiquitous computing systems for Sports equipment, in *PerCom*, pp. 79–86, IEEE Computer Society, 2007. ISBN:0-7695-2787-6.

20

Building Blocks of the Internet of Things: State of the Art and Beyond

Alexandru Serbanati, Carlo Maria Medaglia and Ugo Biader Ceipidor
CATTID- "Sapienza" University of Rome
Italy

1. Introduction

ICT has simplified and automated many tasks in the industry and services sector. Computers can monitor and control physical devices from very small to very large scales: they are needed in order to produce semiconductor wafers and can help operating ships, airplanes or manufacturing devices. Until some years ago though, these solutions were monolithic and thus application specific.

In the field of monitoring and control, the wide adoption of modular design patterns and standardization, together with the improvements in communication technologies, paved the way to the diffusion of single component products that could be integrated as building blocks for ever more complex applications. An array of embedded devices and autoID technologies are now available as well as off-the-shelf platforms (ref Oracle, IBM, Arduino, Arch Rock, Sensinode) which can be used and customized for addressing specific purposes.

One of the biggest paradigms behind this trend is the Internet of Things (IoT) which foresees a world permeated with embedded smart devices, often called "smart objects", inter-connected through the Internet[1]. These devices should help blending together the digital and the physical world by providing Things with "identities and virtual personalities" (European Technology Platform on Smart Systems Integration [EPoSS], 2008) and by providing pervasive sensing and actuation features.

This scenario is very challenging as not all the building blocks of the IoT are yet in place. Standardization efforts are essential and have only recently been made and a reference architecture is still missing. Other researches on this topic nowadays focus on hardware and software issues such as energy harvesting, efficient cryptography, interoperability, communication protocols and semantics. The advent of IoT will also raise social, governance, privacy and security issues.

This work provides a historical and conceptual introduction to the IoT topic. In the second part of the chapter, a wide perspective on the aforementioned issues is provided. The work also outlines key aspects in the process of moving from the current state of the art of IoT, where objects have digital identities, towards a network of objects having digital personalities and being able to interact with each other and with the environment. In the last part, a selection of the possible impacts of the IoT is analyzed.

---

[1] A better definition of the phrase "Internet of Things" will be provided in the next Section.

2. Evolution of a vision

The concept of Internet of Things was originally coined by Kevin Ashton of the MIT Auto-ID Center to describe the possibility of using RFID tags in supply chains as pointers to Internet databases which contained information about the objects to which the tags were attached. The concepts heralded in the presentation made by Ashton in 1998, were soon realized in practice with the birth of the EPCglobal, a joint venture aiming to produce standards from the Auto-ID Center, which eventually created the EPC suite of standards and the homonymous architecture framework (Armenio et al., 2007).

The phrase maintained this meaning (Meloan, 2003), untill 2004, when, for the first time a world where "everyday objects [had] the ability to connect to a data network" was conceived (Gershenfeld et al., 2004). Innovative concepts such as the extreme device heterogeneity and IP-based, narrow-waist protocol stack were for the first time introduced for what was also called Internet0.

In the last years the hype surrounding the IoT grew in proportions. In the last years, quite a few definitions have been given and we will analyse them briefly in order to provide a better definition of the Internet of Things phrase.

In the final report of the Coordination and Support Action (CSA) for Global RFID-related Activities and Standardisation [CASAGRAS] project (CASAGRAS, 2009) the reader can find a compiled list of definitions which capture different aspects of and meanings given to the concept of Internet of Things:

Initial CASAGRAS definition: *"A global network infrastructure, linking physical and virtual objects through the exploitation of data capture and communication capabilities. This infrastructure includes existing and evolving Internet and network developments. It will offer specific object-identification, sensor and connection capability as the basis for the development of independent cooperative services and applications. These will be characterised by a high degree of autonomous data capture, event transfer, network connectivity and interoperability"*, Anthony Furness, European Centre of Excellence for AIDC

The CASAGRAS definition was given in the first part of year 2009, and was then confirmed in the final report of the project. In this definition the IoT is first and foremost a network infrastructure. This is coherent with the semantic meaning of the phrase which assumes that the IoT builds upon the existing Internet communication infrastructure. The definition is also focused on connection and automatic identification and data collection technologies that will be leveraged for integrating the objects in the IoT.

SAP definition: *"A world where physical objects are seamlessly integrated into the information network, and where the physical objects can become active participants in business processes. Services are available to interact with these 'smart objects' over the Internet, query and change their state and any information associated with them, taking into account security and privacy issues."* Stephan Haller, SAP AG

We would like to note here the focus on the physical objects which are in the center of the attention as main participants of the IoT. They are described as active participants in the business processes. Besides, the IoT here is more a vision than a global network, as the word "world" would suggest. Also the idea of using services as communication interfaces for IoT is explicited. Services will soon become one of the most popular tools to broaden the basis of communication interoperability in the IoT vision. Security and privacy, though not related to the definition of IoT, are also highlighted as critical issues (see Section 5.3).

Future Internet Assembly/Real World Internet definition: *The IoT concept was initially based around enabling technologies such as Radio Frequency Identification (RFID) or wireless sensor and actuator networks (WSAN), but nowadays spawns a wide variety of devices with different computing and communication capabilities – generically termed networked embedded devices (NED). [...] More recent ideas have driven the IoT towards an all encompassing vision to integrate the real world into the Internet [...].*

More recent definitions seem to emphasize communication capabilities, and to assign a certain degree of intelligence to the objects (EPoSS, 2008; cited in Botterman, 2009).

"a world-wide network of interconnected objects uniquely addressable, based on standard communication protocols."

"Things having identities and virtual personalities operating in smart spaces using intelligent interfaces to connect and communicate within social, environmental, and user contexts."

In conclusion, we can thus identify two different meanings (and thus definitions) of the phrase: the IoT network and the IoT paradigm. First and foremost, the Internet of Things is a global network, an extension of the current Internet to new types of devices - mainly constrained devices for WSANs and auto-ID readers -, aiming at providing the communication infrastructure for the implementation of the Internet of Things paradigm. The Internet of Things paradigm, on the other hand, refers to the vision of connecting the digital and the physical world in a new worldwide augmented continuum where users, either humans or physical objects (the things of the Internet of Things), could cooperate to fulfill their respective goals.

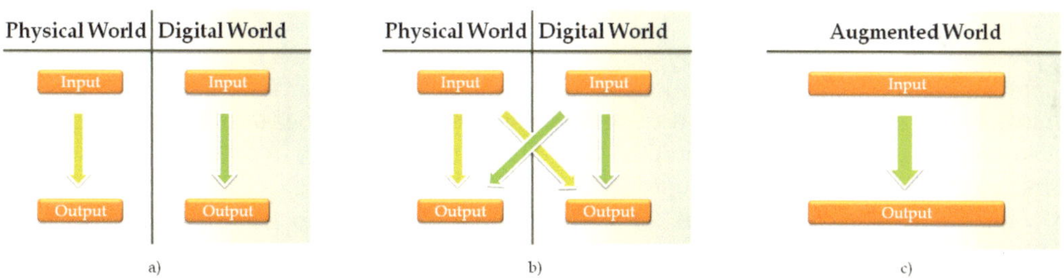

Fig. 1. The paradigm of IoT: from the current situation where digital and physical environments are uncoupled (a), to one where physical and digital world can interact (b) and finally to one where physical and digital worlds are merged sinergically in an augmented world (c).

In order to realize the IoT paradigm, the following features will be gradually developed and integrated in or on top of the Internet of Things network infrastructure, slowly transforming it into an infrastructure for providing global services for interacting with the physical world:

- object identification and presence detection
- autonomous data capture
- autoID-to-resource association
- interoperability between different communication technologies
- event transfer
- service-based interaction between objects
- semantic based communication between objects
- cooperation between autonomous objects.

3. A model for the Internet of Things

The aim of this section is to provide insight on the actors and components of the Internet of Things and how they will interact. We will provide our definition on the concepts we deem essential in the Internet of Things as previously defined in Section 2 .What is expressed in the following paragraphs has been heavily influenced by the fruitful interaction with our partners in the IoT-A project.

The generic IoT scenario can be identified with that of a generic *User* that needs to interact with a (possibly remote) *Physical Entity* of the physical world. In this short description we have already introduced the two key actors of the IoT. The *User* is a human person or a software agent[2] that has a goal, for the completion of which the interaction with the physical environment has to be performed through the mediation of the IoT. The *Physical Entity* is a discrete, identifiable part of the physical environment that can be of interest to the User for the completion of his goal. *Physical Entities* can be almost any object or environment, from humans or animals to cars, from store or logistic chain items to computers, from electronic appliances to closed or open environments.

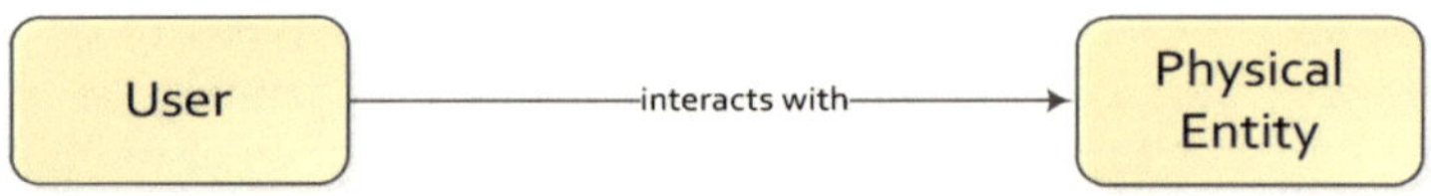

Fig. 2. Basic abstraction of the IoT interaction

In the digital world *Digital Entities* are software entities which can be agents that have autonomous goals, can be services or wimple coherent data entries. Some *Digital Entities* can also interact with other *Digital Entities* or with *Users* in order to fulfill their goal. Indeed, *Digital Entities* can be viewed as *Users* in the IoT context. A *Physical Entity* can be represented in the digital world by a *Digital Entity* which is in fact its *Digital Proxy*. There are many kinds of digital representations of *Physical Entities* that we can imagine: 3D models, avatars, objects (or instances of a class in an object-oriented programming language) and even a social network account could be viewed as such. However, in the IoT context, *Digital Proxies* have two fundamental properties:

- they are *Digital Entities* that are bi-univocally associated to the *Physical Entity* they represent. Each *Digital Proxy* must have one and only one ID that identifies the represented object. The association between the *Digital Proxy* and the *Physical Entity* must be established automatically
- they are a synchronized representation of a given set of aspects (or properties) of the *Physical Entity*. This means that relevant digital parameters representing the characteristics of the *Physical Entity* can be updated upon any change of the former. In the same way, changes that affect the *Digital Proxy* could manifest on the *Physical Entity* in the physical world.

While there are different definitions of smart objects in literature (Kortuem et al., 2009), we define a *Smart Object* as the extension of a *Physical Entity* with its associated *Digital Proxy*. We have chosen this definition as, in our opinion, what is important in our opinion is the

---

[2] We prefer, wherever it is possible, not to introduce a distinction between the world of constrained devices and the one of full function devices. Some authors refer to the IoT as a concept related only to constrained devices. We prefer to stick to the previously provided definition, where the IoT is conceived as an extension of the Internet, thus including it and all the related concepts and components.

In this case for example, the 'software agent' can equally be one residing on a server, on an autonomous constrained device or running on the mobile phone.

synergy between the *Physical Entity* and the *Digital Proxy*, and not the specific technologies which enable it. Moreover, while the concept of "interest" is relevant in the IoT context (you only interact with what you are interested in) the term "Entity of Interest" (Haller, 2010) focuses too much attention on this concept and doesn't provide any insight on its role in the IoT domain. This term was an alternative to Entity in (Sensei, 2008), which in turn we view as an unnecessary abstraction that can also be misleading. For these reasons we have preferred the term *Smart Object*, which, even if not perfect (a person might be a *Smart Object*), is widely used in literature.

Indeed, what we deem essential in our vision of IoT though, is that any changes in the properties of a *Smart Object* have to be represented in both the physical and digital world. This is what actually enables everyday objects to become part of the digital processes.

This is usually obtained by embedding into, attaching to or simply placing in close vicinity of the *Physical Entity* one or more ICT devices which provide the technological interface for interacting with or gaining information about the *Physical Entity*, actually enhancing it and allowing it to be part of the digital world. These devices can be homogeneous as in the case of Body Area Network nodes or heterogeneous as in the case of RFID *Tag* and *Reader*. A *Device* thus mediates the interactions between *Physical Entities* (that have no projections in the digital world) and *Digital Proxies* (which have no projections in the physical world) extending both.

From a functional point of view, Device has three subtypes:

- *Sensors* can provide information about the *Physical Entity* they monitor. Information in this context ranges from the identity to measures of the physical state of the *Physical Entity*. The identity can be inherently bound to that of the device, as in the case of embedded devices, or it can be derived from observation of the object's features or attached *Tags*. Embedded *Sensors* are attached or otherwise embedded in the physical structure of the *Physical Entity* in order to enhance and provide direct connection to other *Smart Objects* or to the network. . Thus they also identify the Physical Entity. *Sensors* can also be external devices with onboard sensors and complex software which usually observe a specific environment in which they can identify and monitor *Physical Entities*, through the use of complex algorithms and software training techniques. The most common example of this category are face recognition systems which use the optical spectrum. *Sensors* can also be readers (see *Tags* below).
- *Tags* are used by specialized *Sensor* devices usually called readers in order to support the identification process. This process can be optical as in the case of barcodes and QRcode, or it can be RF-based as in the case of microwave car plate recognition systems and RFID.
- *Actuators* can modify the physical state of the *Physical Entity*. *Actuators* can move (translate, rotate, ...) simple *Physical Entities* or activate/deactivate functionalities of more complex ones.

It is also interesting to note that, as everyday objects can be logically grouped together to form a composite object and as complex objects can be divided in components, the same is also true for the *Digital Entities* and *Smart Objects* which can be logically grouped in a structured , often hierarchical way. As previously said, *Smart Objects* have projections in both the digital and physical world plane. Users that need to interact with them must do so through the use of *Resources*. *Resources*[3] are digital, identifiable components that implement different capabilities, and are associated to *Digital Entities*, specifically to *Digital Proxies* in the case of IoT. More than one *Resource* may be associated to one *Digital Proxy* and thus to one *Smart Object*. Five general classes of capabilities can be identified and provided through *Resources*:

[3] In this work we depart from the original and abstract meaning of the term (Berners-Lee, 1998) which we consider closer to the definition of *Entity of Interest*.

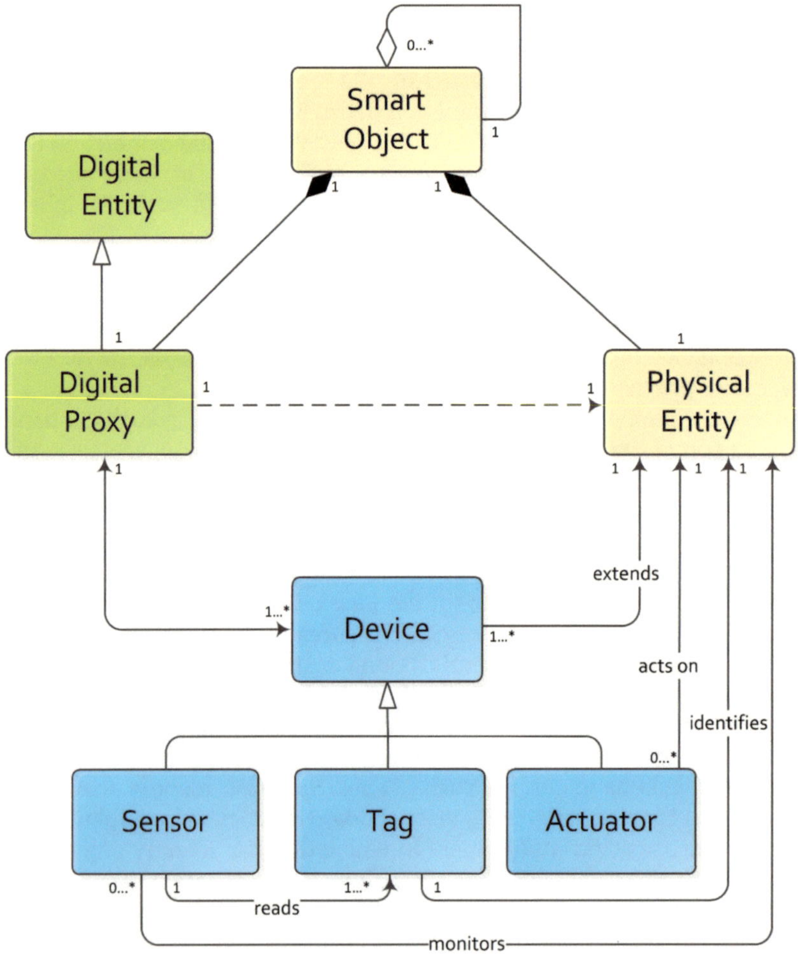

Fig. 3. Conceptual model of a Smart Object

- retrieval of physical properties of the associated *Physical Entity* captured through *Sensors;*
- modification of physical properties of associated *Physical Entity* through the use of *Actuators;*
- retrieval of digital properties of the associated *Digital Proxy;*
- modification of digital properties of the associated *Digital Proxy;*
- usage of complex hardware or software services provided by the associated *Smart Object*[4].

In order to provide interoperability, as they can be heterogeneous and implementations can be highly dependent on the underlying hardware of the *Device,* actual access to *Resources* is provided as *Services.*

---

[4] The use of remote processing capabilities for computation intensive operations (e.g. the resolution and lookup processes) or the usage of specific hardware (e.g. printers or projectors) are good examples of this kind of *Resources.*

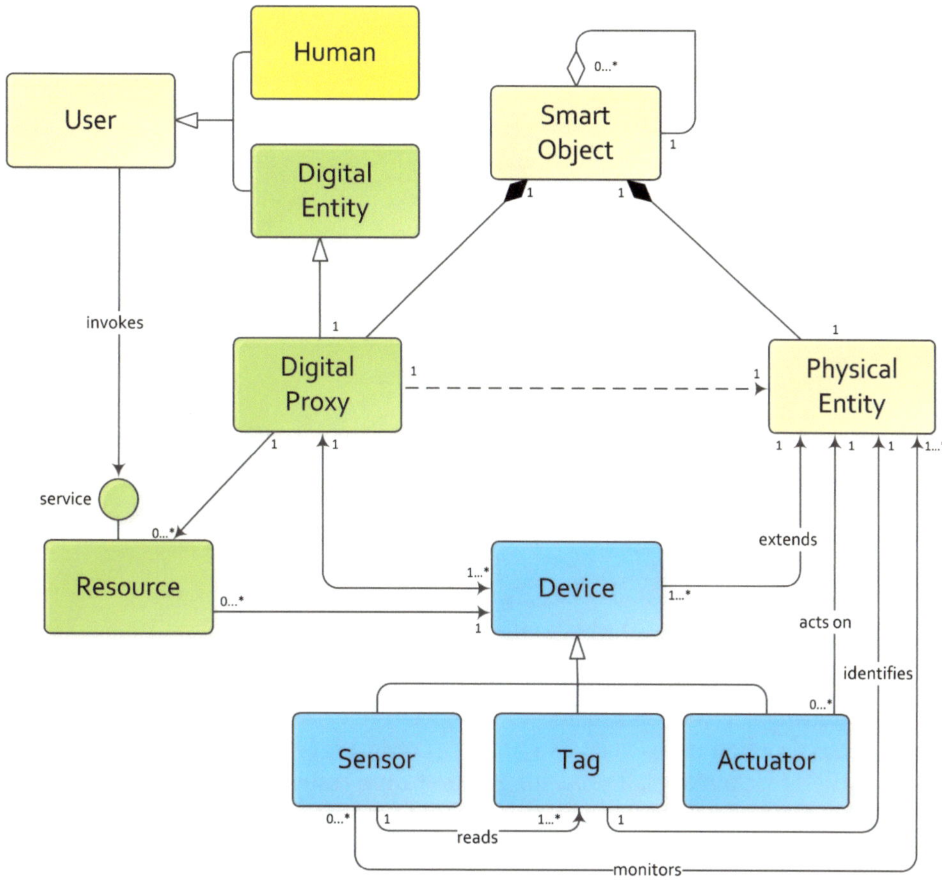

Fig. 4. Proposed Internet of Things reference model

The associations between *Smart Objects* and *Resources* (i.e. their identity) and the locations (i.e. network addresses) of the relative *Services* is either recorded in the *Smart Object* itself or can be stored (along with a small amount of auxiliary information) in what we call *Resolution Service*, an infrastructural component of the Internet of Things. The *Resolution Service* is conceived as a registry-based provider of the essential resolution service. Its task is very similar to that of current DNS or ONS service: it takes as input the ID of a *Smart Object* or *Resource* and provides as output the network addresses of the *Services* associated to it.
In the same way, a semantic description of the *Resources* and the ID of the associated *Virtual Proxy* is recorded in what we define the *Lookup Service*. This is similar to nowadays semantic search engines in that it accepts an input query and provides a relevance-ordered set of IDs, identifying *Resources* that might be useful to the *User*, according to the semantic query provided by the *User*.
Both the resolution and the lookup services can be provided as *Services*.

4. Identification, data collection and communication

The IoT vision had its base in the automatic identification (autoID). For the first time, ICT systems could assign an identity to common objects and soon these were able to become –

passive - part of automated, computer-managed processes. Such processes initially aimed at shadowing physical processes by monitoring them through the use of autoID.

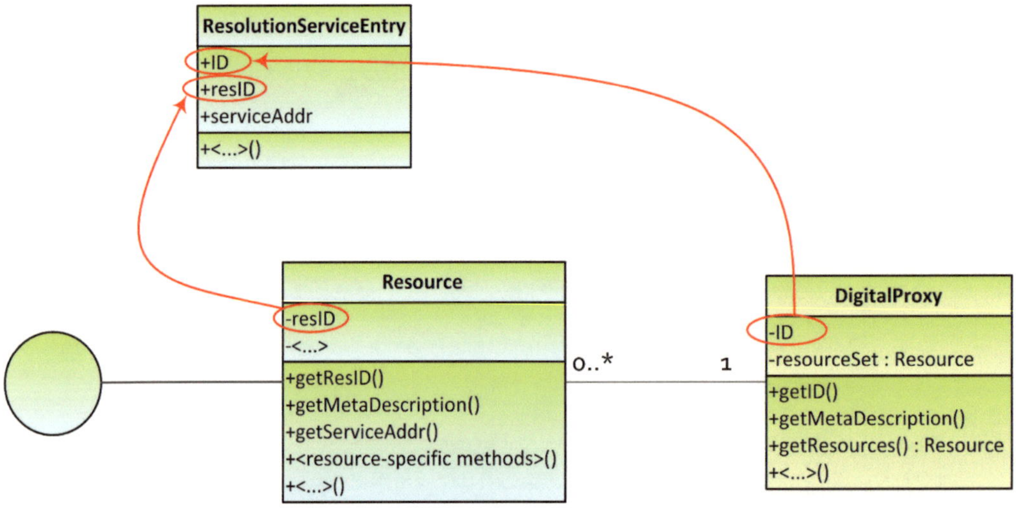

Fig. 5. Representation of the resolution registry

In the beginning, barcodes provided the first means of identifying items through optical labels. Barcodes eventually evolved, also thanks to the spread of camera-integrating mobile phones, to bi-dimensional optical codes such as QRcode (Denso Wave, n.d.). In the meanwhile, the well known RFID technology allowed for the first time real-world objects to be efficiently integrated in the digital processes, making in this way the first step towards the convergence and integration of digital and real world as the IoT paradigm proclaims. A relative small form-factor and low price together with the limited need of maintenance made this technology a good solution for specific supply chain and asset management solutions (Bose & Pal, 2005).

Unfortunately though, the RFID technology has its limits. Designed for identification, it can only provide information about presence and it also brings along a set of privacy and security issues. Semi-passive RFID tags can provide readings from battery powered sensors, but communication is still one-way and objects are not connected.

Sometimes passive RFID is also erroneously thought of as an authentication technology. This is a misconception, albeit common. The option of using RFID for authentication purposes should be thoroughly investigated prior to adoption and could prove even dangerous if system designers believe that RFID could provide a secure way of identifying things (Lehtonem et al., 2009).

When it comes to data collection, networks provide the most powerful solution. Bidirectional communication, enabling constant monitoring as well as command actuation, an always-available connection and higher data-rates sound definitely appealing. Wireless networks on the other hand prove to be a good solution because they need no physical infrastructure for operating and the deployment process is easier. And so, Wireless Sensor Networks (WSN) were born and provided new performance levels that were needed in some fields of data collection.

WSNs are made of a number of network (usually WPAN or LR-WPAN) nodes that often have automatically (re-)configuration capabilities and provide a wireless communication channel for the data gathered by onboard sensors. A user or a central business logic can get the collected data through a special node, usually the coordinator of the network, that acts as gateway. A good knowledge base on WSNs can be found in (Akyldiz & Vuran, 2010). Bidirectional communication is also useful for requesting real-time data and commanding actuators. Hence the phrase Wireless Sensor and Actuator Network[5] thereafter WSAN) was coined. Bidirectional communication is also useful for reprogramming devices directly on the field (Karlof et al., 2004).

These technologies paved the way to a whole new set of applications thanks to their ease of deployment. With almost no need for a physical network infrastructure, WSANs attracted a lot of interest from application designers aiming to employ them in fields ranging from home and industrial automation (Sleman & Moeller, 2008), smart metering (Kistler et al., 2008), to precision agriculture (Xuemei et al., 2008), environmental monitoring and healthcare (Yang & Yacoub, 2006).

These applications though are just the top of a submerged iceberg when it comes to the possibilities provided by embedding sensor and actuators in the environment. The real revolution will take place when embedded devices will be able to provide and access resources through the Internet. This, together with the use of semantics will also uncover the untapped potential of context-awareness and autonomous decision making.

The first steps towards this vision have already been taken. As the IP protocol is the cornerstone of the Internet and, as the IoT will be an extension of the current Internet, many have proposed to use IP, and in particular IPv6, as the shared narrow-waist of IoT-capable protocol stacks (Vasseur, & Dunkels, 2010). Indeed, the perspective of having 50 to 100 billion devices by 2020 (Sundmaeker et al., 2009) can be even viewed as one of the drivers of the adoption of the IPv6.

In this context, the work of the 6LoWPAN group in providing an adaptation layer between IPv6 NWK layer and the MAC layer of IEEE 802.15.4 is worth mentioning (Bormann et al., 2009). The adaptation was needed because of the different purposes of the IPv6 and of the IEEE 802.15.4 standard for Low Rate WPANs (LR-WPANs). The former was based on the existing features of IPv4, and was designed for the Internet while, at design time, LR-WPANs were required to optimize energy consumption. Thus the work had to deal with the typical limitations of constrained devices.

One of the greatest issues was that the LR-WPAN PHY layer packet length of 127 bytes. This forced the workgroup to rely on the compression for the 40 bytes IPv6 header in order to achieve larger application-level payloads and thus greater efficiency in communication, which lead to RFC4944 (Montenegro et al., 2007). The reasons behind this choice can be understood considering that the MAC header has a maximum length of 25-bytes, that the possible overhead due to the MAC layer security can take up to 21 bytes and that fragmentation support in upper layers can reduce even more the actual application payload.

The potential of having small - though constrained devices - to the Internet has been readily perceived by the actors of the embedded devices market. For example, alongside the interest focused from the academic environment, it is relevant that all embedded platforms previously cited already provide support to 6LoWPAN. Contiki and Tiny OS, two of the major operating systems for embedded devices, also provide modules for 6LoWPAN.

[5] While in literature the term 'WSN' is much more used, we prefer to use 'WSAN' because, limiting the functional definition of such network to sensing doesn't fit with the IoT scenario, where the interaction with the real world is bidirectional

Communication capabilities are essential for achieving other features that have been associated to the Internet of Things. Cooperation among *Smart Objects* and the auspicated context-awareness are the most relevant. In order for devices to exchange meaningful data a proper support at service and application layer level is essential.

5. The missing building blocks

The IoT paradigm is a visionary one. Currently there are more questions than answers and many challenges need to be taken into account. Some building blocks, such as autoID technologies, WSANs and basic IP-based communication are (almost) available, yet others are still needed and obstacles pave the path to the advent of IoT. Nonetheless, this vision, unlike many others, is in the realm of possibility and the sheer momentum of the effort it focuses might lead to its success.

This section lists and analyzes the most relevant technological and scientific missing building blocks. Many of these topics have been discussed in the frame of the Internet of Things Architecture [IoT-A] project (IoT-A, 2010)[6], which aims at bridging many of these gaps.

A governance framework is also considered to be necessary, yet missing, and the relative issues will be depicted in the relative sub-section.

5.1 Interoperability

The paramount challenge at the moment seems to be interoperability. This issue has many facets, some of which are tightly intertwined to technical aspects. Even though there are many other challenges for the IoT, one of the most important requirement to keep in mind when addressing them is that they need to be solved in a common way for interoperability's sake. We have identified the following topics on which efforts from the research and stakeholder community in creating inter-operable solutions and towards standardization are most needed:

- reference architecture and protocol suites
- identification schemes
- routing and addressing
- resource resolution and lookup
- semantics

Though not strictly related to standardization, governance and intellectual property management also have to be addressed jointly and in an international frame. In this case though, it's not the research or stakeholder community that has to make efforts and take decisions, but the international entities that will be responsible of the management of the infrastructure of the IoT.

5.2 An architecture and a reference conceptual framework

Despite the interest in the topic and the huge amount of scientific papers, books and workshops about the Internet of Things, there is a manifested lack of consensus on some concepts and definitions related to the IoT.

As seen in Section 2, there is a certain degree of misalignment even in the definition of the Internet of Things and this also extends to other concepts used in this context. This

[6] See http://www.iot-a.eu

misalignment translates in the fact that the set of expected capabilities of the IoT is not the same throughout the scientific community. For example, it is not clear whether the ability of co-located objects to interact must be necessarily mediated by the central infrastructure services or could be realized by local service discovery processes.
Also, there's much uncertainty on the functional components of the IoT. Depending on the required features of the IoT, new infrastructure services will be needed. In Section 2, we have proposed the definition of *Lookup* and *Resolution Services*, but many other may be needed to cope with security and privacy issues for example. Such services also raise the problem of scalability from three perspectives:

- number of devices requesting a service from the IoT infrastructure.
- number of *Resource* entries in the registry of an infrastructure service on which to perform the search
- client device resources (bandwidth, battery, processing power, which decrease going towards the periphery of the IoT network)

In this context, the fact that there is no reference architecture for the IoT is almost a consequence. To our best knowledge, there is very limited literature on the topic yet (Tsiatsis et al., 2010; Vazquez et al., 2010). The IoT-A project (IoT-A, 2010), as the name suggests, will address thoroughly this issue in its three years' course.

5.3 Privacy and security

Privacy and security, or the lack of, also pose a significant challenge for the correct deployment of the IoT concept. Clearly, the peripheral part of the IoT is the most vulnerable one. Here, networks of constrained devices and data-collection systems, generally characterized by very limited resources, aim to collect and transport sensible and sometimes critical data.
More and more often, such systems rely on wireless communication, which has greatly improved the ease of deployment of data-collection systems, overcoming physical limitations related to the weaving of cables needed for the communication infrastructure. From a security point of view though, wireless systems (such as today's WNANs or RFID systems) have an intrinsic downside: they use a shared physical medium for communication. To share the air as physical medium means that attackers can easily and anonymously obtain access to packets sent over the air from far away and with minimum costs. Access to data is then a simple matter if this is not encrypted. Moreover, as there is no physical authentication, malicious users can inject forged packets at Link Layer level, disrupting the network and possibly compromising any functionality of the upper layers.
Though many solutions for improving passive RFID security have been proposed in the scientific community, very few standards actually implement relevant security features (Oertel et al., 2005). The general problem is that passive RFID tags provide a very limited and vulnerable memory storage as well as minimal processing capabilities. These aspects limit in turn the flexibility of the security features, so that, at the best of our searches to date, it is impossible to secure (provide at least authentication, confidentiality and freshness) the typical IoT scenarios where RFID tags can move around and interact with different readers, pertaining to different security domains.
For what concerns peripheral networks, in order to provide confidentiality, integrity and authentication features, security frameworks (Casado et al., 2009; Karlof et al., 2004; Luk et al., 2007) can be used. These frameworks work at Link Layer level in order to protect the functionalities of the higher layers. On the downside though, they introduce a relatively consistent communication and processing overhead to achieve their goal. Authentication in particular is essential in order to deny packet forging and avoid replay attacks.

Even for these systems, there is another common issue: in such systems there is no trusted actor (i.e. device) by default. The process of defining a trusted actor and sharing the "secret", or key, subsequently used for authentication or encryption, is a critical and vulnerable one. Because of its utter importance, it also has to be done in a safe environment which generally means connecting to the devices physically (by cable) or wirelessly in a safe environment. Moreover, in such systems, keys are usually network-wide because of memory constraints, which means that compromising one node, might compromise the whole network/system. Also, such keys cannot be as long as the standard length for unconstrained devices because of the limited computational power.

As pointed out in (Vazquez et al., 2010), smart objects with communication capabilities usually use a gateway in order to connect to the Internet. This gateway usually is at the edge between the domain of constrained and unconstrained devices and usually having less constraints than peripheral devices. It is interesting to note that these devices are also on the border of two domains characterized by different security capabilities. It is thus reasonable to delegate to these devices the task of providing the needed security scalability for providing end-to-end security features. In Figure 5 we describe three possible scenarios for what concerns authentication scalability. We consider that gateways can authenticate all traffic incoming from the Internet side with a standard length key and that, in the most demanding scenario, they, at the same time authenticate all the outgoing from the WSAN. These scenarios can be easily adapted for the confidentiality and integrity features.

A more complex scenario though configures in the case of nomadic nodes that use unfamiliar networks to connect to the Internet and thus were not pre-configured. In this case, there is also the issue of having nodes that do not trust the gateway by default. A Certification Authority (thereafter CA) could be used to provide mutual trust between the gateway and he mobile node, but this might prove risky as the access to the CA is provided by the (un-trusted) gateway.

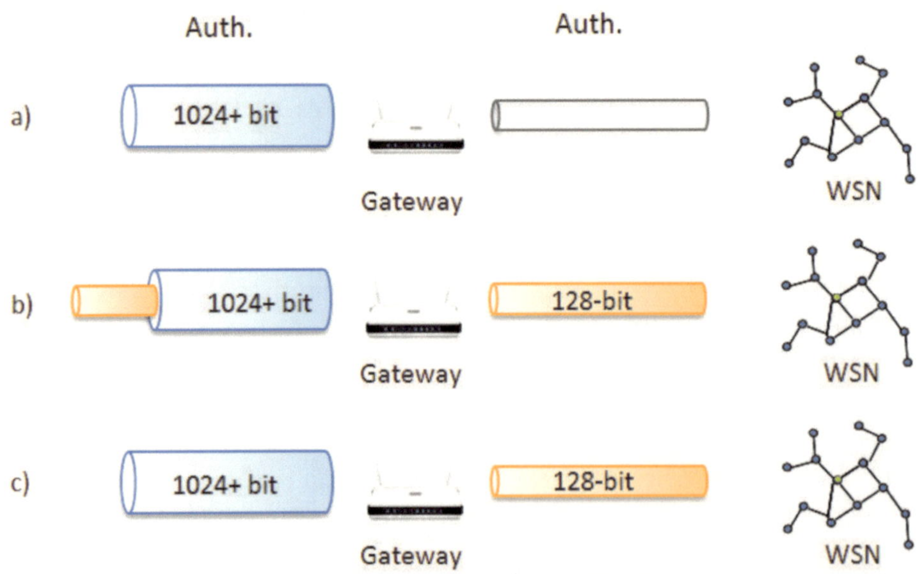

Fig. 6. Security scalability scenarios: only the gateway (and thus the source network) can be authenticated, b) tunnelling and c) active scaling of features

In conclusion, in current WSAN-based IoT-like systems a) an Internet-wide security framework implementation for granting Link Layer security features is not feasible, b) long keys cannot be used for cryptography/signing on embedded devices, c) security of mobile constrained devices is even more problematic and d) we suggest the use of gateways as tools for scaling the security features between the core area of the IoT network and the peripheral part.

5.4 Governance

A specific regulatory frame that takes into account private, business and public needs for the IoT is needed. Also, governing bodies have to adopt a shared strategy for managing, maintaining a global, international and possibly critical infrastructure such as the Internet of Things.

Defining and enforcing policies for such a network is also an important and delicate issue due to its characteristics (physical pervasiveness, trans-national reach, transport of sensitive data, ...) and the criticality of the potential implications.

If today's Internet has stretched the previous definitions of intellectual property, the IoT scenario will likely challenge the old definitions. For example, imagine a future environment, a private ground that is publicly accessible, such as a supermarket. Many people move through this space and the devices they wear continuously collect data in the open environment. This data might be made accessible to users all over the world through the IoT. But whose property is this data? Does it belong to the environment's owner, to the sensor's owner, or to the collector of the data? And in the case that the sensor simply traces other devices? How will the user be informed that some data regarding him has been acquired? Actually will he be informed? How will privacy rules be enforced in such an environment?

Many questions, but few answers. It is certain though that the Internet of Things will introduce a whole new set of security and privacy issues and that users shall be able to understand and manage the security and privacy features of their devices in order to benefit from the deployment of IoT. This is of the utmost importance and we believe that it is one of the main priorities of the governance to design and supervise this process.

6. Implications

6.1 Social implications

Pervasive technologies might have a consistent and positive impact on society but they also have the power to be very disruptive. For these reasons, the design and adoption of the Internet of Things shall be performed taking into account the implications on society that we can foresee and limiting, where it can be done, the consequences we cannot predict.

For example, the digital divide is one of the issues which we can foretell that will be accentuated by the adoption of the IoT, if not correctly designed. In first stance, the digital culture is not homogeneously distributed across the territory: people living in densely populated areas usually are more used to technology and adapt easier to new IT developments. IoT will be a great challenge for all users because the interaction paradigm will be completely new to them. In order just to realize how difficult it will be, one could think of how the users will manage the privacy and security features of the 3 to 8, display-less devices which have been forecasted belong to their personal space.

In second stance, while deploying stand-alone WSAN solutions in remote areas is relatively easy, taking the Internet of Things to rural areas will be very difficult due to the fact that a proper infrastructure and maintenance will be needed. The advent of IoT without a properly

established and pervasive infrastructure for connecting to the Internet could accentuate the division between less and more urbanized areas.
Another underestimated impact of the IoT in on education. The acceleration of the information flow and the handily availability of information in what will be an augmented reality will drastically change the way people learn things. If young people manage to master the new interaction paradigms that will characterize IoT, their relation to the physical world will also be drastically changed and we do not know how these changes will manifest later on.

6.2 Economic implications

The IoT will doubtlessly have an impact on economy. While in the first phase we expect only a limited, vertical impact, with the wider adoption of IoT-enabled solutions, the benefit derived from adopting the IoT paradigm will increase in a typically exponential way.
The first impact we foresee is the improvement of process efficiency in all economic sectors thanks to the adoption of large-scale automation. In second stance, brand new services for private, public and business users will be designed and developed.
There are though many open questions related to the economic implications of large-scale adoption of IoT. Such questions involve all scales from enterprise-level to an international level:

- what will the underlying business models look like? When will the ROI rate be high enough to sustain the spontaneous adoption of the IoT paradigm by mainstream enterprises in industry and agriculture?
- will excessive automation change the economic model of countries? Will it have a negative impact on society?
- as the time of adoption of the IoT by developed countries will come sooner than in in-development countries, will this accentuate the gap between them? Or could this even help the economy of such countries?

While we strongly believe that it's important to keep these questions in mind while designing the Internet of Things, we also believe that, due the expected highly-accelerated rate of development and adoption, having accurate long-term forecasts in the IoT scenario is very difficult.

6.3 Environmental impact

Having such a large amount of "things" integrating electronic circuitry and components might have a significant impact on the environment. First of all, the shear amount of hardly recyclable or even hazardous materials that will be introduced in the environment could represent a serious pollution danger. Moreover objects integrating electronic devices that are disposed of at the end of their life cycle will be difficult to treat, let alone to recycle, which again can increase pollution. Thus, new materials and recycling techniques for objects incorporating electronic devices should be developed.

7. Conclusions

From a technical point of view, the Internet of Things is an interesting evolutionary path in the Age of Information, likely making the newborn smart objects replace humans as the main consumers of Internet-transported data in a 10 years' time span. It also has the potential to radically change our life and the relationship with the environment. We have provided a reference model for the Internet of Things and highlighted some of the most

critical missing building blocks needed for its advent. We also looked into some impacts of this advent that apparently have been neglected till now.
One important concept we would like to disseminate through this article is that designing the IoT in all its aspects is very important. The results of this design phase will affect human life in almost all aspects and at all territorial scopes.

8. Acknowledgment

The authors would like to thank the European Commission for their timely commitment to the research on the Internet of Things topic and, in particular, for the sponsorship of the IoT-A project in the frame of FP7. We would also like to acknowledge the fruitful discussions with the partners in the IoT-A frame, which were the seeds of many of the ideas presented herein.

9. References

Akyldiz, I., Vuran, M.C. (2010), *Wireless Sensor Networks*, John Wiley & Sons Inc., 978-0-470-03601-3 (H/B), New York, NY, USA.

Arch Rock, (2009) Available from: http://www.archrock.com Note: as Arch Rock has been acquired by Cisco as of Sept. 20, 2010, some resources on the site might not be accessible.

Arduino, Available from: http://www.arduino.cc

Armenio, F., et al. (2007) EPCglobal Architecture Framework v1.2, available http://www.gs1.org/sites/default/files/docs/architecture/architecture_1_2-framework-20070910.pdf

Berners-Lee, T., Fielding, R., Irvine, U. C., Masinter, L., (1998) RFC2396: Uniform Resource Identifier (URI): General Syntax, last accessed A 2011, available from http://tools.ietf.org/html/rfc2396

Bormann, C., Mulligan, G., Arkko, J., Townsley, M., (2009) IPv6 over Low power WPAN (6LoWPAN), last accessed March 2011, available from http://www.ietf.org/html.charters/6lowpan-charter.html

Bose, I., Pal, R., (2005) Auto-ID: managing anything, anywhere, anytime in the supply chain, *Communications of the ACM*, 48(8), pp.100-106.

Botterman, M., (2009) Internet of Things: an early reality of the Future Internet. Workshop Report, European Commission Information Society and Media.

Casado, L., Tsigas P. (2009) ContikiSec: A Secure Network Layer for Wireless Sensor Networks under the Contiki Operating System, *Lecture Notes in Computer Science*, vol. 5838, pp. 133-147, Springer Berlin / Heidelberg.

Coordination and Support Action (CSA) for Global RFID-related Activities and Standardisation [CASAGRAS] (2009) Final report, Available from http://www.rfidglobal.eu

European Technology Platform on Smart Systems Integration, [EPoSS], (2008) Internet of things in 2020, report of *Beyond RFID - the Internet of Things*, joint EU-EPoSS workshop, Brussels (BE), February 2008, last accessed March 2011, available from http://www.smart-systems-integration.org/public/internet-of-things

Gershenfeld N., Krikorian R., Cohen D., (2004) The Internet of Things, *in Scientific American*, October 2004, pp 76-81.

Haller, S., The Things in the Internet of Things, *Proceedings of Internet of Things Conference 2010*, Tokyo, 2010

IBM, IBM Mote Runner, last accessed March 2011, available from: http://www.sensinode.com/

Internet of Things Architecture [IoT-A], 2010, available from http://www.iot-a.eu

Karlof, C., Sastry, N., Wagner, D. (2004) TinySec: a link layer security architecture for wireless sensor networks. *Proceedings of 2nd International Conference on Embedded Networked Sensor Systems*, SenSys 2004, Baltimore, MD, USA, November 3-5, 2004, pp. 162-175, ACM.

Kistler, R., Knauth, S., Klapproth, A., (2008) EnerBee - An Example of an Advanced Metering Infrastructure Based on ZigBee, EuZDC.

Kortuem, G., Kawsar, F., Fitton, D., Sundramoorthy, V. (2009) Smart objects as building blocks for the Internet of things, Internet Computing, IEEE Journal of, Vol.14, pp44–51.

Lehtonen, M., Ruhanen, A., Michahelles, F. and Fleisch, E., Serialized TID numbers - A headache or a blessing for RFID crackers?, *Proceedings of RFID, IEEE International Conference on*, Orlando (FL), April 2009.

Luk, M., Mezzour, G., Perrig, A., Gligor, V., (2007) MiniSec: a secure sensor network communication architecture. *Proceedings of 6th International Conference on Information processing in Sensor Networks.*

Meloan, S., (2003) Toward a Global "Internet of Things", Available from http://java.sun.com/developer/technicalArticles/Ecommerce/rfid/.

Montenegro, G., Kushalnagar, N., Hui, J., Culler, D. (2007) IPv6 over IEEE 802.15.4", RFC 4944.

Oertel, B., Wölk, M., Hilty, L., (2005) Security Aspects and Prospective Applications of RFID Systems, Federal Office for Information Security, last accessed March 2011, available from https://www.bsi.bund.de/SharedDocs/Downloads/EN/BSI/Publications/Studies/RFID/RIKCHA_englisch_pdf.pdf .

Oracle, Oracle Sun SPOT, Available from: http://www.sunspotworld.com/.

SENSEI, (2008) SENSEI Reference Architecture, Retrieved from: http://www.ict-sensei.org/index.php?option=com_docman&task=doc_download&gid=44&Itemid=49

Sleman, A., Moeller, R., (2008) Integration of Wireless Sensor Network Services into other Home and Industrial networks; using Device Profile for Web Services (DPWS), *Information and Communication Technologies: From Theory to Applications 2008*, pp. 1-5.

Sundmaeker, H., Guillemin, P., Friess, P., Woelfflé, S., (2009) Vision and Challenges for Realising the Internet of Things, last accessed March 2011, available from http://docbox.etsi.org/tispan/open/IoT/CERP-IOT_Clusterbook_2009.pdf.

Tsiatsis, V., Gluhak, A., Bauge, T., Montagut, F., Bernat, J., Bauer, M., Villalonga, C., Barnaghi, P., Krco, S., (2010), The SENSEI Real World Internet Architecture, *in Towards the Future Internet - Emerging Trends from European Research*, pp 247-256, 2010, IOS Press, 978-1-60750-538-9

Vasseur, J.P., Dunkels, A. (2010) *Interconnecting Smart Objects with IP: The Next Internet*, Morgan Kaufmann, 978-0-12375-165-2.

Vazquez, J. I., Ruiz-de-Garibay, J, Eguiluz, X. , Doamo, I., Rentería, S., Ayerbe, A., (2010) Communication Architectures and Experiences for Web-Connected Physical Smart Object, *in Pervasive Computing and Communications Workshops (PERCOM Workshops), 2010 8th IEEE International Conference on*, 978-1-4244-6605-4 , Mannheim, March 29 - April 2 2010.

Xuemei, L., Yuyan, D., Lixing, D. (2008) Study on Precision Agriculture Monitoring Framework Based on WSN, International Conference on Anti-counterfeiting Security and Identification, pp. 182-185.

Yang, G.Z., Yacoub, M., (2006) *Body Sensor Networks*, Springler Berlin / Heidelberg, 978-1-84628-272-0, London, UK.

21

RFID Security and Privacy

Michel Arnaud
University Paris Ouest Nanterre la Défense
France

1. Introduction

The European Commission has published in May 2009 a recommendation *"on the implementation of privacy and data protection principles in applications supported by radio-frequency identification"*, which is designed to provide *"guidance to Member States on the design and operation of RFID applications in a lawful, ethical and socially and politically acceptable way, respecting the right to privacy and ensuring protection of personal data."* This recommendation requires RFID operators to conduct a "Privacy and Data Protection Impact Assessment" before any RFID application is deployed, and make its results available to the competent authority. The RFID recommendation is also designed to promote *"information and transparency on RFID use"*, in particular through the development of *"a common European sign developed by European Standardisation Organisations, with the support of concerned stakeholders"*, designed *"to inform individuals of the presence of readers"*.

RFID PIA (Privacy and Impact Assessment) process aims to reach several objectives:

- to favor "privacy by design" by helping data controllers to address privacy and data protection before a product or service is deployed,
- to help data controllers to address privacy and data protection risks in a comprehensive manner, an opportunity to reduce legal uncertainty and avoid loss of trust from consumers,
- to help data controllers and data protection authorities to gain more insight into the privacy and data protection aspects of RFID applications.

The industry has proposed a RFID PIA framework which classifies a RFID application into 4 possible levels:

- Level 0: applications that do not process personal data and where tags are only manipulated by users, and which are rightly excluded from conducting a PIA.
- Level 1: applications where no personal data is processed, yet tags are carried by individuals.
- Level 2: applications which process personal data but where tags themselves do not contain personal data.
- Level 3: applications where tags contain personal data.

If the RFID application level is determined to be 1 or above, the RFID operator is required to conduct a four part analysis of the application, with a level of detail that is proportionate to identified privacy and data protection implications. The first part is used to describe the RFID application. The second part allows highlighting control and security measures. The third part addresses user information and rights. The final part of the proposed PIA framework requires the RFID operator to conclude whether or not the RFID application is

ready for deployment. As a result of the PIA process, the RFID operator will produce a PIA report that will be made available to the competent authority.
For the industry, only levels 2 and 3 are to be submitted to a PIA because it considers that information contained in a level 0 tag are not personal. However level 1 rises concerns of Article 29 Working Party because tagged items carried by a person contain unique identifiers that could be read remotely. In turn, these unique identifiers could be used to recognize that particular person who will be tracked by a third party without her knowledge. When a unique (or multiple identifiers) is associated to a person, it falls in the definition of personal data set forth in Directive 95/46/EC, regardless of the fact that the "social identity" (name, address, etc.) of the person remains unknown (i.e. she is "identifiable" but not necessarily "identified"). Additionally, the unique number contained in a tag can also serve as a means to remotely identify items carried by a person, which in turn may reveal information about social status, health, or more. Even in those cases where a tag contains solely a number that is unique within a particular context and without additional personal data, care must be taken to address potential privacy and security issues if this tag is going to be carried by persons. The Working Party has urged the industry to fully address this issue, by clearly mentioning it as part of a revised risk assessment approach for level 1.
This chapter will address issues of protecting privacy of RFID tag carriers in a "privacy by design" model which is described below on four different layers: legal aspects, policy services, technical specifications and security services. The idea is to provide easy-to-use tools to accept or not to be tracked at PIA level 1. In case of a negative decision, tags have to be deactivated. Authentication techniques are to be used to protect user identity for PIA levels 2 and 3. Security measures have also to be taken to protect personal information on RFID tags against information leak which could lead to identity theft.

2. Legal framework

Personal data is any information relating to an identified or identifiable natural person ('data subject'). An identifiable person is one who can be identified, directly or indirectly, in particular by reference to an identification number or to one or more factors specific to their physical, physiological, mental, economic, cultural or social identity. Personal data exist in many digital forms and are included in browsers as certificates; mobile phones are generally related to an individual; home, appliances and clothing may include technology (e.g. smart metering, Internet of Things and RFID) which represent owner or user's identity; social networking sites reflect personal information in great detail including : digital information stored in databases, video, pictures, documents, files, notebooks, invoices, medical records, RFID, ID cards, passports, cookies, flash objects, eID middleware, biometric identifiers (e.g. fingerprints, DNA, etc.).
Basic principles of Directive 95-46 of the European Parliament include the following regarding protected data: fairly and lawfully processed, for limited, adequate, relevant and not excessive purposes, accurate and up to date, not kept for longer than necessary, processed in line with individual's rights, secure in processing, storage and transfer, not transferred to other countries without an adequate level of protection.

2.1 Identity management

The concept of "identity management" is not well defined with reference to currently available international standards, although there is relevant work in ISO/IEC JTC1/SC27/WG5 *"Identity management and privacy technologies"*.

An individual during its lifetime may have many multiple different personae, i.e. names, depending on the roles that it has or qualifies for. For example, at the time of marriage an individual may acquire and use a new (legal) persona. Consequently, an individual may have multiple legally recognized names (LRNs), recognized individual names (RINs), recognized individual identities (riis) at the same time (and so used in various business transactions). Examples include a persona which an individual assigns to himself and is one which also serves as an identifier such as an e-mail address (on a hotmail or gmail account, Facebook, Twitter, as an "avatar", etc.).

A recognized individual name is any persona associated with a role of an individual which is recognized as having legal status, i.e., if a legally recognized name (LRN) is recognized in a jurisdictional domain and accepted in compliance with the registration corresponding schema. Associated with a registered individual name is (usually) a registration number of the document attesting that the RIN has legal status of some kind. A registration authority shall assign a unique identifier to each of its registered members including and especially identifying where the member is acting as an individual. This unique identifier has the properties and behaviors of an ID code in the coded domain used to support management and maintenance of the registration authority schema.

From an eBusiness perspective, one often does not need to distinguish whether the entity which is party to a business transaction is a "natural person" or "legal person", or an "individual" or "organization", etc. Credit worthiness, ability to pay, secure payment, etc., of a "person", as a buyer, is often a more important criterion for doing business with the person in the role of seller based applications, business (including e-commerce, e-government, e-health, etc.). This is particularly so when modeling Open-edi scenarios and scenario components from an internal constraints perspective only. In much of consumer trade, a buyer can remain anonymous vis-à-vis a seller by presenting a money token in which a seller has 100% trust (e.g., cash).

Privacy protection requirements have made "anonymity" an external constraint matter which needs to be supported. At times it is desired that an individual can establish a long-term relationship (including a reputation, trust relationship, etc.), with some other person, without the individual's actual identity being disclosed. For convenience, it may be useful for the individual, or the other party concerned, to establish a unique (new) persona, identifier, token, etc., known as "pseudonym" with the other person. Pseudonymization is recognized as an important method for privacy protection of personal information. Pseudonymization techniques, mechanisms and services may be used within an organization or public administration, within a jurisdictional domain as a whole or across jurisdictional domains for transborder data flows.

The following set of rules summarizes privacy protection requirements which apply. A buyer (and its agent(s)) or third party (or any other party to the business transaction), shall not retain any personal information on the individual as the buyer for any time longer than is consented to by the individual for post-actualization purposes unless external constraints of the applicable jurisdictional domain requires retention of such personal information for a longer period.

2.2 Good practices

Good practices have been defined within the CEN/ISSS Workshop on Data Protection and Privacy (WS/DPP). Organizations should appoint a person who periodically checks whether notified information is still complete, accurate and up-to-date, or whether grounds

for exemption are still valid. The principal purpose of having notification and a public register is transparency and openness. It is a basic principle of data protection that the public should know who is carrying out the processing of personal information as well as other details about processing. Notification, therefore, serves the interests of individuals by helping them understand how personal information is being processed by data controllers.
Data subject has the right of access, rectification, erasure, blocking and objection to retention. Data controller should respect these rights. Under Section 3 of the Directive, data subjects have the right to find out, free of charge, if any entity (either an individual or an organization) holds information about them. They might also request a description of the information and inquire about the purpose(s) for holding their information.
Anyone having access to the organization's documents, media, computers or information systems is responsible for complying with the information security policy and all other associated documentation that is applicable to it. The information security policy will preserve an appropriate level of confidentiality, integrity, availability, lawful purpose. Support contractors who have access to sensitive information in paper, electronic or other format should sign a written agreement stating they will comply and adhere to organization's policies to keep information secure. Their compliance should be monitored to verify they adhere to these obligations.

2.3 PIA framework for RFID

A privacy impact assessment (PIA) enables organizations to anticipate and address likely data protection impacts of proposed initiatives and foresee problems. This process reflects measures taken to protect privacy of individuals about whom sensitive data are kept and addresses legal obligation to use appropriate security measures. Systems should be designed to avoid unnecessary privacy intrusion and with privacy-by-design features implemented to reduce possibility or effects of a security incident.
Individuals responsible for data protection (including their processing service provider) should be identified in the security policy. These documents identify roles, individual responsibilities, incident handling and reporting practices that have been put in place to protect personal data and their processing with appropriate technical and organizational measures to ensure, that at all times, integrity, confidentiality and availability of personal/sensitive data.
The PIA Framework for RFID of January 12, 2011 explains key concepts, internal procedures and classification criteria for RFID applications. For these criteria the PIA Framework provides a two phases approach. The initial analysis phase is used to determine if a PIA of RFID application is required. The decision, to which level an application belongs, has to be made after working through a decision tree where level 1 implies a small scale PIA while levels 2 and 3 require a full scale PIA. If an application is designed according to level 0 which means that no private data are concerned, there is no privacy threat given and further documentation is not needed. Level 2 applications may have controls to protect back-end data while level 3 applications may have controls to protect both back-end data and tag data. For level 1 applications, required controls and corresponding documentation in the PIA report are simplified.
The objective of the risk assessment phase is to document how risks are pro-actively mitigated through technical and organizational controls. The PIA process requires any RFID application operator to:

1. Describe the RFID application;
2. Identify and list how the RFID application under review could threaten privacy and estimate the magnitude and likelihood of those risks;
3. Document current and proposed technical and organizational controls to mitigate identified risks;
4. Document the resolution (results of the analysis) regarding the application.

The risk assessment requires evaluating the applicable risks from a privacy perspective. The RFID operator should consider:

a. The significance of a risk and the likelihood of its occurrence.
b. The magnitude of the impact should the risk occur.

The resulting risk level can then be classified as low, medium or high. A prime risk is that RFID tags could be used for profiling and/or tracking of individuals. In this case RFID tag's information - in particular its identifier(s) - would be used to re-identify a particular individual. Retailers who pass RFID tags on to customers without automatically deactivating or removing them at checkout *may* unintentionally enable this risk. A key question, though, is whether this risk is likely and actually materializes into an *undismissable* risk or not.

According to recommendation, retailers should deactivate or remove at the point of sale, tags used in their application unless consumers, after being informed of the policy in accordance with this framework, give their consent to keep tags operational. Retailers are not required to deactivate or remove tags if the PIA report concludes that tags that are used in a retail application and would remain operational after the point of sale do not represent a likely threat to privacy or protection of personal data.

The RFID operator should use categories below to indicate privacy and data protection implications of the RFID application:

- Ready for deployment: the RFID application as described provides for suitable practices, controls, and accountability.
- Not ready for deployment: the RFID application is not approved for operations in its current state. A specific corrective action plan has to be developed, and a new privacy impact assessment has to be performed and documented to determine if the application has reached an approvable state.

The PIA Framework provides only a generic scheme for the PIA and has to be complemented by more detailed schemes like roles, security targets, classes and templates reflecting the special aspects of industry-specific and individual applications.

2.4 Technical guidelines as templates for PIA

The approach of the European Commission suggests using so-called templates as extensions to the Framework document in order to reach the level of detail that is necessary to conduct a complete application-specific Privacy Impact Assessment. Such templates are specific to an application area and should provide a detailed guidance for the creation of a PIA report. This puts the "Technical Guidelines for the Secure Use of RFID" (TG RFID) into perspective which have been issued by Germany's Federal Office for Information Security (BSI). In 2007 the BSI launched this project which aims at providing technical recommendations for RFID systems that ensure secure implementations and protection of personal data but nevertheless support RFID operators' and service providers' business needs. The BSI achieved a consensus between supporters and critics. TG RFID are accepted by relevant parties and are now available for application areas: public transport, event ticketing, NFC-ticketing, retail & logistics and employee cards. First implementations proved practicality and viability of this approach.

A major goal of development for the TG RFID is to find a consensus and to gain acceptance of all relevant stakeholders. Therefore the BSI installed an intense review and alignment process and invited experts and relevant stakeholders from specific application area to participate. Representatives of RFID operators, service providers, customers, Data Protection Agencies (DPAs) and also critics of RFID have had the opportunity to comment early versions of the document and take part in review and alignment sessions. In this process, security goals, potential threats, security measures and especially remaining risks were identified, discussed and described. This process provided information on potential impact and risks of RFID applications and generated transparency that is necessary to build trust and acceptance. So far Technical Guidelines for five application areas have been created. In all cases a consensus including acceptance from participating DPAs was achieved.

Unfortunately, TG RFID for logistics and retail have not been piloted so far, because progress with RFID in this sector is far behind former projections by retail stakeholders. RFID tags are actually mostly used on pallets and cartons. Products in supermarkets shelves are still only marked with traditional bar codes or with GS1 data bar, except cases like Gillette razors. Whereas in the sectors of ticketing, NFC (13.56MHz) and employee cards (125 kHz HID) a great progress with RFID is on its way.

TG RFID provide patterns for application specific templates which can be efficiently set up as required by PIA Framework.

Stakeholders of an application have individual and sometimes diverging requirements for a technical guideline. Data Protection Agencies (DPAs) want to protect data and privacy of citizens, customers and employees. TG RFID address their objectives by a detailed description of all relevant threats, appropriate safeguards and potentially remaining risks. Operators are focused on their business objectives. Their intention is on practicality, acceptance of their customers and a cost efficient and future proof solution. Balance between objectives of both parties is achieved by a scalable definition of safeguards. Minor threats are mitigated by simple, low-cost safeguards. Strong and costly controls are only applied in case of high protection demand and severe threats. This approach makes sure that cost of security measures and impact on usability are reduced to what is necessary.

Interoperability is an imperative for RFID implementations. Operators need to cooperate with business partners and customers want to use services from multiple service providers and across borders. This requires standardized and interoperable technical interfaces and security measures. In addition, comparability of security levels is of major importance. Operators can only cooperate if they can trust partner's system implementation. This includes a certain level of data protection, privacy and as well information security and safety. TG RFID support these fundamental requirements by two dedicated features:

i. TG RFID include not only an assessment of privacy and data protection. In addition, a risk analysis and documentation of information security and safety is provided. The latter is mandatory to cover business requirements of operators.
ii. Risk assessment methodology and documentation of results comply with worldwide standard ISO27005. This makes it easy to compare PIA and security assessment reports of different implementations and systems.

Operators will refrain from investing in RFID applications if they can't determine the cost of security measures and their potential impact on services and usability. Both aspects have major influence on the overall business case. TG RFID define appropriate technical safeguards for specific scenarios of an application. This information builds a solid base for cost calculations and tenders. This feature of TG RFID counters a major roadblock for introduction of RFID.

The European Commission has identified lacking confidence in legal situation for RFID-implementations as one major roadblock for the broad adoption of RFID. Use of TG RFID is not mandatory in a legal sense. Nevertheless they will provide a solid basis for legal judgments of RFID applications because they are accepted by all stakeholders and represent the current state-of-the-art for implementations of RFID.

2.5 Description of structure and security methodology of TG RFID

TG RFID are created for specific application areas and consist of three major parts: description of the application area, assessments and recommendations. A detailed but generic description of all service and business models of an application area is given in the first part. This is the foundation for assessments and recommendations and covers role models, services, products, business processes, use cases and any other information that may be relevant for security and privacy assessments. In order to ensure practicality and usability for all service providers and operators, this part is done in close cooperation and alignment with experts from the application domain.

The assessment part is based on description of application area and specific security targets. It covers all three domains of information security: security, privacy and safety. Security targets are defined and aligned with all stakeholders. Methodology of risk assessment is compliant with ISO 27005. Results of assessment are a list of relevant threats, appropriate safeguards that can mitigate these threats and a description of remaining risks.

The third part of guidelines document provides recommendations on how to implement a RFID-system in an appropriate way. Based on example scenarios from the application domain it is shown how findings of risk assessment are transformed into specific safeguards that should be applied to the relevant system components. This provides a clear and economically viable guidance for the design of system.

Organizations must be able to demonstrate that they have implemented a data protection management system (DPMS) using appropriate technology (PETs) and operational protective measures (OPMs) to protect personal data. PIAs incorporate tests of PETS and OPMs to prove data protection principles are met by the system. All personnel within the organization have a responsibility to ensure that they take steps to safeguard security of information that they are entrusted with and to use OPMs and PETs as established policy.

3. Privacy framework models

3.1 OASIS Privacy Management Reference Model (PMRM)

OASIS Privacy Management Reference Model (PMRM) Technical committee aims at achieving a standard-based framework that will help business process engineers, IT analysts, architects, and developers implement privacy and security policies in their operations. PMRM picks up where broad privacy policies leave off. Most policies describe fair information practices and principles but offer little insight into actual implementation. PMRM provides a guideline or template for developing operational solutions to privacy issues. It also serves as an analytical tool for assessing the completeness of proposed solutions and as the basis for establishing categories and groupings of privacy management controls.

This model is based on a service-based approach, describing them in three categories:

- core policy services : agreements (with options and permissions), control (with policies and data management),
- presentation and lifecycle services : interaction (manages data/preferences/notice), agent (software that carries out processes), usage (data use, aggregation, anonymization), access (individual review/updates to personal information),
- privacy assurance services : certification (credentials, trusted processes), audit (independent, verifiable accountability), validation (checks accuracy of personal information), enforcement (including redress for violations)

Personal information is stored in a container accessed by an agent (at entry point) for specific processing which must abide to privacy rules (referred to as agreement and control procedures). Assurance service guarantees conformity to these rules which can be a simple validation or a certification, leading eventually to an audit and an enforcement procedure.

Each use case invokes a sequence of service calls. Each service call executes a sequence of functions: define (operational requirements), select (input, process, and output) data and parameters, input (data and parameter values in accordance with select), process (data and parameter values within functions), output (data, parameter values and actions), link to other services, secure with appropriate security functions.

3.2 Open identity exchange trust framework

In the context of digital identity systems, a trust framework is a certification program that enables a party who accepts a digital identity credential (called the relying party) to trust the identity, security, and privacy policies of the party who issues the credential (called the identity service provider) and vice versa. In the Open Identity Trust Framework (OITF) model, an open identity trust framework provider can administer any trust framework that meets: 1) the principles of openness, and 2) any additional requirements imposed by the Trust Framework Provider (TFP).

The rules of every trust framework are defined for a particular set of participants in online (and possibly offline) interactions. The Open Identity Trust Framework Model defines six standard trust framework roles (in addition to the trust framework provider role played by OIX):

1. Users
2. Identity service providers
3. Relying parties
4. Assessors
5. Auditors
6. Dispute resolution service providers
7. In addition, OIX has defined a seventh role, special assessor, which is an assessor responsible for assessing the qualifications of other assessors.

As defined in the Open Identity Trust Framework Model, a level of assurance (LOA) is a unit of measure for the degree of confidence a relying party can have in assertions for an identity credential from an identity provider. A level of protection (LOP) is a unit of measure for the degree of confidence: a) an identity provider can have in the protection provided by a relying party for the identity information disclosed in an identity credential, or b) a user can have in the protection provided by an identity provider and/or a relying party for the identity information disclosed in an identity credential.

3.3 Technical profiles

A technical profile is a specification of requirements for use of a specific technology, RFID in our case, in order to achieve technical interoperability in exchange of digital identity credentials that is consistent with associated LOA or LOP. Once an OIX trust framework is accepted for listing in the OIX Listing Service, participants may apply for certification.

For RFID open identity trust technical profile, four main functions have to be taken into consideration to provide appropriate tools for agents: anonymization and pseudonymization facilities, attributes management tools, identity management tools, security management tools.

4. Conclusion

All TG RFID follow a common security concept. Whereas RFID Recommendation is primarily directed towards privacy and data protection, TG RFID cover all three security domains: safety, security and privacy. Furthermore, TG RFID provide detailed guidance how to carry out all detailed work PIA Framework leaves out, because it is understood as a high level document more for senior management and non-IT people. TG RFID are written for IT experts who are responsible for designing systems, investigating threats and weaknesses and providing for the right protection provisions. Definition of generic controls and proposition of scenario-specific safeguards are carried out as a joint approach. This reflects the fact that threats for privacy are often threats to information security as well. Vice versa certain safeguards can counter threats for privacy and information security. The approach of TGs optimizes the impact of safeguards and minimizes cost of security and privacy and complements PIA Framework.

TG RFID provide guidance and information that will enable operators to conduct a PIA and minimize efforts for completing the report. Major parts of the PIA can simply be covered by referencing appropriate chapters as templates and selecting particular services, processes and scenarios mentioned in the guideline. This will work out in most cases because TG are describing all known eventualities of an application area. The operator's application will normally be a subset of what is documented. Furthermore TG RFID provide detailed patterns to develop templates as required by the PIA Framework. All this brings quality of compliance statement to a level that can be trusted by all parties that will deal with RFID-based systems.

5. Glossary

Individual anonymity. The state of not knowing the identity or not having any recording of personal information on or about an individual.

Anonymization process. Whereby the association between a set of recorded information (SRI) and an identifiable individual is removed.

Information Security. Preservation of the confidentiality, integrity and availability of information.

Monitor. Carrying out an activity for the purpose of detecting, observing, copying or recording the location, movement, activities, or state of an individual.

Personal Data. Any information relating to an identified or identifiable natural person ("data subject"); an identifiable person is one who can be identified, directly or indirectly, in particular by reference to an identification number or to one or more factors specific to his physical, physiological, mental, economic, cultural or social identity.

RFID Application. An application that processes data through the use of tags and readers, and which is supported by a back-end system and a networked communication infrastructure.

RFID Application Operator. The natural or legal person, public authority, agency, or any other body, which, alone or jointly with others, determines the purposes and means of operating an Application, including controllers of personal data using an RFID Application.

Radio Frequency Identification (RFID). The use of electromagnetic radiating waves or reactive field coupling in the radio frequency portion of the spectrum to communicate to or from a tag through a variety of modulation and encoding schemes to uniquely read the identity of a radio frequency tag or other data stored on it.

RFID Reader. A fixed or mobile data capture and identification device using a radio frequency electromagnetic wave or reactive field coupling to stimulate and effect a modulated data response from a tag or group of tags.

RFID Tag or 'tag'. An RFID device having the ability to produce a radio signal or an RFID device which re-couples, back-scatters or reflects (depending on the type of device) and modulates a carrier signal received from a reader or writer.

RFID Tag Information or information on the RFID Tag. The information contained in an RFID Tag and transmitted when the RFID Tag is queried by an RFID Reader.

User. Specifically, an RFID Application User, i.e., a person (or other entity, such as a legal entity) who directly interacts with one or more components of an RFID Application (e.g., back-end system, communications infrastructure, RFID Tag) for the purposes of operating an RFID Application or exercising one or more of its functions.

6. References

Privacy and Data Protection Impact Assessment Framework for RFID Applications, 12 January 2011, http://www.statewatch.org/news/2011/feb/eu-art-29-wp-rfid-opinion-annex.pdf

Industry Proposal Privacy and Data Protection Impact Assessment Framework for RFID Applications, March 31 2010 http://ec.europa.eu/justice/policies/privacy/docs/wpdocs/2010/wp175_annex_en.pdf

Opinion 5/2010 on the Industry Proposal for a Privacy and Data Protection Impact Assessment Framework for RFID Applications Article 29 Data Protection Working Party 00066/10/EN WP 175, July 13, 2010 http://www.statewatch.org/news/2011/feb/eu-art-29-wp-rfid-opinion.pdf

Technical Guidelines RFID as Templates for the PIA-Framework, Federal Office for Information Security (BSI) Bonn, Germany, 2010. https://www.bsi.bund.de/SharedDocs/Downloads/EN/BSI/Publications/TechGuidelines/TG03126/TG_RFID_Templates_for_PIA_Framework_pdf.pdf?__blob=publicationFile

Digital Agenda: new guidelines to address privacy concerns over use of smart tags April 6, 2011

http://europa.eu/rapid/pressReleasesAction.do?reference=IP/11/418&format=HTML&aged=0&language=EN&guiLanguage=fr

22

The Ethics of RFID Technology

Joël Schlatter and Fouad Chiadmi
University Hospital of Jean Verdier
France

'Everyone's right to life shall be protected by law. No one shall be deprived of his life intentionally save in the execution of sentence following his conviction of a crime for which this penalty is provided by law. Deprivation of life shall not be regarded as inflicted in infringement of this Article when it results from the uses of force which is no more than absolutely necessary: (a) in defence of any person from unlawful violence; (b) in order to effect lawful arrest to prevent the escape of a person lawfully detained; (c) in action lawfully taken for the purpose of quelling riot or insurrection.'
(Korff, 2006)

1. Introduction

In the Parisian subway, a passenger quickly passes through the terminal of the subway without showing a ticket. The terminal indicates by a sound and a visual sign that the passage is allowed and the door is automatically unbolted. The passenger passes his bag at a distance of a few centimetres above the terminal, containing a Navigo chart integrating an RFID chip. The personal information is transmitted to the data processing unit in real-time and analyzed. The passage of the person is thus traced.
When passing through US customs, the traveller shows his passport to the American official who passes it on to a reader. Thus, it immediately obtains all information of the passenger. On the chip all information found on the old passport is recorded and by this technique police controls are more easily facilitated and false passports detected.
In Rotterdam or Barcelona, VIP visitors of Baja Beach Clubs are identified by a subcutaneous chip developed by the company Applied Digital Solutions (ADS). This chip is also used as a credit card without contact. ADS developed the VeriChip, a subcutaneous chip the size of a grain of rice, allowing applications of security access to buildings, such as for the identification of patients in hospitals or security against abduction in South America.
With these examples, we see *de facto* that the RFID technology applied to humans or objects immediately generates many questions about the right to a private life because of its somewhat suspect nature, in particular, when recording personal data. Doubts about the justification of the collection of data or on the storage of data and on the protection against access by third parties are revealed. If these fears prove to be relevant, it will be necessary for us either to espouse refusal of the technology in the name of respect for freedom, or, in the name of a defence of inescapable progress, to find the form of political society where the definition of freedom would not inevitably include a right to anonymity. It will thus be

necessary for us to see, on the basis of the fear of the individuals towards technology, if one can plan to reconcile freedom and technology, in a society where humans see guaranteed freedom and where the use of constantly evolving technology does not dehumanize. Hobbes proposed that fear is caused by the threat of violence and thus the state is created with the role of protecting individuals. But he points out to us another passion associated with this fear which is the ceaseless desire to enjoy life and secure more pleasure, and the role of the state must also therefore be to protect and guarantee peace.
The fear of individuals of this type of technology, but also the desire to have it, are expressed openly and thus the question that should be posited with regard to politics is: does RFID go against a political society where the individuals are free or does RFID better contribute to the maintenance of political society?

2. RFID Technology and instinctive fears

In defence of their freedom, individuals are afraid of being constantly supervised, card-indexed or tracked without their consent. The ultimate fear is that of the shrouded arrival of a society where there would be no more freedom, in other words, the imposition of a totalitarian state. One could certainly interpret the fear of RFID technology and this concern with regard to losing freedom as simple paranoia, but it should not be forgotten that this fear also has historical precedent. The totalitarian societies of the 20th century did not hesitate to categorize individuals according to their race, religious affiliation, politics etc. and then to exterminate some of them, according to the profiles thus drawn up. In the concentration camps, as it was the case in Auschwitz, numbers were tattooed on the skins of prisoners, contributing to the process of dehumanization. One can understand that the loss of freedom represents a fear for the 21st century. In this direction, the pure and simple refusal of RFID technology could be an obvious and radical solution to preserving freedom. The technological advancement would increase the danger, and the capacity to harm, of a totalitarian society. Therefore, the question should be asked: does the application of this technology carry in it the germ of the advent of a totalitarian society?
On the basis of the postulate that we live in a liberal society, where the freedom of each individual is preserved and does not harm that of others, we can suppose that the fear of citizens to losing their freedom because of a technological advance would be related to the disappearance of a political society which is only capable of guaranteeing this freedom. It is necessary for us to determine whether or not RFID is compatible with a certain idea of liberalism, but in addition, it is necessary to keep in mind that liberalism can also mean the freedom of business or 'free trade'.
The second human passion exposed by Hobbes is fear. From it arises the request for safety, since humans have the desire to preserve themselves. It is thus fear, a motive of human action, which leads to the centralised capacity of the state in Hobbes' design. Fear thus generates mistrust. In this design, RFID allows for data collection to be in the hands of the state, meaning access to the data collected by the society is possible when important issues are called upon. The state does not itself need to collect this data since industry takes care of this in part, but it can of course decide to impose identity RFID cards upon citizens, in which case it would have the techniques to control at any moment each citizen individually. The modern individual expects the state to protect, which gives to the state the obligation to do everything to anticipate the worst. However, RFID facilitates forecasts by providing considerable amounts of reliable information, fast. The fear of terrorism constitutes a

sufficient motive to encourage the state to use the data given by RFID technology to operate a better monitoring system, as any individual potentially represents a threat to others. Of course that raises the question of the respect due to private life, but this requirement is not at the heart of Hobbes' design of the state.

RFID technology makes possible, in theory, the identification of all things in the world but at the same time it causes in many people the fear of permanent monitoring. The critical approaches as to the employment of RFID technology relate mainly to this threat to the private life. Action groups, such as CASPIAN, led by Katherine Albrecht, categorize RFID chips as being 'spy chips'.

But *a contrario*, RFID can release a human being from daily constraints and thus enable him to fulfil his desires more easily. Today, patients voluntarily agree to have a chip placed in the arm so that there is no error of attribution on their medical file. Tomorrow, we will certainly agree to be chipped because that will give us privilege, of rights or accessibilities, thus RFID technology is ambivalent, tearing the individual between the desire to facilitate existence and the fear of a loss of freedom, in particular in the private sphere.

3. RFID Technology and interconnection of all things in the world

It is necessary to widen the narrow field of RFID, to position it in the scope of broader development, that of the 'Internet of things'. After a report by the ITU (International Telecommunication Union, 2005) the Internet of things represents a technological revolution which gathers developments in various fields, such as the techniques of communication without contact. From now on the question of the things will relate to those which think, which communicate between each other and react with their medium. 'Indeed, with the benefit of integrated information processing, industrial products and everyday objects will take on smart characteristics and capabilities. (...) eventually, even particles as small as dust might be tagged and networked. Such developments will turn the merely static objects of today into newly dynamic things, embedding intelligence in our environment and stimulating the creation of innovative products and entirely new services.' This revolution is not science fiction. This development is considered inevitable. RFID then plays a part in broader developments, potentially with the wireless connectivity of the things, allowing communication and interaction. The various applications of RFID contribute with other developments, such as mobile telephony, Bluetooth and GPS (Global Positioning System) with the interconnection of all the things in the world. Such technologies will contribute to the realization of futuristic visions, such as the automated house which allows the connection of the house with the outside world without direct contact.

Kenneth F. Fishkin and Jay Lundell outlined the possibilities of a system of observation and warning for the elderly. In their project they imagine the following situation. A man, Chester, is 90 years old and lives alone in a house. His daughter, Molly, lives a few kilometres away. The researchers propose to equip all the objects in the house with RFID markers. Chester himself carries a reader in his clothing. Thus it is possible to rebuild his entire timetable and by processing the data also recover Chester's entire manner of living. The revolution in the interconnection of all the things in the world is not a unilateral development, but a term which gathers multifarious developments. Thus, the Internet of things will have a broad impact on many processes of life that characterize our daily existence and potentially influencing our behaviour.

But is the passage of information using RFID technology protected? Is the coding of the information transmitted by the chip placing it in the private life of people? The first disadvantage concerning subcutaneous chips is that they are not protected. If the computer age has taught us one thing it is that systems and the data are always less protected than is said and although the chips, such as VeriChip, are marketed for access control, they lack protection. In a recent publication in the Journal of the American Medical Informatics Association (JAMIA), Ari Juels showed that VeriChip has no more protection than a code bar and that it was very simple to build an object able to scan a VeriChip (Juels, 2006). The vulnerability of RFID chips does not protect us from those who would deploy them as sedentary tools and the data which will compose this type of chip will never be completely protected or confidential. That means that there are certain applications which these technologies will never be able to carry out. It should be understood that the asset of these technologies, and in particular of this one, is not to make the world a safer place. The goal, if there is one, is to simplify the use of it, to return the world to a more malleable and less formalized place. The issue is not to make safe, control, sign and certify, but to create a new place for the abstract one which already composes our relations with the world, the objects and others. The risk, and there is one, is that more and more applications have only one access key, too fragile to be worthy of confidence.

4. RFID Technology in relation to man

The technology brought upheaval to our world and invaded all fields. The general tendency is then to declare oneself for or against the technology. The technology plays a great part in the constitution of our lives. The relations between the human and the technology are not neutral since they enable us to increase the standard of living when comfort has become unceasingly important. But humans fear losing control, to be worked and controlled by the technology. The seizure of humans by the technology is understood, like hybridization or progressive fusion of human and technology. Hybridization takes the form of a progressive delegation of the body's functions, like intellectual functions and the determination of human behaviour. RFID technology thus tends to influence us in multiple ways. Generally this influence functions while intervening in the relations which we constitute with others and by which our lives are structured. Engagement in the ethics of the technology gives the required authority to conclude hybridization. It is by the imposition of the laws that developments described as intolerable by the majority of people could be born. But a whole field of hybridization escapes legislation. By hybridization with technology, the human integrates new forms of use of the technology and brings about novel modes of life. In addition, these practices create new forms of existence. It is not a question of freedom or autonomy against the technology, but of an attitude toward practices which do not want to only undergo a shaping by the techniques, but aims at establishing an autonomy which consists in shaping of oneself. It is not a question either of a categorical denunciation of RFID, on the basis of universal principles of autonomy which would prevent conscious engagement in relations of constraints. It is about an attitude of recognition and acceptance of RFID technology. Such is the message of the ethics of constituting oneself through the use of the technology.

Let us take the Navigo system in the Parisian subway as an example. It is clear that the Navigo system imposes behaviours on the users. All the technical devices in the subway, such as the terminals, direct the behaviour of people. This development is not understood as an imposition, but rather as a progressive engagement in structured procedures. The introduction of the Navigo system increased the sale of yearly subscriptions, which indicates that people gradually wish to benefit from the Navigo master key and accept integration with the system.

5. RFID technology and human desire

Accordingly, RFID allows the human passions expressed in the state of nature to be appeased. If RFID allows the collection of information on the purchases of individuals, this can potentially be made known to companies. With information on the type of goods individuals want to enjoy, when, at what frequency etc., companies can then do everything to ensure the individual has all he wants, even before he formulates his desire, since RFID allows for envisaging desires to come. It can also better stimulate this desire by 'profiling' the customer in order to create publicities which cause him to desire buying and to consume more intensely. RFID is a tool which returns the work of modern companies even faster, more precisely and more completely: human desire being insatiable by definition, the companies continue to produce their products and produce them as long as this desire is maintained. The information furnished by RFID should allow for even more important marketing, made more reliable as it is supported on precise, real data. Since the state is what allows the development of economic activity, from this point of view the state could not be opposed to the use of RFID, in that it contributes to the search of pleasure.

What about those who do not wish to consume, or do not consume according to the discounted forecasts, or who refuse to allow their private information to be gathered on a database? They will not be able to benefit from the advantages granted to the remainder of the consumers. Under the increased pressure of publicities, presented as more individual and instrumentalized, is there a real risk that social links transform under this impulse into a 'common bond' which would be that of the same desire formatted by industry? But the use of the data collected in any time and any place by RFID could well lead to a gradual standardization of the desire, i.e. part of passions. From this point of view, could RFID represent the efficient cause of a turning in the design of the social link? The individuals 'would thus be linked' by a social link rather than understood 'to wish for the same type of things'. This social link born of a common, or rather similar, desire would be then a 'fossilized' bond because it is registered in foreseeable diagrams. *De facto,* all those who refuse to see their data collected or do not consume in a foreseeable way, would then be excluded from this social link.

6. Discussion

RFID allows tracking and identifying things equipped with an RFID marker. This technology allows for the dream of tracing all the things and people in the world, but represents at the same time a fear of total monitoring. RFID poses problems with the private life. This time it is not the publication of images and facts on the personal life, but the collection and recording of the data for reasons of statistics and management. With the RFID technology, a new concept of private life appears. The daily use of the products equipped with RFID functions often returns a benefit of the specific services in exchange for the collection of information on the user. The arbitration of the state is then essential to control the exchanges of information emitted by the RFID by the constitution of the laws on the protection of information. One thus needs new distinctions between acceptable and unacceptable forms of tracing. The state has two complementary roles to play: to facilitate trade by the contribution of the RFID technique and to guarantee the safety of the information exchanges. Indeed, the use of RFID techniques in daily life involves a fear of insecurity insofar as the actions of people can be known in real-time (purchases, displacements, communications…). The private sphere then becomes permanently public. With RFID, humans lose the control of their own data and lifestyle choices. It involves itself in a process of hybridization with RFID techniques, implying an increase in its capacity on nature, but also dependence on the technique. The human being becomes a virtual being, a being of information to be collected in one way for thousands of companies. The exchanged information is instrumentalized and the desires of the individuals are standardized. The RFID technique creates a tension between the human need for safety and the need for freedom. But, the individual must determine what is more essential for him: fear of death, willingness to conserve life or fear of domination and the desire to be free. The ethical approach must be concerned here with the subjectification of people through the use of the technology.

7. References

Korff, D. 2006. The right to life. Guide to the implementation of Article 2 of the European Convention on Human Rights Strasbourg. Council of Europe, Human Right Handbooks n° 8.

Hobbes, T. 2010. Leviathan: However the Matter, Form and Power of the Commonwealth, Ecclesiastical and Civil Library. Ed. Yale University Near. ISBN: 0300118384.

Graafstra, A. 2006. RFID Toys: Cool Projects for Home, Office and Entertainment. ED. John Wiley & Sons Inc. ISBN: 0471771961.

Juels, A. 2006. The Practice of Informatics: Technology Evaluation: The Security Implications of VeriChip Cloning. JAMIA; 13:601 - 607 DOI:10.1197/jamia.M2143.

International Telecommunication Union. 2005. ITU Internet Carryforwards 2005: The Internet of Things. 7th edition, Geneva, www.itu.int/internetofthings.